Chilling Injury
of
Horticultural Crops

Editor

Chien Yi Wang, Ph.D.

Research Horticulturist
Horticultural Crops Quality Laboratory
U.S. Department of Agriculture
Beltsville, Maryland

CRC Press, Inc.
Boca Raton, Florida

Library of Congress Cataloging-in-Publication Data

Chilling injury of horticultural crops / editor, Chien Yi Wang.
    p. cm.
  Includes bibliographical references.
  ISBN 0-8493-5736-5
  1. Horticultural crops—Physiology. 2. Horticultural crops-
-Postharvest technology—Diseases and injuries. 3. Plants, Effect
of cold on. I. Wang, Chien Yi.
  SB314.5.C48 1990
  632'.11--dc20
                                89-70893
                                     CIP

  Direct all inquires to CRC Press, Inc., 2000 Corporate Blvd., N.W., Boca Raton, Florida, 33431.

1990 by CRC Press, Inc.

International Standard Book Number 0-8493-5736-5

Library of Congress Card Number 89-70893
Printed in the United States

PREFACE

Susceptibility to chilling injury prevents the cultivation of many crops in regions where temperatures can drop below the optimal growth temperatures for individual plant species. Chilling injury also limits the extended storage of many horticultural commodities. These commodities develop symptoms of injury after exposure to chilling temperatures. The mishandling of such crops can result in tremendous losses. The postharvest losses resulting from chilling injury are probably greater than what has been recognized because most symptoms of chilling injury are not apparent until after the commodities have been transferred from cold storage to warmer temperatures in the markets, and a substantial portion of market losses ascribed to infection by pathogens or ripening disorders may actually result from chilling injury. With the great diversity of horticultural crops, it is difficult to make generalized statements concerning the symptoms, responses, and characteristics of the damage caused by chilling. Exposure of sensitive commodities to chilling temperatures may not always lead to development of the typical chilling injury symptoms. However, the disruption of normal physiological functions by chilling stress may lead to conditions, such as enhancement of ethylene production and metabolic imbalances, which accelerate the deterioration and senescence of these commodities. A better understanding of the basic mechanisms which underlie the injury and a greater comprehension of the physiological and biochemical responses of plants to chilling stress will help us design better cultural practices and develop more effective techniques for the alleviation of chilling injury.

This book brings together discussions on various aspects of chilling injury, including five introductory chapters on the description of chilling injury; three chapters on the contributing factors and assessment of chilling injury; seven chapters on the biochemical changes, molecular basis, and concepts of chilling injury; and three chapters on the prevention and reduction of chilling injury. The purpose of this book is to provide readers with a historical background as well as current knowledge and theories about various aspects of chilling injury. All the chapters are written by scientists who are experienced and actively engaged in chilling injury research. The extensive collection of references at the end of each chapter also offers readers an abundance of resources for further detailed study. It is my hope that this book will provide readers with a better understanding of chilling injury problems and will stimulate ideas for further research.

I wish to thank all the authors for their tremendous efforts in fulfilling the difficult task of researching, compiling, and writing the assigned chapters and for sharing their knowledge and thoughts with the readers. It is only through this kind of collective effort that a book covering a wide range of chilling injury research in depth is made possible. I wish also to express my sincere appreciation to Marguerite Benedict, former Technical Editor of *Agricultural Research Service,* and George Kramer and Bruce Whitaker of the Horticultural Crops Quality Laboratory, U.S. Department of Agriculture for their careful reading and comments, and finally, to my wife, Shiow Ying, for her valuable suggestions and continuous encouragement.

Chien Yi Wang

THE EDITOR

Chien Yi Wang, Ph.D., is a Research Horticulturist in the Horticultural Crops Quality Laboratory at the U.S. Department of Agriculture, Beltsville, Maryland.

Dr. Wang received his B.S. degree in Horticulture from National Taiwan University, Taipei, Taiwan, in 1964 and his Ph.D. degree in Horticulture and Plant Physiology from Oregon State University, Corvallis, Oregon, in 1969. From 1969 to 1976, he worked at the Mid-Columbia Experiment Station, Hood River, Oregon and conducted research on production, harvesting, handling, storage and ripening of tree fruits. Since 1976, he has been a Research Horticulturist at the Beltsville Agricultural Research Center, Agricultural Research Service, U.S. Department of Agriculture. His current research interests include chilling injury, controlled atmosphere storage, ripening, and packaging of fruits and vegetables. His research has led to numerous publications in scientific journals and book chapters, as well as to many presentations at national, international, and industrial meetings.

Dr. Wang is a member of American Society for Horticultural Science and American Society of Plant Physiologists. He reviews manuscripts frequently for various scientific journals and has served in the review panels to evaluate research proposals for several organizations including the Agency for International Development and the USDA Special Grants Program for Tropical and Subtropical Agriculture. He has been the recipient of an award from the National Academy of Science and the International Advisory Panel to serve as an advisor in the Chinese University Development Program. He has been selected as an expert in food preservation by the United Nations Industrial Development Organization. He has also been invited to write an article for the *World Book Encyclopedia*. Recently, he was appointed by the Office of International Cooperation and Development to serve as a Technical Advisor in a research project on the reduction of chilling injury in fruits.

CONTRIBUTORS

Kazuhiro Abe, Ph.D.
College of Agriculture
University of Osaka Prefecture
Sakai, Osaka, Japan

William J. Bramlage, Ph.D.
Professor
Department of Plant and Soil Sciences
University of Massachusetts
Amherst, Massachusetts

R. W. Breidenbach, Ph.D.
Professor
Plant Growth Laboratory
University of California
Davis, California

Charles R. Caldwell, Ph.D.
Plant Physiologist
Plant Stress Laboratory
USDA/ARS/BARC-W
Beltsville, Maryland

Roger J. Field, Ph.D.
Professor
Department of Plant Science
Lincoln University
Canterbury, New Zealand

Charles F. Forney, Ph.D.
Plant Physiologist
Horticultural Crops Research Laboratory
U.S. Department of Agriculture
Fresno, California

John A. Greaves
Plant Physiologist
Research Department
Garst Seed Company
ICI Seeds
Slater, Iowa

T. T. Hatton, Ph.D.
Research Leader (retired)
Horticultural Research Laboratory
USDA/ARS
Orlando, Florida

Robert C. Herner, Ph.D.
Professor
Department of Horticulture
Michigan State University
East Lansing, Michigan

Werner J. Lipton, Ph.D.
Plant Physiologist (retired)
Horticultural Crops Research Laboratory
U.S. Department of Agriculture
Fresno, California

James M. Lyons, Ph.D.
Professor
Department of Vegetable Crops
University of California
Davis, California

Shimon Meir, Ph.D.
Researcher
Department of Fruit and Vegetable
 Storage
The Volcani Center
Bet Dagan, Israel

Leonard L. Morris, Ph.D.
Professor Emeritus
Department of Vegetable Crops
Mann Laboratory
University of California
Davis, California

Norio Murata, Ph.D.
Professor
National Institute for Basic Biology
Okazaki, Japan

Takao Murata, Dr.
Professor
Department of Horticulture
Shizuoka University
Shizuoka City, Japan

Ikuo Nishida, Ph.D.
National Institute for Basic Biology
Okazaki, Japan

Glenda R. Orr, M.Sc.
CSIRO Division of Horticulture
Sydney Laboratory
School of Biological Sciences
Macquarie University
North Ryde, Australia

Brian D. Patterson, Ph.D.
Principal Research Scientist
Postharvest Laboratory
CSIRO Division of Horticulture
North Ryde, Australia

Robert E. Paull, Ph.D.
Professor
Department of Plant Molecular
 Physiology
University of Hawaii
Honolulu, Hawaii

Albert C. Purvis, Ph.D.
Department of Horticulture
University of Georgia
Coastal Plain Experiment Station
Tifton, Georgia

John K. Raison, Ph.D.
Chief Research Scientist
CSIRO Division of Food Research
School of Biological Sciences
Macquarie University
North Ryde, Australia

Michael S. Reid, Ph.D.
Department of Environmental Horticulture
University of California
Davis, California

Mikal E. Saltveit, Jr., Ph.D.,
Department of Vegetable Crops
Mann Laboratory
University of California
Davis, California

Philip R. van Hasselt, Ph.D.
Department of Plant Biology
University of Groningen
Haren, The Netherlands

Chien Yi Wang, Ph. D.
Research Horticulturist
Horticultural Crops Quality Laboratory
U.S. Department of Agriculture
Beltsville, Maryland

John M. Wilson, Ph.D.
Lecturer
School of Plant Biology
University College of North Wales
Bangor, Wales

TABLE OF CONTENTS

Part I. Introduction and Description of
Chilling Injury

Chapter 1

OVERVIEW ON CHILLING INJURY OF HORTICULTURAL CROPS

M. E. Saltveit, Jr. and L. L. Morris

TABLE OF CONTENTS

I. INTRODUCTION

Temperature management is the predominant method used to maintain the quality of harvested horticultural crops. Reductions in temperature can substantially reduce the velocity of many metabolic processes which lead to the natural deterioration and loss of quality of horticultural products. For certain commodities, however, the commonly accepted practice of quickly cooling freshly harvested material to low, nonfreezing temperatures can actually shorten their market life because of their susceptibility to a disorder called chilling injury.

Those vascular plants that evolved in warm tropical and subtropical environments, where the ambient temperatures narrowly fluctuate around 20° C, suffer a physiological dysfunction if exposed to temperatures above the freezing point of the tissue, but below around 10 to 15°C. This disorder, commonly called chilling injury, applies to many economically important agronomic and horticultural crops and severely limits not only the minimum temperatures to which they should be exposed after harvest, but also the geographical areas and seasons in which these sensitive crops can be grown. While much practical information is available on the expression and prevention of symptom development, the primary changes caused by chilling temperatures and their physiological transduction into commonly recognized symptoms of chilling injury remains unresolved. A better understanding of the cause and prevention of chilling injury is, therefore, both scientifically interesting and economically important.

A large body of literature exists on the practical and scientific study of chilling injury of horticultural crops, and is referenced in several recent reviews,[1-5] books,[6-9] and symposia.[10-19] This introduction will draw upon these references to provide a brief overview of the history, terminology, symptoms, causes, and technologies associated with chilling injury of horticultural crops.

II. HISTORY

Chilling injury was recognized as a factor that limited plant growth by persons involved in plant production long before it became a subject of study by plant physiologists. One cannot ascribe a date, or even a period of time, when the phenomenon was first recognized. However, the maintenance of tropical plants in heated structures in temperate regions during ancient times represented an early recognition of the existence of chilling injury and a method to avoid it.

For many years, persons in the produce industry have recognized that some harvested horticultural commodities respond unfavorably to chilling temperatures. Bananas, citrus, sweet potatoes, and tomatoes are important horticultural crops that can suffer serious losses from chilling injury under conditions of distribution and marketing. It was a common procedure in the early part of this century for technicians to accompany rail cars of imported bananas being shipped into colder regions of the U.S. They would remove ice, add heat, or ventilate the rail cars when needed to avoid chilling the fruit. Similarly, before thermostatically controlled, mechanically refrigerated cars were available, tomato shippers manipulated icing and ventilation of ice-refrigerated rail cars during transportation to avoid chilling temperatures. For many years it has been the practice to store sweet potatoes in nonrefrigerated (and often heated) structures. The necessity of harvesting sweet potatoes before the onset of cold weather has long been recognized. For home storage, the desirability of storing pumpkins and squash in heated areas of the house rather than in cold root cellars has been recognized since colonial times.

The widely held belief that produce removed from cold storage deteriorates more quickly than freshly harvested produce is no doubt partially based on the observation that chilling-sensitive commodities rapidly break down after transfer from cold storage to warm, nonchilling temperatures. There are many other examples of the early recognition of chilling injury as an important factor in plant growth and quality, and of the means to prevent or reduce the severity of chilling injury symptoms.

Molisch[20] cited an article by Bierkander that was published in 1778 and is the earliest published scientific work regarding chilling injury that has come to our attention. In that article, Bierkander is cited as reporting that several susceptible species of plants were killed when exposed to temperatures just above freezing. In another article cited by Molisch,[20] Hardy reported in 1854 that some examples of young, woody plants introduced from tropical regions were killed by exposure to temperatures in the range of 1 to 5°C. In 1865, Sachs[21] found that exposure to chilling temperatures for one to a few days resulted in the death of plants of tropical and subtropical origin.

Sachs[22] had previously reported in 1864 that the minimum temperature for protoplasmic streaming in certain chilling-susceptible plants approximated that of the upper limit of their chilling range. This suggested the possible involvement of cessation of streaming in the series of events resulting in chilling injury. This idea was later pursued by Lewis[23] and Wheaton[24] and is of current interest to investigators since it may have a direct connection with chilling-induced membrane changes.[25]

Molisch[20] studied the effect of a water-saturated atmosphere during the chilling of susceptible leaves and plants. He found that preventing water loss did not prevent death from chilling. His studies followed those of Sachs,[21] who proposed that the chilling-induced leaf wilting observed in many chilling-sensitive plants was the result of an inability of the roots to absorb and transport sufficient water from the cold soil to the leaves. Death was proposed to result from a transpirational deficit and dehydration. While reduced water uptake and transport has been reported for about ten species of chilling-sensitive plants to date, water stress probably does not account for the rapid death of many sensitive species. Molisch[20] attributed the development of brown spots on some of the tropical plants that were exposed to 1 to 4°C to the accumulation of toxic materials. Molisch[26] is often credited with having suggested the term ''chilling injury'' (Erkaltung) as opposed to ''freezing injury'' (Erfrieren).

Scientific investigations into chilling injury of plants, animals, and lower organisms continued during the first half of this century, but at a relatively slow rate. A monograph by Belehradek in 1935,[27] which seems to have escaped the attention of many investigators, presents an excellent survey of the early work on chilling injury of biological systems. Over 2 decades later, Belehradek published a comprehensive review on the physiological effects of temperatures.[28]

While the term ''chilling'' has been used in horticulturally related sciences for many years in conjunction with the exposure of deciduous fruit trees to near freezing temperatures and the breaking of dormancy, it has only been applied to the development of injury symptoms on plants of tropical origin for a comparatively short time. Some early studies of ''low-temperature injury'' focused on agronomic crops.[29-31] Sellschop and Salmon[32] present an excellent article on chilling injury that deals with a wide range of chilling-sensitive crops and of treatments to reduce the symptoms of chilling injury. The first research article in the *Proceedings of the American Society for Horticultural Science* on ''low-temperature injury'' that used the term ''chilling injury'' was in 1938.[33] The *Journal of the American Society for Plant Physiology* published their first article on ''low-temperature injury'' in 1942.[34] Chilling injury was discussed in two early articles in the *Annual Review of Plant Physiology*.[35,36] The first use of the term ''chilling injury'' in the indexes to *Biological Abstracts* occurred in 1959.

In the past 2 decades, the level of interest, the focus of effort, and the volume of work on the physiological basis of chilling injury has increased dramatically. The International Seminar on Low Temperature Stress in Crop Plants, held in 1979,[10] the American Society for Horticultural Science Symposium on the Chilling Injury of Horticultural Crops, held in

1981,[14] and the biannual Gordon Conference on Temperature Stress in Plants are evidence of the increased interest in this field.

III. TERMINOLOGY

The physical and/or physiological changes induced by exposure to low temperatures, together with the subsequent expression of characteristic symptoms, are commonly combined in the term "chilling injury". For clarity, it is desirable to distinguish between the inclusive term "chilling injury" and its components. The terms "chilling", "chilling treatment", or "chilling exposure" refer to the treatment itself, which involves both the duration and temperature of exposure. The temperature treatment is commonly thought to cause a primary event that is followed by a series of subsequent physiological and physical events which give rise to symptoms. The exact type and extent of symptom development varies with the species, the severity of the chilling injury, and the environmental conditions during and following chilling (see also Chapter 6).

The terms "chilling damage" and "chilling breakdown" can be used as synonyms of "chilling injury". The term "chilling stress" applies to the action of the low temperature that results in "chilling injury". The term "low-temperature injury" and other combinations where "low temperature" is substituted for "chilling" should be used as more general terms to include both chilling injury and freezing injury.

Raison and Lyons[37] have proposed definitions for terms applied to this general topic. We believe the above terminology is compatible with their definitions. They emphasize the importance of making a distinction between "chilling-sensitive" and "chilling-resistant" or "chilling-tolerant" tissue. They propose that sensitivity to chilling refers to the threshold temperature below which the primary event of chilling injury occurs. In contrast, "chilling resistance" or "chilling tolerance" would refer to the ability of the plant to resist or tolerate changes caused by the primary event. Thus, a factor in the environment could retard symptom development and, hence, increase the tolerance of the tissue to chilling injury without actually changing the sensitivity of the tissue to chilling temperatures. Since the primary event for each species may be somewhat time dependent as well as temperature dependent, a clear distinction between tolerance and sensitivity may be difficult to maintain in the future when refinements in techniques allow closer observation of the molecular effects of temperature on the primary event.

Controversy still exists as to whether certain low-temperature disorders that occur in temperate zone plants should be considered as symptoms of chilling injury. Examples include the disorders that some cultivars of apple, peach, and plum exhibit after prolonged low-temperature storage and the somewhat abbreviated storage life of asparagus at temperatures near 0°C. The question arises as to whether the term "chilling injury" should be limited to the phenomenon that has a threshold temperature in the range of 10 to 15°C, as shown by species of tropical origin.

The following short reiteration of some of the more important concepts discussed above may be helpful. *Chilling* is the act of exposing plant material to a critical or threshold low temperature, i.e., the *chilling temperature,* which is generally thought to cause a *primary response* in *sensitive plants.* This primary event is the initial, rapid response to the chilling temperature. The primary response causes a *dysfunction,* i.e., an impaired functioning of the tissue, which is reversible if the tissue is returned to a nonchilling temperature after a brief period of chilling, but which becomes irreversible after a longer period of time at the chilling temperature. The cause of the primary event and the dysfunction may be one and the same, e.g., an alteration in the fluidity of a critical membrane, or a change in the activation energy of a critical enzymatic reaction. After some time, the dysfunction results in the appearance of horticulturally important symptoms. *Chilling-tolerant* or *chilling-re-*

TABLE 1
Some Examples of Commercial Crops that are
Susceptible[a] to Chilling Injury

African violet	Corn[b]	Pineapple
Apples[c]	Cucumber	Plums[c]
Asparagus	Eggplant	Poinsettia
Avocado	Ginger	Pumpkin
Banana	Mango	Rice
Basil	Melons[d]	Sorghum
Beans, French	Okra	Soybean
Beans, Mung	Orchids	Squash
Cassava	Papaya	Sugar cane
Citrus	Peaches[c]	Sweet potato
Coffee	Peanut	Taro (and other aroids)
Coleus	Pepper	Tomato
		Tradescantia

[a] Susceptibility depends on the plant part and cultivar tested.
[b] Only the plant is sensitive.
[c] Only fruit of certain cultivars are chilling sensitive.
[d] Probably all types of melons are sensitive.

sistant plants are sensitive plants which are able to resist developing symptoms of chilling injury. Most horticultural treatments for chilling increase the tolerance of the material to develop symptoms of chilling injury, but not its sensitivity to chilling temperatures.

IV. SUSCEPTIBLE CROPS

Although the emphasis of this book is on harvested horticultural commodities, it should be remembered that chilling-sensitive plants indigenous to the tropics are subject to injury at all stages of their development,[15-18] except as dormant, dry seeds.[15,18] In contrast, only specific organs from plants indigenous to temperate areas, such as certain deciduous fruit and young asparagus shoots, are injured by prolonged exposure to chilling temperatures.[17] In both tissues, the level of tolerance to chilling can change with maturation and development. For example, immature fruit are thought to be less tolerant of injury than ripe fruit. Symptoms may also differ in their type and extent as the commodity matures (see also Chapters 2 and 3).

Chilling injury can occur during germination of seeds, storage of transplants, growth of the crop in the field, storage, transportation and market distribution, and holding in the home refrigerator. The nature and severity of symptoms will be a function of the species, the variety, the part of the plant, and its maturity, as well as the severity and duration of the exposure. Some examples of economically important crops that are susceptible to chilling injury are listed in Table 1. As can be seen, the examples range over many families of vascular plants. The ambient environment of the commodity during and following chilling will also influence symptom development. In addition, exposure of the commodity to other stresses or injuries can influence the nature and severity of symptom development.

V. SYMPTOMS OF CHILLING INJURY

There are no easily measured physiological or visual changes that are unique to chilling-injured plant tissue.[15] Often the symptoms that develop are merely exaggerations of the effects of physical injury or other physiological stresses. For example, chilling often increases the occurrence of senescence and decay after chilling and the rate of water loss both during and after chilling. Similar symptoms can be induced by factors other than chilling temperatures, (e.g., mechanical injury and contaminated handling facilities, and low relative humidity). However, several commonly occurring symptoms are often used as indicators of the severity of chilling injury. Some of the more commonly recognized symptoms of chilling injury are listed below in order of their response time.

1. **Cellular changes** — These include changes in membrane structure and composition, cessation of protoplasmic streaming, and plasmolysis of cells and an increased rate of leakage from the cells occurring in a variety of tissues (see also Chapter 12).

2. **Altered metabolism** — Abnormal increases in the rate of production of carbon dioxide and ethylene occur sometimes during chilling, and very often after the chilled tissue is placed at warmer, nonchilling temperatures (see also Chapters 14 and 15). Increased levels of the products of anaerobic respiration and other abnormal metabolites have also been reported in chilled tissue.

3. **Reduced plant growth and death** — Field-grown crops and ornamental plants are often stunted in their growth or killed by chilling temperatures. Wilted leaves and development of necrotic areas often occur after chilling.

4. **Surface lesions** — These include surface pitting, large sunken areas, and discoloration of the surface (see also Chapter 5). These symptoms are aggravated by mechanical damage and environmental conditions. The dull appearance of chilled bananas is a common example.

5. **Water-soaking of the tissue** — Cellular breakdown and loss of membrane integrity result in the exudation of cellular fluids into the intracellular spaces.

6. **Internal discoloration** — The pulp, vascular strands, and seeds often turn brown. Vascular browning of avocado fruit is a good example.

7. **Accelerated senescence** — This occurs when the loss of chlorophyll in light (i.e., photooxidation) and in darkness, and the loss of cellular integrity proceed at a faster pace.

8. **Increased susceptibility to decay** — The preceding changes provide a favorable medium for the growth of pathogens. The microorganisms successfully attacking chilled tissue are often weak parasites that do not grow readily on healthy tissue. *Alternaria* rot of chilled tomato fruit is a good example.

9. **Failure to ripen normally** — Fruits which are harvested at a mature but unripe stage are unable to progress through the normal ripening sequence if chilled. Even if they ripen, there is a loss of the development of the characteristic flavor and aroma, and often the development of off-flavors occurs. Common examples are chilled mature-green bananas and tomatoes.

10. **Loss of vigor** — Stored propagules such as sweet potatoes loose their ability to sprout and grow after chilling.

The development or expression of many of these chilling symptoms is often mediated by current environmental factors. For example, crops such as corn which are only marginally sensitive to chilling in the dark, suffer severe injury if exposed to sunlight while at the chilling temperature. At the other extreme, many symptoms of chilling injury do not appear during chilling, but only appear after subsequent holding at warm, nonchilling temperatures. In the case of water loss-related symptoms, such as those discussed above, the water stress usually imposed at higher holding temperatures can accelerate development of the symptoms. Pitting, as a symptom of chilling injury, can be effectively inhibited by reducing transpiration. However, the increased susceptibility to disease or rapid senescence is unaffected by such treatments.

Chilling injury is not translocatable. When chilling was localized to one half of an intact cucumber, the symptoms were confined to the chilled portion.[38] Localized regions of injury, as in the development of pits or discolorations, could indicate distinct regions of tissue that are more susceptible than the surrounding tissue.

Since there are no symptoms specific to chilling injury, it is difficult to diagnose chilling injury or to quantify it based solely on a few observed symptoms. As analytical techniques improve, scientists will be able to make better objective measurements of the primary changes taking place during chilling. For example, developments such as refined calorimeters and use of infrared spectrophotometers promise to allow closer observation of some of the first events associated with chilling.

VI. DEVELOPMENT OF SYMPTOMS

Several factors contribute to the development of chilling injury symptoms. They include intrinsic qualities of the tissue, (such as species, cultivar, growing conditions, age, maturity, previous exposures to stress, and plant part) and extrinsic qualities of the environment (such as temperature, duration of exposure, whether exposure is continuous or interrupted, relative humidity, composition of the atmosphere, and postharvest treatments). Many of these factors will be discussed in the following section on preventing chilling injury symptoms.

There is a time and temperature interaction in the development of chilling symptoms. In general, greater injury occurs in plants exposed to lower chilling temperatures for longer periods of time. Below a threshold temperature, there is a curvilinear relationship between degree-hours and the extent of injury.[38] After a critical number of degree-hours has accumulated, each additional hour of chilling contributes less to the overall severity of the injury. The initial event in chilling is presumed to be relatively rapid in comparison to the time required for sensitive tissue to develop gross symptoms. The intervening time is necessary for changes caused by the primary event to be transduced into permanent damage that causes the observed symptoms. In some cases, however, intermediate temperatures have caused greater chilling injury than either temperatures immediately below the critical chilling temperature or temperatures near zero.

The presence of a lag period during which no permanent damage appears to occur in exposed tissue has a number of important consequences. This means that most tissues can recover from short periods of exposure to chilling temperatures that are within the lag period, and that interruption of the exposure to chilling temperatures with brief periods at nonchilling temperatures before permanent or irreversible damage occurs can reduce or prevent development of chilling injury symptoms.

Although sensitive tissue can recover from brief, subinjurious exposures to chilling temperatures, the severity of chilling injury symptoms are usually cumulative if insufficient time is allowed for the tissue to overcome the effects of the chilling exposure, or if the exposures are at an injurious level. For example, exposure to chilling temperatures in the field or during subsequent handling increases the susceptibility of the tissue to additional injury if exposed to further chilling temperatures after harvest.[39]

Symptoms may appear during the exposure to the chilling temperature, but their appearance more usually occurs after removal to a warmer, nonchilling temperature. In fact, if the material is to be used immediately after removal from storage, an acceptable product can be obtained after a period of storage at chilling temperatures that would normally result in development of severe damage during subsequent holding at higher temperatures.

Since many symptoms of chilling injury involve alterations of normal ripening, it is commonly thought that ripe fruit are more resistant to chilling injury than unripe fruit. In fact, ripe fruit may appear more chilling resistant simply because they cannot exhibit alterations in the already accomplished processes of ripening.

VII. BASIS OF CHILLING INJURY

Many theories have been proposed to explain the mechanism of chilling injury (see also Chapter 9). The sensitivity to chilling ranges over a wide spectrum of temperatures and exposure times from plants indigenous to the tropics that are injured by a few hours at temperatures below 15°C, to plants indigenous to temperate regions that are injured only after many weeks at temperatures near 0°C. While it may not be reasonable to ascribe similar physiological causes of chilling injury to plants that exhibit these extremes in temperature sensitivity, the large number of economically important horticultural crops that share a similar threshold temperature for chilling injury suggests that a common temperature-dependent physical event is involved. The similar symptoms shown by these species suggests that there might also be some common pathways leading to injury. While the observation that there are several types of chilling injury over the entire range of chilling-sensitive plants that are symptomatically distinguishable speaks more to differences in the subsequent development of symptoms in the various plants than to differences in the primary event. While the primary event may vary with the species, it is certain that the resulting events in symptom development will surely vary with the species and with the environment. It may not be practical to propose a single model that is applicable to all plants that show chilling injury.

It is difficult to find any selective advantage that would result in resistant plants developing a specific sensor to low temperatures. Rather, it is more likely that the chilling-sensitive progenitors of the chilling resistant species of the present developed resistance or tolerance to chilling temperatures as they radiated from tropical regions into temperate areas. In contrast to the evolutionary shift toward chilling resistance in many temperate species, breeders have inadvertently produced varieties of many horticultural species, e.g., tomato and potato, that are more chilling sensitive than their wild progenitors. It is probably incorrect to think that there is a specific low-temperature sensor that is responsible for sensing and transducing a signal for chilling injury. It is much more likely that a temperature-induced alteration in either membrane properties or enzyme activities results in chilling injury.

One of the earliest theories related to the cause of chilling injury proposed that the accumulation of toxic levels of metabolites was caused by a temperature induced imbalances in metabolism.[40] Nelson[40] suggested that surface pitting was associated with the products of abnormal respiration. In fact, accumulation of toxic compounds as the result of abnormal metabolism was the only theory discussed at length in an early review by Pentzer and Heinze.[36] While certain preventative measures, such as intermittent warming, appear to work because they allow dissipation of accumulated toxic compounds, no compound has been isolated from chilled tissue that fulfills the criteria of the toxic material. In fact, the failure of chilled tissue to produce an essential metabolite could just as well be proposed as a cause of chilling injury.

The possibility that differences in the lipids and fatty acids of chilling-sensitive plants could contribute to their susceptibility has been suggested by many workers (see also Chapter 11). Lewis and Morris[41] suggested that if lipids of the membrane are affected by low temperatures, then permeability could be altered by chilling. Furthermore, many of the symptoms of chilling stress suggest that membrane permeability and/or fluidity could be involved.[42,43]

The fact that many symptoms of chilling injury, such as pitting, water soaking, and increased ion leakage, appear to arise from a common cause, i.e., failure of the cellular membranes to maintain compartmentalization of the cellular contents, was noticed quite early. Fluids that leaked from the cell could form water-soaked areas, reduce normal gas diffusion, provide a medium for the growth of pathogens, and cause wilting, desiccation, and necrosis of the affected tissue. Evaporation of the water from these areas, especially if they are near the epidermis, could cause the epidermis to develop sunken areas and pits.

Both mechanical injuries to the epidermis and low relative humidities in the surrounding atmosphere exaggerates the development of these surface abnormalities.

Other symptoms of chilling appear to arise from abnormal metabolism. If the activation energy of a few key enzymes have large temperature coefficients near the chilling temperature, and their enzyme activities change at different rates, then at some critical temperature there could be a crossover of control, and metabolism could shift to the unregulated production of inappropriate compounds that could accumulate to toxic levels. Such changes could also occur during temperature-induced phase changes in critical membranes if the activity of the enzymes were tightly coupled with the affected membranes.

A model suggested by Lyons[1] incorporated previous proposals by Lyons and Raison,[44] Raison et al.,[45] Singer,[46] Levitt,[47] and Singer and Nicolson[48] and brought together the ideas of a phase change of the membrane lipids and the resulting alterations in the metabolism of the chilled cells (see also Chapter 9). This model proposes the existence of a primary event that results in a series of secondary events which, in turn, results in the symptoms of chilling injury.[49] It is compatible with many of the empirical observation and experimental data collected over the years. Although this model is currently considered to have many critical shortcomings, it has been the predominant hypothesis over the past 15 years and has stimulated a great deal of interest and research in the area of chilling injury. Recent alterations to this evolving proposal is that only a small portion of the membrane actually undergoes the phase change which results in both changes in cellular membrane permeability and the activation energy of membrane-bound enzymes. However, it is still not clear how a change in a small fraction of the lipids could cause the pronounced changes associated with chilling.

Other problems with this theory include the fact that the degree of fatty acid unsaturation in bulk lipids rarely correlates with membrane fluidity or the level of chilling sensitivity. Also, membrane fluidity has been calculated from equations which assume a homogenous lipid environment, a condition not usually found in biological membranes. Neither does the hypothesis take into account the pronounced effects of membrane-associated sterols and proteins on the physical properties of the membrane.[50] The emphasis has shifted to studying the effect of other membrane components and the unsaturation of fatty acids in specific membranes and lipid fractions on chilling (see also Chapters 10 and 11).

Procedural problems include the fact that the electron spin resonance probes used to measure fluidity can preferentially reside in a specific fraction of the membrane or cause aggregation of specific components of the membrane itself. In many cases, differences are not found in Arrhenius plots of membrane phase transition behavior between chilling-sensitive and chilling-resistant plants.

Katz and Reinhold[51] recognized that if phase changes were the direct cause of chilling-induced leakage, then changes in membrane permeability should be rapid and leakage should be detectable within a few minutes of chilling. This has only been observed in a blue-green alga in which good evidence exists for chilling-induced membrane changes.[52] In higher plants, however, increased leakage is only measurable after hours to days of exposure to chilling temperatures.

Other currently proposed theories involve energy balance (i.e., levels of adenosine triphosphate) or altered levels of abscisic acid[5] or calcium.[4] The physiological cause of chilling injury awaits definitive research and resolution.

VIII. PREVENTION OR ALLEVIATION OF CHILLING INJURY SYMPTOMS

Many investigations have focused on treatments or procedures designed to increase tolerance to chilling injury or to reduce the severity of the resulting symptoms.[53-56] From a practical view, an important consideration is to recognize the potential danger from chilling and then adopt measures to avoid or minimize the injury. The first avenue of protection would be to prevent exposure to chilling temperatures.

To facilitate research, it would be helpful to make a distinction between reducing the development of symptoms after chilling and reducing the sensitivity of the tissue to chilling. However, since there is currently no accepted measure of chilling sensitivity other than a measure of the degree of subsequent development of various symptoms after chilling, it is impossible to clearly distinguish between the two processes.

Production of chilling-sensitive crops should be limited to areas and seasons that pose minimum risk of exposure to chilling temperatures. Rapidly growing or maturing cultivars can also be used to avoid those periods when chilling temperatures are likely to occur. The use of transplants rather than field seeding may be used to avoid early-season injury to chilling sensitive seedlings. The use of protective structures or materials can help avoid low temperature at the start and at the end of the growing season. Harvesting at a desirable stage of maturity can reduce the effect of chilling temperatures, especially for fruits that are picked mature-green and ripened after harvest. Fall harvests should be terminated before chilling injury has become serious. Chilling temperatures should be avoided at all steps in the transportation and distribution chain.

If chilling cannot be avoided, treatments should be developed to either increase the tolerance of the tissue before chilling or to reduce the development of injury symptoms after chilling. Conditioning, the holding of sensitive tissue slightly above the critical chilling temperature for a period of time, appears to increase the tolerance of the tissue to chilling[53] (see also Chapter 17). In contrast, subsequent development of some symptoms can be reduced after chilling by modifying the environment in which they develop. As one would expect, the symptom of increased water loss can be diminished by either reducing the vapor pressure deficit between the commodity and surrounding air or by increasing the resistance to gas diffusion from the commodity. In the first instance, the commodity can be held in high relative humidity created either through the use of packaging or humidification of the entire storage facility. The resistance to water loss can be increased by waxing or wrapping the commodity in semipermeable films.

The use of films is gaining increased popularity, and films could be used to produce altered oxygen and carbon dioxide levels that may be beneficial. Development of chilling injury symptoms can be reduced in some crops by reduced oxygen or elevated carbon dioxide concentrations in the atmosphere before or during exposure to chilling temperatures (see also Chapter 16). Limited studies with hypobaric, or low-pressure treatments have indicated they are of limited value. Immersion in calcium or potassium salt solutions before chilling has been reported to reduce chilling symptoms. It may be desirable to combine different treatments to alleviate chilling injury. For example, Ilker[56] reported favorable results when okra was given a dip treatment in 1% $CaCl_2$ or KCl and exposed to 0°C for 4 d with 10% CO_2 in the atmosphere.

Intermittent warming treatments, in which the chilling period is interrupted by periods at nonchilling temperatures, were formulated to allow dissipation of toxic materials that were thought to accumulate during chilling. For example, cucumber fruits develop symptoms of chilling injury after 3 d at 2.5°C. However, these fruits showed no symptoms of chilling injury when held for 13 d at 2.5°C if the fruits were warmed to 12.5° for 18 h every 3 d. This is a very successful treatment, even though the physiological basis for its effectiveness is unknown. Intermittent warming may become a practical treatment to prolong the storage of chilling-sensitive crops with improvements in materials handling technology that would facilitate movement of the commodity between temperatures (see also Chapter 18).

IX. CONCLUSIONS

The importance of chilling injury in reducing the quality of horticultural crops will increase with increased shipments of horticultural commodities from the tropical and sub-tropical countries and with the increased use of mechanical refrigeration in tropical countries. The increased use of mixed loads in which commodities with different temperature sensitivities are enclosed in the same space will also contribute to the growing importance of chilling injury. As agricultural production is displaced into marginal areas by encroaching urbanization and the need to expand production, limitations imposed by chilling temperatures at the beginning and end of the growing season will become increasingly important.

Further study is needed to understand the basis of chilling sensitivity, not only on the practical side to prevent and reduce chilling injury in commerce, but also in the scientific area of molecular biology, biochemistry, and physiology. Since the response of sensitive tissue is a function of both the length of exposure and the temperature, a considerable body of information will be needed to formulate proper temperature management schedules for the commercial handling of sensitive crops. This will especially be true for new cultivars, new crops, and sensitive crops grown and handled under new conditions, e.g., the introduction of new tropical crops into the temperate zone. Since chilling is just another stress which is altered by and will alter the response of tissue to other stresses, new methods of fumigation, controlled and modified atmospheres, and heat treatments to control pathogens in international trade may significantly change the response of sensitive crops to low-temperature treatments.

The possibility for a genetic solution to alleviate chilling injury either with traditional breeding or genetic engineering techniques can not be discounted, but it should not detract from the value of basic physiological research. The disorder appears to be quite complex and refractory to simple solutions. A more complete understanding of the physiological basis of chilling injury will be necessary to determine which genes are responsible for chilling resistance, and how these genes should be incorporated into commercially important horticultural crops to increase their chilling resistance. In contrast, unintentional genetic modification could result in increased susceptibility of new cultivars to chilling injury.

Elucidation of the cause of chilling injury and resolution of the problem of chilling sensitivity promise to continue to be exciting scientific endeavors.

ACKNOWLEDGMENT

The authors are indebted to Dr. Michael G. Guye for his helpful suggestions and critical review of this manuscript.

REFERENCES

1. **Lyons, J. M.,** Chilling injury in plants, *Annu. Rev. Plant Physiol.,* 24, 445, 1973.
2. **Uritani, I.,** Temperature stress in edible plant tissue after harvest, in *Postharvest Biology and Biotechnology,* Hultin, H. O. and Milner, M., Eds., Food and Nutrition Press, Westport, CT, 1978, 136.
3. **Graham, D. and Patterson, B. D.,** Responses of plants to low, non-freezing temperatures: proteins, metabolism and acclimation, *Annu. Rev. Plant Physiol.,* 33, 347, 1982.
4. **Minorsky, P. V.,** An heuristic hypothesis of chilling injury in plants: a role for calcium as the primary physiological transducer of injury, *Plant Cell Environ.,* 8, 75, 1985.
5. **Markhart, A. H.,** Chilling injury: a review of possible causes, *HortScience,* 21, 1329, 1986.
6. **Ryall, A. L. and Lipton, W. J.,** *Handling, Transportation and Storage of Fruits and Vegetables,* Vol. 1, 2nd ed., AVI Publishing, Westport, CT, 1979.
7. **Pantastico, E. B., Mattoo, A. K., Murata, T., and Ogata, K.,** Chilling injury, in *Postharvest Physiology, Handling and Utilization of Tropical and Subtropical Fruits and Vegetables,* Pantastico, E. B., Ed., AVI Publishing, Westport, CT, 1975, 339.

8. **Levitt, J.,** *Responses of Plants to Environmental Stress,* Vol. 1, Academic Press, New York, 1980.
9. **Lyons, J. M. and Breidenbach, R. W.,** Chilling injury, in *Postharvest Physiology of Vegetables,* Weichmann, J., Ed., Marcel Dekker, New York, 1987.
10. **Lyons, J. M., Graham, D., and Raison, J. K., Eds.,** *Low Temperature Stress in Crop Plants: The Role of the Membrane,* Academic Press, New York, 1979.
11. **Lyons, J. M., Raison, J. K., and Steponkus, P. L.,** The plant membrane in response to low temperature: an overview, in *Low Temperature Stress in Crop Plants: The Role of the Membrane,* Lyons, J. M., Graham, D., and Raison, J. K., Eds., Academic Press, New York, 1979, 1.
12. **Patterson, B. D., Graham, D., and Paull, R.,** The role of the membrane, in *Low Temperature Stress in Crop Plants: The Role of the Membrane,* Lyons, J. M., Graham, D., and Raison, J. K., Eds., Academic Press, New York, 1979, 25.
13. **Wade, N. L.,** Physiology of cool-storage disorders of fruit and vegetables, in *Low Temperature Stress in Crop Plants: The Role of the Membrane,* Lyons, J. M., Graham, D., and Raison, J. K., Eds., Academic Press, New York, 1979, 81.
14. **Watada, A. E.,** Introduction to the symposium, *HortScience,* 17, 160, 1982.
15. **Morris, L. L.,** Chilling injury of horticultural crops: an overview, *HortScience,* 17, 161, 1982.
16. **Couey, H. M.,** Chilling injury of crops of tropical and subtropical origin, *HortScience,* 17, 162, 1982.
17. **Bramlage, W. J.,** Chilling injury of crops of temperate origin, *HortScience,* 17, 165, 1982.
18. **Wolk, W. D. and Herner, R. C.,** Chilling injury of germinating seeds and seedlings, *HortScience,* 17, 169, 1982.
19. **Wang, C. Y.,** Physiological and biochemical responses of plants to chilling stress, *HortScience,* 17, 173, 1982.
20. **Molisch, H.,** Das Erfriern von Pflanzen bei Temperaturen uber dem Eispunkt, *Sitzungber. Akad. Wiss. Wien, Math. Naturwiss. Kl.,* 105, 1, 1896.
21. **Sachs, J.,** Handbuch der Experimental — Physiologie der Pflanzen, in *Handbuch der Physiologischen Botanik,* Vol. IV, W. Hofmeister, Leipzig, 1865.
22. **Sachs, J.,** Uber die obere Tempertatur-grenze der Vegetation, *Flora,* 22, 5, 1864.
23. **Lewis, D. A.,** Protoplasmic streaming in plants sensitive and insensitive to chilling temperatures, *Science,* 124, 75, 1956.
24. **Wheaton, T. A.,** Physiological Comparison of Plants Sensitive and Insensitive to Chilling Temperatures, Ph.D. thesis, University of California, Davis, 1963.
25. **Woods, C. M., Reid, M. S., and Patterson, B. D.,** Response to chilling stress in plant cells. I. Changes in cyclosis and cytoplasmic structure, *Protoplasma,* 121, 8, 1984.
26. **Molisch, H.,** *Untersuchungen uber das Erfieren der Pflanzen,* Fisher, Jena, 1897, 1.
27. **Belehradek, J.,** Temperature and living matter, *Protoplasms Monographien,* Borntraeger, Berlin, 1935, 117.
28. **Belehradek, J.,** Physiological aspects of heat and cold, *Annu. Rev. Physiol.,* 19, 59, 1957.
29. **Marcarelli, B.,** La produzione risicola 1917 nel vercellese in relazione ai principali faltori meteorologici. *G. Risic.,* 8, 7, 1918.
30. **Faris, J. A.,** Cold chlorosis in sugar cane, *Phytopathology,* 16, 885, 1926.
31. **Collins, J. L.,** A low temperature type of albinism in barley, *J. Hered.,* 18, 331, 1927.
32. **Sellschop, J. P. F. and Salmon, S. C.,** The influence of chilling, above the freezing point, on certain crop plants, *J. Agric. Res.,* 37, 315, 1928.
33. **Morris, L. L. and Platenius, H.,** Low temperature injury to certain vegetables, *Proc. Am. Soc. Hortic. Sci.,* 36, 609, 1938.
34. **Platenius, H.,** Effect of temperature on the respiration rate and respiratory quotient of some vegetables, *Plant Physiol.,* 17, 179, 1942.
35. **Biale, J. B.,** Postharvest physiology and biochemistry of fruits, *Annu. Rev. Plant Physiol.,* 1, 183, 1950.
36. **Pentzer, W. T. and Heinze, P. H.,** Postharvest physiology of fruits and vegetables, *Annu. Rev. Plant Physiol.,* 5, 205, 1954.
37. **Raison, J. K. and Lyons, J. M.,** Chilling injury: a plea for uniform terminology, *Plant Cell Environ.,* 9, 685, 1986.
38. **Eaks, I. L. and Morris, L. L.,** Deterioration of cucumbers at chilling and non-chilling temperatures, *Proc. Am. Soc. Hortic. Sci.,* 69, 388, 1957.
39. **Kader, A. A., Lyons, J. M., and Morris, L. L.,** Postharvest responses of vegetables to pre-harvest field temperature, *HortScience,* 9, 523, 1974.
40. **Nelson, R.,** Storage and transportational diseases of vegetables due to suboxidation, *Mich. Agric. Exp. Stn. Tech. Bull.,* 81, 1926.
41. **Lewis, D. A. and Morris, L. L.,** Effects of chilling storage on respiration and deterioration of several sweet potato varieties, *Proc. Am. Soc. Hortic. Sci.,* 68, 421, 1956.
42. **Wolfe, J.,** Chilling injury in plants — the role of membrane lipid fluidity, *Plant Cell Environ.,* 1, 241, 1978.

43. **Wright, M.,** The effect of chilling on ethylene production, membrane permeability and water loss of leaves of *Phaseolus vulgaris, Planta,* 120, 63, 1974.
44. **Lyons, J. M. and Raison, J. K.,** Oxidative activity of mitochondria isolated from plant tissue sensitive and resistant to chilling injury, *Plant Physiol.,* 45, 386, 1970.
45. **Raison, J. K., Lyons, J. M., and Thomson, W. W.,** The influence of membranes on the temperature-induced changes in the kinetics of some respiratory enzymes of mitochondria, *Arch. Biochem. Biophys.,* 142, 83, 1971.
46. **Singer, S. J.,** in *Structure and Function of Biological Membranes,* Rothfield, L. I., Ed., Academic Press, New York, 1971.
47. **Levitt, J.,** *Responses of Plants to Environmental Stress,* Academic Press, New York, 1972.
48. **Singer, S. J. and Nicolson, G. L.,** The fluid mosaic model of the structure of cell membranes, *Science,* 175, 720, 1972.
49. **Raison, J. K.,** The influence of temperature-induced phase changes on the kinetics of respiratory and other membrane-associated enzyme systems, *Bioenergetics,* 4, 285, 1973.
50. **Guye, M. G.,** Sterol composition in relation to chill-sensitivity in *Phaseolus* spp., *J. Exp. Bot.,* 39, 1091, 1988.
51. **Katz, S. and Reinhold, L.,** Changes in the electrical conductivity of Coleus tissue as a response to chilling temperatures, *Isr. J. Bot.,* 13, 105, 1965.
52. **Ono, T. A. and Murata, N.,** Chilling susceptibility of the blue-green alga *Anacystis nidulans.* II. Stimulation of the passive permeability of cytoplasmic membrane at chilling temperatures, *Plant Physiol.,* 67, 182, 1981.
53. **Wheaton, T. A. and Morris, L. L.,** Modification of chilling sensitivity by temperature conditioning, *Proc. Am. Soc. Hortic. Sci.,* 91, 529, 1967.
54. **Kader, A. A. and Morris, L. L.,** Amelioration of chilling injury symptoms on tomato fruits, *HortScience,* 10 (Abstr.), 324, 1975.
55. **Lyons, J. M. and Breidenbach, R. W.,** Strategies for altering chilling sensitivity as a limiting factor in crop production, in *Stress Physiology in Crop Plants,* Mussell, H. and Staples, R. C., Eds., John Wiley & Sons, New York, 1979, 179.
56. **Ilker, Y.,** Physiological Manifestation of Chilling Injury and its Alleviation in Okra Fruit (*Abelmoschus esculentus* (L) Moench), Ph.D. thesis, University of California, Davis, 1976.

Chapter 2

CHILLING INJURY OF CROPS OF TROPICAL AND SUBTROPICAL ORIGIN

Robert E. Paull

TABLE OF CONTENTS

I. INTRODUCTION

Temperature limits plant growth and development. The growth and development of plants that evolved in the tropics and, to a lesser extent, subtropics are severely limited by temperatures between 0°C and up to 10 to 12°C. The injury that occurs between 0°C and 10 to 12°C is referred to as chilling injury. This injury occurs under two situations. The first is when chilling-sensitive crops are planted in temperate areas.[1] An example of this was the chilling injury of young cotton plants exposed to unseasonably cool weather with reported losses of up to $60 million in 1980.[2] Other examples include limitations to maize growth,[3] French beans,[4] and tomato seed germination.[5]

The second and more studied aspect of chilling injury is the postharvest storage temperature response of commercial tropical and subtropical crops. Chilling injury limits the postharvest life of these commodities. Low temperatures, <10 to 12°C for tropical and <7°C for subtropical crops, cannot be used without caution to extend postharvest life. The extent of postharvest loss due to chilling injury is difficult to determine since chilling injury predisposes all commodities to pathogenic attack, especially saprophytes, and disrupts ripening of climacteric fruit.

A number of reviews have appeared in the last 10 years dealing with basic[6-9] and applied aspects[10] of chilling injury. This review will briefly consider postharvest aspects of chilling injury of tropical and subtropical fruits and vegetables, including symptom development, the time-temperature response of chilling injury, and the response to chilling temperatures.

II. SYMPTOM DEVELOPMENT

Symptoms of chilling injury generally develop after removal from the chilling temperature to nonchilling temperatures. The development of symptoms is very slow, and the symptoms are similar to those which occur for other stresses and injury. The chilling injury symptoms have been described by many authors, and a summary has been given by Morris.[11]

The symptoms most commonly seen are surface lesions. These lesions take the form of pitting (eggplant),[12] scald (papaya, citrus),[13,14] and large sunken areas (peppers).[15] Internal brown discoloration of pulp (pineapple, taro),[16,17] vascular strands (avocado),[18] and seeds (eggplant, tomato)[12,19] is also common. Water soaking of the tissue coupled with wilting and desiccation (ung choi)[20] is a common symptom in leaves.

The above visual symptoms frequently occur with dysfunction of a number of developmental processes. The most commonly reported is a failure of fruits to ripen (banana, mango)[21,22] and acceleration of senescence (ung choi).[20] These last symptoms often lead to a reduction of postharvest life (Figure 1) and increased susceptibility to decay.[23,24] The organisms involved in this decay are usually not found in healthy tissue.[11]

Leaves of many species, such as *Phaseolus vulgaris,* show rapid wilting within a few hours of chilling. Initially, this effect is rapidly reversible. Sunken pits, which turn into brown necrotic areas, later appear. The extreme is the rapid leaf symptoms in *Episcia reptans,* which is completely killed in as little as 4 h at 5°C.[2] Light[25,26] and water stress[27] are important in determining the speed of chilling injury symptom development. However, under postharvest storage conditions, commodities are not stored at chilling temperature in the light and attempts are made to minimize water stress. Therefore, much of the chilling research on leaf tissue is not directly relevant to postharvest conditions.

The postharvest storage life of tropical and subtropical crops does not gradually increase with decreasing storage temperature (Figure 2). The postharvest storage life of tropical and subtropical crops reaches a maximum at around 10 to 12°C and 4 to 7°C, respectively. Higher temperatures allow continued senescence and ripening with an expected Q_{10} near 2. Below the 10 to 12°C or 4 to 7° threshold, storage life decreases due to the development

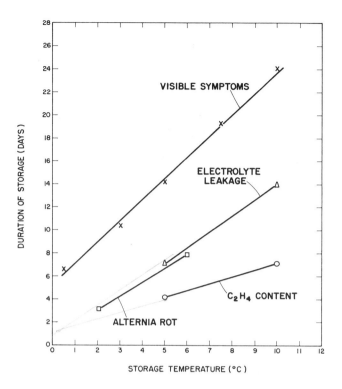

FIGURE 1. The time and temperature relationship of incipient chilling injury symptom development in papaya fruit. The data were taken from various sources: electrolyte leakage and fruit cavity ethylene accumulation,[30] Alternaria rot,[23] and visible symptom development.[13] Above the lines, injury symptoms become more severe.

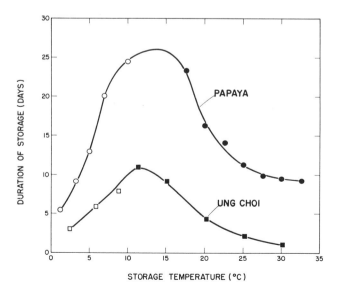

FIGURE 2. The time and temperature relationship of incipient chilling injury for ung choi (□), papaya fruit (○), and ung choi leaf yellowing (■) and papaya fruit ripening (●) at various temperatures. (Redrawn from Hirata, K., Chachin, K., and Inata, T., *J. Jpn. Soc. Hortic. Sci.,* 55, 516, 1987; Chen, N. M. and Paull, R. E., *J. Am. Soc. Hortic. Sci.,* 111, 639, 1986; and Paull, R. E., Unpublished data, 1987.)

of chilling injury (Figure 2), but ripening and senescence are severely restricted. This pattern explains why most storage recommendations for tropical commodities are just in the range where chilling injury occurs. For example, papaya can be stored for about 25 d at 10°C and still have up to a week of life at 20°C without developing chilling injury. Commercially, papaya is shipped at 7 to 10°C. These patterns are in direct contrast to the increasing storage life found for temperate commodities stored at decreasing temperatures to near 0°C.[28]

III. TIME-TEMPERATURE RESPONSE

The response of plant tissue to chilling injury has frequently been separated into primary and secondary responses. This separation is based on the perception of a series of events involving multiple events cascading-from the primary event after a certain level[7] of chilling injury. Two distinct possibilities are suggested as the primary event — lipid changes and protein changes. Changes in either lipids or proteins as the primary event in chilling injury are thought to occur instantaneously when the tissue is subjected to chilling temperatures. The lipid hypothesis has received the most attention after the original proposal was put forward in 1965.[7-9,29] These changes may or may not cause secondary responses or damage depending upon temperature, exposure time, and tissue sensitivity.

The sequence of events leading to visible symptoms after primary sensing of chilling temperatures clearly vary with tissue. Physiological symptoms of chilling injury in papaya include increased ethylene production,[30] electrolyte leakage,[30] and Alternaria rot,[23] all of which occur before the appearance of visible symptoms (Figure 1). At 10°C, increased ethylene production occurs after 7 d, electrolyte leakage after 14 d, and visible symptoms occur after 24 d.

Similar time sequences of physiological and visible symptom developments can be determined from the published literature for tomatoes, sweet potatoes, mung beans, and avocado.[9] The actual sequence of events leading to visible symptoms are unknown. The relationship between the physiological symptoms of chilling injury — ethylene production and electrolyte leakage (Figure 1) — and the visible symptoms of chilling injury in papaya is unknown. The sequence is probably not disruption of ethylene synthesis, causing excessive electrolyte leakage, followed by the appearance of visible symptoms. The time-temperature for incipient visible symptoms of chilling injury development is the point at which irreversible dysfunction occurs; prior to this stage injury is fully or partially reversible. Information on the time-temperature response to incipient chilling injury for tropical and subtropical fruits and vegetables is limited. This information is essential for the development of optimum postharvest handling and disinfestation procedures.

IV. STAGE OF TISSUE DEVELOPMENT

The stage of fruit ripeness and age of leaf tissue at the time of low-temperature storage significantly influences the plant sensitivity to visible chilling injury symptom development. The preclimacteric stage is generally more sensitive than the postclimacteric stage for avocado, papaya, honeydew melon, tomato, and mango (Table 1). The extreme sensitivity of bananas to chilling does not allow a difference to be determined. However, banana fruit in the less mature apical hands are less susceptible to chilling injury than the more mature, distally borne fruit.[31,32] Avocado increases in chilling injury sensitivity up to the climacteric, then sensitivity declines.[33]

A study of "Marsh" grapefruit chilling injury symptom development has been carried out in Florida.[34,35] Fruit stored at 5.5°C were susceptible at the beginning and end of the season, with the midseason fruit being more tolerant.[34,36] The midwinter resistance period to chilling injury in Florida is possibly associated with preharvest water stress and field temperature and not to fruit "maturity".[37]

TABLE 1

Time Required for Chilling Injury Symptoms to Develop for Various Fruits and Vegetables at Different Stages of Development Stored Between 0 and 7.5°C

Crop	Stage of ripeness at chilling	Criteria	Storage temperature (°C)	Days to symptoms	Ref.
Avocado	Preclimacteric	Visible symptoms	2	31	33
	Climacteric	Visible symptoms	2	19	33
	Postclimacteric	Visible symptoms	2	43	33
Banana	Mature green	Ripening	5	5	63
		Visible symptoms	1	<1	60, 219
	Ripe	Visible symptoms	2	1	24
Eggplant	Unripe (5)[a]	Internal browning	1	5	89
	Normal (15)	Pitting	1	8	89
		Internal browning	1	7	89
	Overripe (25)	Pitting	1	5	89
		Internal browning	1	12	89
Grapefruit[b]	Green	Visible symptoms	4.5	20	34
	Yellow	Visible symptoms	4.5	50	34
Honeydew melon	Mature	Visible symptoms	5	ca. 16	115
	Ripe	Visible symptoms	5	>19	115
Mango	Color breaking	Visible symptoms	1	< 7	220
	Ripe	Visible symptoms	2	7—14	220
Papaya	Color Break	Visible symptoms	2	9	13
	60% ripe	Visible symptoms	2	18	13
Pineapple	Base turning	Internal browning	7.5	3	16
	Full yellow	Internal browning	7.5	7	16
Tomato	Mature green	Visible symptoms	0	10	221
	Ripe	Visible symptoms	0	>30	222
Ung choi[c]	Young leaves	K$^+$ leakage	1	2	20
		Visible symptoms	1	2	20
	Mature leaves	K$^+$ leakage	1	>5	20

[a] Days from flower in parentheses.
[b] Early season fruit. Yellow fruit degreened 48 h with ethylene.
[c] *Ipomoea aquatica* Forsk.

V. COMMODITY RESPONSES

The symptom and time-temperature response for tropical and subtropical crops are described below. There is some difficulty in deciding with some crops whether they are tropical or subtropical. The citrus are included as subtropical.

A. TROPICAL CROPS

Annona spp. (*Annona cherimola x A. squamosa,* atemoya, custard apple; *A. squamosa* L., sweetsop, custard apple; *A. muricata* L., soursop) — The Annonas are very susceptible to chilling with injury occurring within 5 d at 4°C.[38] Atemoya appears to be more tolerant of chilling temperatures. The symptoms include skin darkening, failure to ripen with a jelly-like pulp, pulp discoloration, poor flavor and aroma, and an increase in rot. Unripe atemoya can be stored for 1 week at 8°C and up to 2 weeks at 12°C with some loss of quality.[39] The optimum storage conditions are 15 to 16°C with high humidity for about 2 weeks, lower temperatures lead to fruits with poorer flavor and texture.[39,40] Sweetsop ripening was disrupted by storage for 13 d at 15°C, having poor taste and a dark brown appearance.[38,41] Above 15°C the fruit was said to ripen normally. A similar disruption of flavor development has been noticed for soursop fruit stored for 1 d at 10°C.[24]

Avocado (*Persea americana* Mill.) — The common symptoms of chilling injury are dark brown or gray discoloration, especially around the vascular bundles in the mesocarp. This discoloration is first observed at the distal end.[18] Other symptoms include scalding, pitting of the skin, failure of the fruit to soften properly when removed from storage at chilling temperatures, atypical respiratory rate patterns, and reduced ethylene production rates.[42,43] Chilling injury symptoms are reduced by modified atmosphere storage, fruit calcium levels, and stage of ripening.[18,33,44]

The time temperature conditions for the development of chilling injury vary with the cultivar and its ecological origin. Three types are recognized: Mexican, Guatemalan, and West Indian. The Mexican is regarded as being more resistant to cold, with the Guatemalan being the intermediate and the West Indian being most sensitive.[45] The fruit show the same sequence for sensitivity to chilling injury.[46,47] However, many of the commercial varieties are probably crosses. Fuerte is believed to be a seedling of a Guatemalan × Mexican parentage, while Hass is a Guatemalan type.

At a storage temperature of 10°C, Hass and Fuerte continued to ripen and were ripe by the 3rd week. No softening or chilling injury occurred with 5 weeks at 5 or 0°C, but chilling injury occurred in all fruit stored for more than 1 week at 5°C or lower when removed to 20°C. Severity of injury increases with time in storage. Only slight differences were noted in the chilling response of the two varieties,[42] although Hass has been reported to be less sensitive to chilling than Fuerte.[48] West Indian varieties develop chilling injury with mesocarp darkening, but still soften when stored at temperatures less than 10° C for more than 4 d.[47] Much of the variation and conflicting data on the time-temperature response to chilling injury can be traced to preharvest factors.[33,49]

At the preclimacteric rise and climacteric peak, avocados are most sensitive to chilling.[33,50] This could explain the finding that ethylene treatment or ethylene removal prior to storage increases fruit chilling susceptibility.[50,51]

Bael (*Aegle marmelos* (L.) Corr.) — Brown spots on the skin increasing in number and size is the major symptom of chilling injury.[52] This injury is followed by accelerated susceptibility to decay. The first symptoms appear after 11 weeks continuous storage at 3°C. Recommended storage conditions are 9°C at 85 to 90% RH for up to 12 weeks.

Banana (*Musa* spp., AAA, Cavendish & Gros Michel Groups) — Bananas can be stored over a small temperature range (12.5 to 21.5°C). The preclimacteric period is an inverse linear function of temperature, with a temperature coefficient of 1 to 3 d/1°C. Temperature coefficient varies depending on the maturity of fruit at harvest.[53] Over a wider range (15 to 50°C), a log relationship was found, with a doubling of preclimacteric period for a temperature reduction of 6.5°C, agreeing with the above function.[54] Storage life is increased from 20 d at 15°C to about 60 d by storage in sealed polyethylene bags with an ethylene absorbent.[55,56] At 10°C, the fruit can be held for about 20 d.

At temperatures less than 12.5°C, chilling injury symptoms rapidly develop. After 9 d at 10.5°C, injury symptoms appeared,[14,57] although the fruit can be held at 10°C for up to 4 d with no apparent effect on ripening.[58] The chilling injury symptoms are dull skin color, blackening of the skin, failure to ripen, hardening of the central placenta, complete loss of flavor, clear latex, subepidermal brown streaking, appearance of water-soaked areas, and abnormal susceptibility to mechanical damage during handling.[59] Cultivars differ in their sensitivity to chilling injury.[60]

The preclimacteric period is affected by preharvest cultivation practices, by cultivar maturity at harvest, and by seasonal variation. Mechanical damage, sun burn, water stress, and disease, all reduce the postharvest preclimacteric period by their effect on ethylene levels.[31,61-63]

Plantain (*Musa* spp., AAB, ABB, cooking bananas) — The postharvest behavior of plantains appears to be generally similar to desert bananas, with similar chilling injury symptoms.[64] The fruit have a higher optimum ripening temperature.[65] Black spots developed on fruit stored for 12 d at 7.2°C along with a hardening of the fruit associated with a disruption of ripening.[64] Some slight possible chilling damage (black spots) appeared after 19 d at 12.8°C. However, these fruit had become yellow skinned with considerable starch breakdown due to ripening (20 to 10%), and the spots could be due to senescent spotting characteristic of desert bananas.[66] This conclusion is supported by the finding that green plantains could be kept for 55 d under refrigeration at 12.8°C if sealed in polyethylene bags containing an ethylene absorbent to prevent ripening.[64]

Beans (*Phaseolus vulgaris* L., snap or green beans; *P. lunateas,* L., lima beans) — Pitting, occasional rusty-brown diagonal streaks, and spots are the major symptoms of chilling injury, though it is difficult to detect when the fruit surface becomes leathery.[67] Cultivars of snap beans differ in their sensitivity to chilling at 5°C,[68] with shelf life for different cultivars tested varying from 5 to 14 d at 15°C following 2 weeks at 5°C. The pods of lima beans are more sensitive than the snap beans.[69] Chilling injury in snap beans was induced after 2 d at 0.5°C, 4 d at 2.5°C, 12 d at 5°C,[67] and, for lima beans, after 5 d at 5°C.[69] The recommended storage temperature is 4 to 7°C for snap beans stored for 7 to 10 d, and 5 to 6°C for unshelled lima beans and 3 to 4°C for shelled lima beans for 5 and 7 d of storage, respectively.[28]

Breadfruit (*Artocarpus altilis* (Parkins.) Fosb.) — Storage of breadfruit below 12°C leads to symptoms consistent with chilling injury. Symptoms, which include abnormal fruit ripening and fruit color change from green to a dull brown,[70] develop at temperatures less than 7°C after 2 to 3 d storage. At 12°C, storage life can be increased by storing the fruit in a sealed polyethylene bag for about 2 weeks.[70] Partially mature fruit at 13°C, not subjected to harvesting injury, can be stored for about 3 weeks in a sealed polyethylene bag, while fully mature fruit can be stored for 2 to 3 weeks.[71]

Carambola (*Averrhoa carambola* L., five corner, star fruit) — Symptoms of chilling injury have not been described.[72,73] Fruit which were green and at color break when placed into storage could be held for 5 weeks at 0 to 4.5°C, 4 weeks at 7.5 to 10°C, 3 weeks at 15.5°C, and 2 weeks at 21°C.[74,75] However, the authors did not indicate whether the green fruit held for 5 weeks at 0°C were able to develop normal yellow color and flavor after removal from chilling temperatures. Yellow to yellow-orange fruit can be stored in a polyethylene bag at 1°C for 3 weeks without apparent injury.[24] Waxing the fruit only marginally decreased water loss of green-yellow fruit for which a 10°C storage recommendation was made for a 3- to 4-week storage.[76]

Choyote (*Sechium edule* (Jacq.) Swartz, chokos) — After 2 to 3 weeks of storage at 2.5°C or lower, large dull brown areas due to chilling injury develop on the fruit surface. Fruit stored after 4 weeks at 5°C develop pits which rapidly turn brown. Internal flesh darkens within 4 weeks of storage at temperatures less than 10°C.[77,78]

Cucumber (*Cucumis sativus* L.) — The major symptom of chilling injury in this fruit, which is commercially harvested immature, is surface pitting (dark colored, watery blemishes). These symptoms are followed by water soaking associated with tissue collapse and shriveling. Symptoms of injury are not generally obvious under chilling conditions. Extensive decay occurs when chilling injury fruit are returned to ambient conditions.[79] Chilling injury develops if the fruit are held longer than about 2 to 4 d at 0 to 2°C and 8 to 16 d at 0 to 5°C. No visible symptoms occur if stored at 7 to 10°C for 20 to 25 d, but the fruit deteriorate more rapidly than if stored at 13°C.[14,80-82]

Durian (*Durio zibethinus* Murr.) — Sound durian fruit can be stored for up to 3 weeks at 15°C.[83] Storage of up to 35 d at 4 to 5°C and 85 to 90% RH has been reported.[84] Wardlaw[14] had earlier reported on shipments of durian from Java to Holland at 3°C which ripened and had full aroma.

Eggplant (*Solanum melongena* L.) — Surface pitting of the peel along with browning of seeds and vascular bundles occur in eggplant fruit after 4 to 5 d at 1°C. The symptoms become extremely severe when the fruit are transferred to 20°C.[12,85] The browning symptoms are closely related to chlorogenic acid levels and phenylalanine and tyrosine ammonia lyase activities.[86] Fruit harvested during very warm weather can be held about 1 week at 12°C and about 10 d at 10°C.[82] Eggplants exposed to wet conditions for the 3 d before harvest were more chilling sensitive.[87] Differences in recommendations are due in part to different cultivars used[88] and seasonal effects.[89] Injury can be reduced by temperature conditioning.[90-92]

Ginger (*Zingiber officinale* Roscoe) — Chilled fresh ginger rhizome softens and breaks down after about 2 to 3 weeks at 7°C.[93] Surface mold growth frequently occurs with this physiological breakdown. No sprouting occurs in rhizomes stored at temperatures less than 13°C. Hence, 13°C and 65% RH are the recommended storage conditions for up to 6 months, with about 10 to 20% weight loss.[93,94] More mature rhizomes have less weight loss under the same conditions.[93,95,96]

Guava (*Psidium guajava* L.) — Storage of guava at 0 to 10°C extends the life of the fruit to 2 weeks. Storage life is limited by storage rots, flesh breakdown, and darkening. Fruit storage at 0°C reduced rotting; but at 5°C, there was considerable pulp darkening after 3 and 4 weeks.[97] The optimum storage temperature for ripe fruit is 4 to 5°C.[97-99] Green mature fruit can be stored at 8 to 10°C and 85 to 95% RH for 2 to 5 weeks.[100] Weight loss of about 14% was a significant problem.

Jackfruit (*Artocarpus heterophylla* Lam.) — No description of chilling injury has been found. The storage recommendation is 11 to 13°C at 85 to 90% RH for up to 6 weeks.[101] The report states that initial quality and stage of maturity at harvest are important factors affecting storage life.

Lanzones (*Lansium domesticum* Correa, langsat) — Chilling injury occurs after 5 d at 0°C.[102] Symptoms include skin browning, fermented flavor, and softer fruit with a water-soaked appearance.[14] If disease control is obtained with fungicides, fruit can be stored for more than 2 weeks at 15°C in an atmosphere of 5% O_2 and no CO_2.[103] This confirmed an earlier study which showed up to 2 weeks storage was possible at 11 to 12.5°C. Green fruit were more sensitive to chilling injury than either turning or ripe fruit.[104]

Mango (*Mangifera indica* L.) — Chilling injury is manifested by grayish scald-like discoloration of the skin, skin pitting, uneven ripening, reductions in the level of carotenoids, aroma, and flavor during ripening, and susceptibility to fungal rotting.[14,22,105-107] Considerable variation exists between different cultivars in the chilling time-temperature for symptom development. The threshold temperature is 10°C, with maximum exposure varying for Southeast Asian cultivar from 15 to 28 d at 10°C.[108] "Kent" mangoes develop chilling injury in 10 to 16 d at 10°C.[109] There is no difference in chilling susceptibility between mature-green and color-breaking fruit, but ripe fruit are much less sensitive to chilling. Ripe "Haden" fruit can be held at 1.7°C for up to 4 weeks.[22,110] Preharvest factors can significantly effect the time-temperature response to chilling injury symptom development.[22] Decay control is essential in order to determine the threshold of chilling injury. The storage recommendations are 13° for 2 to 3 weeks.[14,28]

Mangosteen (*Garcinia mangostana* L.) — The first symptoms of chilling injury in mangosteen is the hardening and browning of the cortex. This occurs after about 6 weeks at 4 and 8°C. Flesh segments were still acceptable, though with longer storage, the segments became soft, browning occurred and there was a loss in original lustre.[111] After 6 weeks at 4 and 8°C, the segments developed a slight fermented smell and flavor.[112] Other workers have reported storage periods of from 4 to 7 weeks at 4 to 7°C.[14,111,113] The differences lie, in part, with the different standards used to judge storage life.

Melon (*Cucumis melo* L., honeydews, Crenshaws, Persian melons, cantaloupe) — All melons develop chilling injury when stored below 5°C. Chilling injury of honeydew melons is induced by 2.5 weeks of storage at 0 to 2.5°C.[114,115] The injury is characterized by a reddish-tan discoloration of the surface and subsurface, with subsurface discoloration preceding surface symptoms. Sunken lesions develop in honeydews. Pitting, surface decay, and failure to ripen are common symptoms of chilling injury in cantaloupe. Full-slip cantaloupe are more resistant to chilling injury and can be stored for 5 to 14 d at 0 to 2°C.[28] The recommendations for honeydew melons are 7 to 10°C for up to 3 weeks. As with cantaloupes, honeydew melons become less sensitive to chilling injury as the fruit ripen.[114]

Okra (*Hibiscus esculentus* L.) — Severe pitting occurs when okra is held for 3 d at 0°C and 6 d at 1°C. Later symptoms include: surface discoloration (blackening along the ridges), bronzing, and susceptibility to decay.[116] Fruit can be stored for up to 20 d at 12°C without pitting, but toughening and yellowing will occur. The maximum storage is about 10 d at 10°C. Injury can be alleviated by increasing CO_2 levels, intermittent warming, and calcium dip.[117-119]

Papaya (*Carica papaya* L.) — The symptoms of chilling injury include skin scald, hard areas in the pulp around the vascular bundles, water soaking of the tissue, failure to ripen,[13] electrolyte leakage, abnormal ethylene synthesis,[30] and increased susceptibility to fungal decay.[23,120,121] The time-temperature response to incipient chilling injury, mainly scald, is shown in Figure 2. As the fruit ripens, it becomes less sensitive to chilling injury.[13,14,122,123] Maximum storage time is 2 weeks at 7°C and 3 weeks at 10°C, after which chilling injury and decay become increasing problems (Figure 1).

Passion fruit (*Passiflora edulis* Sims. var. *edulis,* purple passion fruit; *P. edulis* Sims. var. *flavicarpa* Degener., golden passion fruit; *P. quadrangularis* L., granadilla) — The two symptoms described for purple passion fruit chilling injury are dark red discoloration of the skin, loss of flavor, and an enhanced susceptibility to mold attack.[14,124,125] This injury occurred after 4 to 5 weeks storage at temperatures less than 10°C. Below 6.5°C, the chilling injury changes were marked, while above 6.5°C loss of flavor and slightly greater mold were the major symptoms after 4 to 5 weeks storage.[126] Just-ripe fruit stored better than partially ripe or more ripened fruit.[100,125] The golden passion fruit and granadilla are possibly more sensitive to chilling injury than the purple passion fruit.[127] The color-turning granadilla fruit reportedly can be stored for 15 d at 7.2°C and still could be ripened at 21°C.[14]

Pepper (*Capsicum annuum* L.) — The characteristic symptoms of chilling injury are sunken, water-soaked sheet pitting,[79] seed discoloration near the calyx, and an increase in Alternaria rot.[128] Pitting developed within a few days at 0 to 2°C and 1 week at 7°C. Storage temperatures recommended are in the range of 7 to 13°C for a maximum of 2 to 3 weeks, if rapidly cooled after harvest.[28,129] Packaging in plastic films and waxing can increase storage life when stored at 8 to 10°C.[15,130]

Pineapple (*Ananas comosus* (L.) Merr.) — The most common symptom of chilling injury is the brown to black discoloration of the flesh. The disorder has been referred to as blackheart and endogenous brown spot, though internal browning (IB) is more commonly used. The early symptom is the development of a dark spot of water-soaked tissue at the base of the fruitlet near the core. The spots coalesce and the tissue finally becomes a dark mass. At a later stage, the skin darkens and the fruit becomes more susceptible to molds.[16] The history of this malady has been reviewed.[131] Symptoms of IB begin to appear in fresh fruit after about 4 d of storage at ambient temperatures following refrigerated storage 1 week at 7°C.[14,132] It occasionally appears in unrefrigerated fruit[132] and may be found at harvest when chilling temperatures occur in the field, as in Australia, Taiwan, and South Africa.[133-135] Aside from a report of injury occurring below 21°C,[136] the critical temperature is thought to be 10 to 12°C.[16] Waxing of fruit[137] and exposure to high temperatures (>32°C) for a short period (24 h) before or after chilling[132] can control storage-related chilling injury symptom development. The usual storage life is about 2 to 4 weeks at 7 to 10°C at the $^1/_2$-ripe stage. After this treatment, the fruit still have 1 week of shelf life at ambient temperatures.

Rambutan (*Nephelium lappaceum* L.) — Temperatures less than 10°C lead to chilling injury synptom to the exocarp and spinterns. The exocarp and spinterns darken and become almost black after 8 d at 12°C, with little apparent affect on eating quality.[138,139] Fruit stored in sealed polyethylene bags remain in good condition for 12 d at 10°C. Chilling injury symptoms began to develop in fruit stored in a sealed polyethylene bag after 6 d at 7°C.[139]

Sweet potato (*Ipomoea batatas* (L.) Lam.) — Storage of sweet potatoes below 12°C leads to internal discoloration from brown to black, internal breakdown, off flavors, increased sucrose levels, hard core when cooked, and increased disease susceptibility. Exposure of 2 weeks to 7°C causes surface pitting and secondary fungal decay. Hard core increases as roots are stored from 1 to 4 d at 1°C, then placed at 21°C for 2 d. No hard core was detected if immediately cooked after chilling. If held for 6 d at 21°C, less hard core appeared.[140-146] Cultivars differ in sensitivity to hard core formation due to a disruption of pectic methylation.[147] Storage recommendations are 13 to 16°C for up to 7 months; at higher temperatures sprouting becomes a problem.[28,82,148]

Sapodilla (*Achras sapota* L., chiku) — Failure to ripen normally is a common symptom of chilling injury.[14,149] Fruit can be held at 4°C for 10 d without serious loss of flavor, and for up to 22 d at 15°C.[14] However, at less than 15°C there was some loss of flavor, though the fruit ripened normally.

Tomato (*Lycopersicon esculentum* Mill) — Chill-injured tomatoes exhibit slow and abnormal ripening, blotchy color development, disease susceptibility, increased rates of respiration and ethylene production, and ion leakage.[11,19,150-152] Green-mature fruit should not be stored for more than 2 weeks at 13°C, a higher temperature of 14 to 16°C is preferable. Light red fruit can be held up to 1 week at 10°C without chilling injury. Firm, ripe fruit are less susceptible to injury and can be held for up to 1 week at 7 to 10°C. Longer storage times for ripe fruit at less than 10°C lead to loss of quality as measured by color loss, firmness, and shelf life.[28] Preharvest temperatures can influence susceptibility to chilling injury.[153]

Taro (*Colocasia esculenta* Schott.) — Internal browning due to chilling injury occurs after storage for 10 d at 4°C.[17] Death of buds and decay are also common symptoms. Corms can be stored at 7°C for up to 3.5 months.[154]

Ung choi (*Ipomoea aquatica* Forsk., Kangkong, water convolvulus, water spinach, and kong xin cai) — The initial symptom of chilling injury is the browning of young, immature leaves and stems. Later, the older leaves show darkening. The symptoms develop within 2 to 4 d of storage (Figure 2) at temperatures less than 9°C.[20,155]

B. SUBTROPICAL CROPS

Chinese jujube (*Zizpyhus jujuba* Mill.) — This subtropical deciduous tree fruit is very cold tolerant, with a very low chilling requirement.[156] Though it originated in northern China,[157] the fruit shows symptoms of chilling injury if stored at 0°C.[158] Whitish-green mature fruit and partially ripe fruit with 20 to 50% brown held at 0°C for 26 d had delayed browning and sheet pitting. The sheet pitting became more pronounced when the fruits were transferred to 20°C. An earlier report suggested that fruit could be kept for 45 d at 0°C without damage, except for skin cracking and wilting.[14] These fruit could have been ripe and, therefore, less subject to chilling injury. The related tropical species, Indian jujube (*Z. mauritiana* Lam.), would be expected to be more susceptible to chilling injury. Fruit can be stored for 30 to 40 d at 3°C in polyethylene bags.[159]

Citrus (*Citrus sinensis* (L.) Osbeck, sweet orange; *C. reticulata* Blanco, Mandarin; *C. paradisi* Macf., grapefruit; *C. grandis* (L.) Osbeck, shaddock, pummelo; *C. limon* (L.) Burm. f., lemon; *C. aurantifolia* (Christm.) Swing., lime; *C. sudachi* Hort., Sudachi) — Rind pitting — also known as pox, chill spotting, brown spot, storage spot, and rind spotting — is the most common symptom of chilling injury in citrus. The disorder consists of clearly demarked sunken spots in the peel. The spots are initially not colored, becoming pink, then finally brown. Frequently the tissue under the pits soften and become infected. The pits are small, but later coalesce, especially in limes.[14,160,161] Lemons also show browning of the albedo or carpellary membranes between segments.[162,163] After very low temperatures (0 to 1°C), superficial reddish- to tan-colored scalding may occur instead of pitting on the most susceptible type of citrus fruit; e.g., brown staining in grapefruit. In grapefruit and oranges, the oil glands may darken.[14,164] These pitted areas become larger with extended storage.

The above pitting is a cosmetic problem not affecting the internal quality of the fruit. Peel pitting apparently makes the fruit more sensitive to mold rot development than brown staining, though not as clearly for grapefruit. A water breakdown has been described for oranges stored at lower temperatures in which the flesh is water logged in appearance and has an off odor.[14,36]

Various citrus fruit show no chilling injury at 12°C. The same cultivars stored at 6°C showed that grapefruit is the most chilling sensitive with pitting and rots, followed by Shamouti orange, then lemon and Valencia oranges.[165] However, the authors clearly state that there were year-to-year variations and significant effects from different harvest areas and harvest dates.

Limes — Injury develops in limes held below 7.5°C. Leathery brown surface pitting initially develops after removal from storage. Fruits stored for 2 to 3 weeks at 4.4°C developed mild pitting within 1 week upon removal to 20°C.[166,167] Fruit held at 10°C did not develop pitting during storage or after removal to 20°C, but began to turn yellow after 2 to 3 weeks at 10°C. Preconditioning for 1 week at 14°C, raising the RH, and storage in a partial vacuum all decrease chilling injury.[168,169]

Year-to-year variation is common, and also day-to-day variation, in optimum storage conditions.[35,170] Waxing increases susceptibility to chilling.[35] Preconditioning and intermittent warming can significantly reduce chilling injury symptom development.[166,171]

Lemons — The chilling injury symptoms develop in California light green lemons after 8 weeks of storage at 0 or 5°C, followed by 1 week at 20°C, and after 12 weeks continously at 0 and 5°C. No symptoms develop after 12 weeks at 12.8°C and 1 week at 20°C.[172] This contrasts with green lemons from Florida which following degreening, showed injury after 14 to 21 d at temperatures below 10°C.[173] The difference probably reflects preharvest differences, degreening treatment effects, and maturity of the fruit at harvest time. Ethylene pretreatments and the presence of ethylene during storage significantly influence the severity of chilling injury and decay,[174,175] while film wrapping, and low O_2 levels can significantly reduce chilling injury.[174-176]

Oranges — Arizona early Valencia fruit (March) develop injury symptoms after 20 weeks at 9 and 3.3°C, while fruit harvested in June were more sensitive, developing chilling injury symptoms after 4 weeks at 9°C with fewer symptoms when stored at 3.3°C for 12 weeks.[177] Florida- and Texas-grown Valencia oranges can be stored with little pitting or decay for 8 to 12 weeks at 1°C.[35] Israeli Valencia showed more pitting at 6°C and less at 2 and 10°C.[165] A similar result was reported for navel oranges, but this finding was regarded as the exception, with more normal occurrence of increasing rind injury at decreasing temperatures.[35] The effect of locality and maturity on storage life limited by pitting and decay was reported much earlier.[178,179]

Grapefruit — In the New York market, chilling injury appears in 0.1 to 0.2% of mainly Florida grapefruit.[180] Wholesale culling was considerably less. Grapefruit has received the most study on chilling injury. Three different time-temperature responses have been described: (1) a very susceptible sample in which pitting is most pronounced at 10°C and less at 4°C and above 14°C after 14 d storage; (2) less susceptible, in which most pitting occurred when stored for 4.5°C after 23 d and less at 1 and 12°C; and (3) least susceptible fruit in which decreasing temperature leads to increasing pitting after 67 d of storage.[179] These fruit were from different areas of South Africa, and similar results have been summarized.[14,181]

The seasonal resistance of grapefruit to chilling injury during storage is related to winter field temperature. The warmer the midwinter period, the more resistant the grapefruit are to chilling injury.[37]

Preharvest weather significantly affects the development of chilling injury, especially in Marsh grapefruit. This relationship, however, is not clear-cut, as fruit maturity must be imposed on this effect. There is a pattern of grapefruit harvested in July to September and April/May to take about 20 d at 4.4°C to develop chilling injury. Grapefruit harvested between November and February could require from 20 to about 60 d at 4.4°C to develop chilling injury. Benomyl sprays, thiabendazole, gibberellin, 2,4-D, and high levels of CO_2 (<20%, except late season grapefruit), and removal of C_2H_4 from the storage rooms can reduce chilling injury, especially when applied postharvest.[167,182-184] Grapefruit from the exterior canopy were more sensitive to chilling injury than fruit from the interior.[185] Early season fruit treated with ethylene has reduced chilling susceptibility, while later harvest fruit receiving the same treatment was more susceptible.[34] Waxing decreases susceptibility to chilling injury in grapefruit.[35] Preconditioning and intermittent warming can significantly reduce chilling injury symptom development in grapefruit.[171,186,187]

Prestorage conditioning at 16°C was better than 21 and 27°C in minimizing chilling injury to grapefruit subsequently stored at 1°C. Fruit preconditioned for 7 d at 16°C did not develop chilling injury during 21 d at 1°C, while unconditioned fruit stored 28 d at 1°C sustained 23% chilling injury.[187]

Pummelo — No description of chilling injury was located. This tree is frost sensitive.[188,189] There are a number of reports of long-term storage (1 to 4 months) at tropical ambient temperatures.[14,189] Cool storage does not appear to be necessary.

Sudachi — Pitting of green fruit occurs after 4 months at 1°C. Conditioning for 2 to 4 d at 10°C decreased pitting. Optimum curing period varied upon time of harvest in a given season.[190]

Litchi (*Litchi chinensis* Sonn) — Litchi have been reported to be chill injured by storage at 4°C.[191] However, the injury reported was skin browning, which is related to desiccation of the skin.[192,193] If water loss is controlled by packing in a plastic film, water loss-related skin browning is controlled.[192,194] Uniform browning of litchi skin under these conditions does not occur until 3 to 4 weeks at 1°C.[24,195,196] The texture and flavor of the flesh were apparently not affected.[197]

Longan (*Euphoria longana* (Lour.) Steud) — There is little research on the storage of this fruit. Initial studies show that longans sealed in polyethylene bags can be stored for up to 5 weeks at 1°C. The pericarp becomes dark brown and dried during storage, with little change in composition and flavor.[24,139]

Loquat (*Eriobotrya japonica* (Thumb.) Lindl) — The nonclimacteric fruit of this subtropical fruit tree are killed by cold and abscise.[198] Lemon-yellow fruit stored for 4 weeks at 0°C, then for 6 d at 23°C showed some injury, with the fruit being less juicy and in poorer condition than similar fruit stored at 2 and 4°C for the same period.[199] Fully ripe fruit could be held at 3°C for 4 weeks without any deterioration. Fruits stored at higher temperatures (7 to 10°C) had about 50% spoilage after 11 d.[200] The above results would explain the 0°C recommended storage temperature when only 0 and 6°C were compared.[201] The last author did find cultivar difference, with one of the three cultivars being unsuitable for storage.

Olives (*Olea europaea* L.) — The symptoms of chilling injury are a slight tannish to brown discoloration in the flesh of the fruit adjacent to or near the pit. This discoloration becomes more intense with time and progresses through the flesh to the skin. Cultivars differ in sensitivity with the most sensitive developing symptoms following 1 week at 0°C and 2 weeks at 2.2°C. The most resistant cultivar studied required 6 to 7 weeks at 2.2°C to produce symptoms.[202-204] Chilling injury can occur either on the tree due to low field temperatures or postharvest.[205]

Oriental persimmon (*Diospyros kaki* L., kaki) — Internal watery breakdown in the flesh occurred in fruit stored for 2 weeks at 2°C, irrespective of postharvest deastringency methods. Softening was very slow at 2 to 5°C following ethanol or carbon dioxide deastringency at 30°C. Watery breakdown is faster at storage temperatures less than 10°C, than at 15°C.[206,207] Gibberellic acid preharvest sprays extends postharvest storage life.[206] Nonastringent oriental persimmons have been reported to have been stored at 0°C in polyethylene bags for several months without injury.[208]

Pomegranate (*Punica granatum* L.) — The external symptoms of chilling injury are surface pitting, skin discoloration, scald, dead skin tissue, and accelerated fungal growth. Internal injury include dead tissue, brown discoloration of the white segments separating the arils, and a pale color of the arils. Fruit kept for 8 weeks at 5°C plus 3 d at 20°C developed only slight brown discoloration of the placental tissue separating the arils. No injury was noted after 8 weeks at 10°C, with decay limiting storage life after 2 weeks. Fruit stored for 5 weeks at 0°C, then 3 d at 20°C, developed both external and internal chilling injury symptoms.[209] These results conflict with the unsupported report that fruit could be stored for up to 7 months at 0 or 4.5°C.[210]

Potato (*Solanum tuberosum* L.) — Potatoes stored at 1°C for 2 to 4 d had increased levels of free sugars.[211] This sugar increase, occurring at temperatures below 3.5°C, is a serious problem for the potato chip and French fry industries, leading to a storage recommendation of 10 to 13°C for a short period.[28] Intermittent warming reduces sugar buildup at low temperatures.[212] Later symptoms of chilling injury include mahogany browning; black heart and hollow heart appear after about 16 weeks. Bluish discoloration and sinking of intact skin areas appear after 15 to 19 weeks at 1°C.[213] Varieties differ in their sensitivity to these latter symptoms of chilling injury.[214] The optimum storage temperature after curing is in the range 3.5 to 6.0°C for up to 7 months,[82] with sprout inhibitors, up to 11 months.[215]

Tamarillo (*Cyphomandra betacea* (Cav.) Sendt., tree tomato) — Fruit stored below 3.5°C develop scald which could be due to chilling injury. Storage rots are a problem above 4.5°C.[216] Fruit can be stored for at least 12 weeks at 3.5°C and still have at least 7 d of shelf life if rots are controlled.

Watermelon (*Citrullus lanatus* (Thumb.) Matsum and Nakai) — The external symptom of chilling injury is primarily brown staining of the rind, followed by rind pitting. Conditioned fruit can be stored for 12 d at 0°C without chilling injury.[217] This work led to a recommendation of prestorage condition for 4 d at 26°C, followed by 7°C for up to 8 d without injury. The conditioning probably also avoided the other reported adverse effect of low-temperature storage: loss of red flesh color intensity at or below 10°C.[218] The conflicting requirements of less than 10°C, with loss of red flesh color and rind brown staining below 7°C after 12 d, has led to various recommendations. Temperatures of 12.8 to 15.6°C were recommended if the storage period was less than 2 weeks.[82] Storage periods longer than 2 weeks should be at 7 to 10°C.[28] The difficulty with making a recommendation is the fact that quality declines with storage, but the fruit are still edible.[82]

REFERENCES

1. **Wolk, W. D. and Herner, R. C.,** Chilling injury of germinating seeds and seedlings, *HortScience*, 17, 169, 1982.
2. **Wilson, J. M.,** The economic importance of chilling injury, *Outlook Agric.*, 14, 197, 1984.

3. **Christiansen, M. N. and St. John, J. B.,** The nature of chilling injury and its resistance in plants, in *Analysis and Improvement of Plant Cold Hardiness,* Olien, C. R. and Smith, M. N., Eds., CRC Press, Boca Raton, FL, 1981, 1.

4. **Dickson, M. H.,** Breeding beans, *Phaseolus vulgaris* L., for improved germination under unfavorable low temperature conditions, *Crop Sci.,* 11, 848, 1971.

5. **Patterson, B. D., Paull, R. E., and Smillie, R. M.,** Chilling resistance in *Lycoperiscon hirsutum* Humb. & Bonpl., a wild tomato with wide altitudinal distribution, *Aust. J. Plant Physiol.,* 5, 609, 1978.

6. **Graham, D. and Patterson, B. D.,** Responses of plants to low nonfreezing temperatures: proteins, metabolism and acclimation, *Annu. Rev. Plant Physiol.,* 33, 347, 1982.

7. **Lyons, J. M.,** Chilling injury in plants, *Annu. Rev. Plant Physiol.,* 24, 445, 1973.

8. **Lyons, J. M., Graham, D., and Raison, J. K., Eds.,** *Low Temperature Stress in Crop Plants: The Role of the Membrane,* Academic Press, New York, 1979.

9. **Wang, C. Y.,** Physiological and biochemical responses of plants to chilling stress, *HortScience,* 17, 173, 1982.

10. **Couey, H. M.,** Chilling injury of crops of tropical and subtropical origin, *HortScience,* 17, 162, 1982.

11. **Morris, L. L.,** Chilling injury of horticultural crops: an overview, *HortScience,* 17, 161, 1982.

12. **McColloch, L. P.,** Chilling injury of eggplant fruits, *U.S. Dep. Agric. Mark. Res. Rep.,* 749, 5, 1966.

13. **Chen, N. M. and Paull, R. E.,** Development and prevention of chilling injury in papaya fruit (*Carica papaya* L.), *J. Am. Soc. Hortic. Sci.,* 111, 639, 1986.

14. **Wardlaw, W. C.,** Tropical fruits and vegetables: an account of their storage and transport, Memoirs No. 7, Imperial College of Tropical Agriculture, Trinidad, West Indies, 1937; reprinted from *Trop. Agric. (Trinidad),* 14, 3, 1937.

15. **Miller, W. R. and Risse, L. A.,** Film wrapping to alleviate chilling injury of bell peppers during cold storage, *HortScience,* 21, 467, 1986.

16. **Paull, R. E. and Rohrbach, K. G.,** Symptom development of chilling injury in pineapple fruit, *J. Am. Soc. Hortic. Sci.,* 110, 100, 1985.

17. **Rhee, J. K. and Iwata, M.,** Histological observations on the chilling injury of taro tubers during cold storage, *J. Jpn. Soc. Hortic. Sci.,* 51, 362, 1982.

18. **Chaplin, G. R. and Scott, K. J.,** Association of calcium in chilling injury susceptibility of stored avocado, *HortScience,* 15, 514, 1980.

19. **Autio, W. R. and Bramlage, W. J.,** Chilling sensitivity of tomato fruit in relation to ripening and senescence, *J. Am. Soc. Hortic. Sci.,* 111, 201, 1986.

20. **Hirata, K., Chachin, K., and Iwata, T.,** Changes of K^+ leakage, free amino acid contents and phenyl propanoid metabolism in water convolvulus (*Ipomoea aquatica* Forsk) with reference to chilling injury, *J. Jpn. Soc. Hortic. Sci.,* 55, 516, 1987.

21. **Wardlaw, C. W. and McGuire, L. P.,** Banana storage, *Trop. Agric. (Trinidad),* 8, 139, 1931.

22. **Hatton, T. T., Reeder, W. F., and Campbell, C. W.,** Ripening and storage of Florida mangos, *U.S. Dep. Agric. Mark. Res. Rep.,* No. 725, 1965.

23. **Sommer, N. F. and Mitchell, F. G.,** Relation of chilling temperatures to postharvest alternaria rot of papaya fruit, *Proc. Trop. Reg. Am. Soc. Hortic. Sci.,* 22, 40, 1978.

24. **Paull, R. E.,** Unpublished data, 1987.

25. **Lasley, S. E., Garber, M. P., and Hodges, C. F.,** After effects of light and chilling temperatures on photosynthesis in excised cucumber cotyledons, *J. Am. Soc. Hortic. Sci.,* 104, 477, 1979.

26. **Van Hasselt, P. R. and van Berlo, H. A. C.,** Photooxidation damage to the photosynthetic apparatus during chilling, *Physiol. Plant.,* 50, 52, 1980.

27. **Wilson, J. M.,** The mechanism of chill- and drought-hardening of *Phaseolus vulgaris* leaves, *New Phytol.,* 76, 257, 1976.

28. **Hardenburg, R. E., Watada, A. E., and Wang, C. Y.,** The commercial storage of fruits, vegetables, and florist and nursery stocks, *U. S. Dep. Agric. Agric. Handb.,* No. 66, 1986.

29. **Bishop, D. G.,** Chilling sensitivity in higher plants: the role of phosphatidylglycerol, *Plant Cell Environ.,* 9, 613, 1986.

30. **Chan, H. T., Sanxter, S., and Couey, H. M.,** Electrolyte leakage and ethylene production induced by chilling injury of papayas, *HortScience,* 20, 1070, 1985.

31. **Marriott, J.,** Bananas physiology and biochemistry of storage and ripening for optimum quality, *CRC Crit. Rev. Food Sci. Nutr.,* 13, 41, 1980.

32. **Marriott, J. C., New, S., Dixon, E. A., and Martin, K. I.,** Factors affecting the preclimacteric period of banana fruit bunches, *Ann. Appl. Biol.,* 93, 91, 1979.

33. **Kosiyachinda, S. and Young, R. E.,** Chilling sensitivity of avocado at different stages of the respiratory climacteric, *J. Am. Soc. Hortic. Sci.,* 101, 665, 1976.

34. **Grierson, W.,** Chilling injury in tropical and subtropical fruits. Effect of harvest date, degreening, delayed storage and peel color on chilling injury of grapefruit, *Proc. Trop. Reg. Am. Soc. Hortic. Sci.,* 18, 66, 1974.

35. **Grierson, W. and Hatton, T. T.,** Factors involved in storage of citrus fruits a new evaluation, *Proc. Int. Soc. Citricult.,* 1, 227, 1977.

36. **Harvey, E. M. and Rygg, G. L.,** Physiological changes in the rind of California oranges during growth and storage, *J. Agric. Res.,* 52, 723, 746, 1936.

37. **Kawada, K., Grierson, W., and Soule, J.,** Seasonal resistance to chilling injury of 'Marsh' grapefruit as related to winter field temperature, *Proc. Fla. State Hortic. Soc.,* 91, 128, 1978.

38. **Broughton, W. J. and Guat, T.,** Storage conditions and ripening of the custard apple *Annona squamosa* L., *Sci. Hortic.,* 10, 73, 1979.

39. **Brown, B. I. and Scott, K. J.,** Cool storage conditions for custard apple fruit (*Annona atemoya* Hort.), *Singapore J. Primary Ind.,* 13, 23, 1985.

40. **Wills, R. B. H., Poi, A., Greenfield, H., and Rigney, C. J.,** Postharvest changes in fruit composition of *Annona atemoya* during ripening and effects of storage temperature on ripening, *HortScience,* 19, 96, 1984.

41. **Smitananda, P.,** A study of the storage temperature requirement of the fruit of Atis, *Annona squamosa* Linn., *Philipp. Agric.,* 26, 425, 1937.

42. **Eaks, I. L.,** Ripening, chilling injury, and respiratory response of 'Hass' and 'Fuerte' avocado fruit at 20°C following chilling, *J. Am. Soc. Hortic. Sci.,* 101, 538, 1976.

43. **Eaks, I. L.,** Effects of chilling on respiration and ethylene production of 'Hass' Avocado fruit at 20°C, *HortScience,* 18, 235, 1983.

44. **Spalding, D. H. and Reeder, W. F.,** Low-oxygen high carbon dioxide controlled atmosphere storage for control of anthracnose and chilling injury of avocado, *Phytopathology,* 65, 458, 1975.

45. **Popenoe, W.,** *Manual of Tropical and Subtropical Fruits,* Hafner Press, New York, 1920.

46. **Gaffney, J. J. and Baird, C. D.,** Susceptibility of West Indian avocados to chilling injury as related to rapid cooling with low temperature air or water, *Proc. Fla. State Hortic. Soc.,* 88, 490, 1976.

47. **Hatton, T. T., Reeder, W. F., and Campbell, C. W.,** Ripening and storage of Florida Avocados, *U.S. Dep. Agric. Agric. Res. Serv. Mark. Res. Rep.,* No. 697, 1965.

48. **Ahmed, E. M. and Barmore, C. R.,** Avocado, in *Tropical and Subtropical Fruits,* Nagy, S. and Shaw, P. E., Eds., AVI Publishing, Westport, CT, 1980, 121.

49. **Zauberman, G., Schiffmann-Nadel, M., and Yanko, U.,** Susceptibility to chilling injury of three avocado cultivars at various stages of ripening, *HortScience,* 8, 511, 1973.

50. **Lee, S. K. and Young, R. E.,** Temperature sensitivity of avocado fruit in relation to C_2H_4 treatment, *J. Am. Soc. Hortic. Sci.,* 109, 689, 1984.

51. **Hatton, T. T. and Reeder, W. F.,** Quality of 'Lula' avocados stored in controlled atmospheres with or without ethylene, *J. Am. Soc. Hortic. Sci.,* 97, 339, 1972.

52. **Roy, S. K. and Singh, R. N.,** Preliminary studies on the storage of Bael fruit, *Prog. Hortic.,* 11(3), 21, 1979.

53. **Peacock, B. C. and Blake, J. R.,** Some effects of non-damaging temperature on the life and respiratory behavior of bananas, *Queensl. J. Agric. Anim. Sci.,* 27, 147, 1970.

54. **Blake, J. R. and Peacock, B. C.,** Effect of temperature on the preclimacteric life of bananas, *Queensl. J. Agric. Anim. Sci.,* 28, 243, 1971.

55. **Scott, K. M., McGlasson, W. B., and Roberts, E. A.,** Potassium permanganate as an ethylene absorbent in polyethylene bags to delay ripening of bananas during storage, *Aust. J. Exp. Agric. Anim. Husb.,* 10, 237, 1970.

56. **Scott, K. J. and Gandanegara, S.,** Effect of temperature on storage life of bananas held in polyethylene bags with ethylene absorbent, *Trop. Agric. (Trinidad),* 51, 23, 1974.

57. **Broughton, W. J. and Wu, K. F.,** Storage conditions and ripening of two cultivars of banana, *Sci. Hortic.,* 10, 83, 1979.

58. **Young, W. J., Bagster, L. S., Hicks, E. W., and Huelin, F. E.,** The ripening and transport of bananas in Australia, *Counc. Sci. Ind. Res. Aust. Bull.,* 64, 1932.

59. **Grierson, E. B., Grierson, W., and Soule, J.,** Chilling injury in tropical fruits. I. Bananas (*Musa paradisiaca* var. *sapientum* cv. lacatan), *Proc. Trop. Reg. Am. Soc. Hortic. Sci.,* 11, 82, 1967.

60. **Aziz, A. B. A., El-Nabawy, S. M., Abdel-Wahab, F. K., and Abdel-Kader, A. S.,** Chilling injury of banana fruit as affected by variety and chilling period, *Egypt. J. Hortic.,* 3, 37, 1976.

61. **Burg, S. P. and Burg, E. A.,** Relationship between ethylene production and ripening in bananas, *Bot. Gaz.,* 126, 200, 1965.

62. **Littmann, M. D.,** Effect of water stress on ethylene production by preclimacteric banana fruit, *Queensl. J. Agric. Anim. Sci.,* 29, 131, 1972.

63. **Pantastico, E. B., Grierson, W., and Soule, J.,** Chilling injury of tropical fruits. I. Bananas (*Musa paradisiaca* var. *Sapientum* cv. Lacatan), *Proc. Trop. Reg. Am. Soc. Hortic. Sci.,* 11, 82, 1968.

64. **Hernandez, I.,** Storage of green plantains, *J. Agric. Univ. P. R.,* 57, 100, 106, 1973.

65. **Sanchez Nieva, F., Hernandez, I., and Buesco de Vinas, C.,** Studies on the ripening of plantains under controlled conditions, *J. Agric. Univ. P. R.,* 54, 517, 1970.

66. **Liu, F. W.,** Ethylene inhibition of senescent spots on ripe bananas, *J. Am. Soc. Hortic. Sci.,* 10, 684, 1976.
67. **Watada, A. E. and Morris, L. L.,** Effect of chilling and non-chilling temperatures on Snap beans fruits, *Proc. Am. Soc. Hortic. Sci.,* 89, 368, 1966.
68. **Watada, A. E. and Morris, L. L.,** Post-harvest behavior of Snap bean cultivars, *Proc. Am. Soc. Hortic. Sci.,* 89, 375, 1966.
69. **McColloch, L. P. and Vaught, C.,** Refrigerated-storage tests with lima beans in the pod, *U.S. Dep. Agric. Agric. Res. Serv. Rep., ARS,* 51-23, 3p, 1968.
70. **Thompson, A. K., Been, B. O., and Perkins, C.,** Storage of fresh breadfruit, *Trop. Agric.,* 51, 407, 1974.
71. **Marriott, J., Perkins, C., and Been, B. O.,** Some factors affecting the storage of fresh breadfruit, *Sci. Hortic.,* 10, 177, 1979.
72. **Lam, P. F. and Wan, C. K.,** Climacteric nature of the carambola (*Averrhoa carambola* L.) fruit, *Pentanika,* 6, 44, 1983.
73. **Oslund, C. R. and Davenport, T. L.,** Ethylene and carbon dioxide in ripening of *Averrhoa carambola, HortScience,* 18, 229, 1983.
74. **Grierson, W. and Vines, H. M.,** Carambolas for potential use in gift fruit shipments, *Proc. Fla. State Hortic. Soc.,* 78, 349, 1965.
75. **Kenney, P. and Hull, L.,** Effects of storage condition on carambola quality, *Proc. Fla. State Hortic. Soc.,* 99, 222, 1986.
76. **Vines, H. M. and Grierson, W.,** Handling and physiological studies with the carambola, *Proc. Fla. State Hortic. Soc.,* 79, 350, 1966.
77. **Littmann, M. D. and Stoler, A.,** Choko storage and disorders, *Queensl. Agric. J.,* 99, 291, 1973.
78. **Littmann, M. D., Stoler, A., and Blake, J. R.,** Chilling disorders in fruit of the Choko, *(Sechium edule), Queensl. J. Agric. Anim. Sci.,* 38, 65, 1981.
79. **Morris, L. L. and Platenius, H.,** Low temperature injury to certain vegetables after harvest, *Proc. Am. Soc. Hortic. Sci.,* 36, 609, 1938.
80. **Abbott, J. A. and Massie, D. R.,** Delayed light emission for early detection of chilling in cucumber and bell pepper fruit, *J. Am. Soc. Hortic. Sci.,* 110, 42, 1985.
81. **Apeland, J.,** Factors affecting the sensitivity of cucumbers to chilling temperatures, *Bull. Int. Inst. Refrig.,* Annex 1, 46, 325, 1966.
82. **Ryall, A. L. and Lipton, W. J.,** *Handling, Transportation and Storage of Fruits and Vegetables,* Vol. 1, 2nd ed., AVI Publishing, Westport, CT, 1979, 425.
83. **Watson, B. J.,** Bombacaceae, in Tropical Tree Fruits for Australia, Page, P. E., Ed., Queensland Department of Primary Industry, 1984, 45.
84. **Mathur, P. B. and Srivastava, H. C.,** Preliminary experiments on the cold storage of durian, *Bull. Cent. Food Technol. Res. Inst. Mysore,* 3, 199, 1954.
85. **Abe, K., Chachin, K., and Ogata, K.,** Chilling injury in eggplant fruits. VI. Relationship between storability and contents of phenolic compounds in some eggplant cutivars, *J. Jpn. Soc. Hortic. Sci.,* 49, 269, 1980.
86. **Kozukue, N., Kozukue, E., and Kishiguchi, M.,** Changes in the contents of phenolic substances, phenylalanine ammonia-lyase (PAL) and tyrosine ammonia-lyase (TAL) accompanying chilling-injury of eggplant fruit, *Sci. Hortic.,* 11, 51, 1979.
87. **Nakamura, R., Fujii, S., Inaba, A., and Ito, T.,** Effects of different soil moisture and fertilizer application during cultivation on chilling sensitivity after harvest in eggplant fruit, *J. Jpn. Soc. Hortic. Sci.,* 55, 490, 1987.
88. **Aubert, S. and Pochard, E.,** Problemes de conservation en frais de l'aubergine (*Solanum melongena* L.) I. Etude bibilographique, *Rev. Hortic.,* 215, 33, 1981.
89. **Abe, K., Chachin, K., and Ogata, K.,** Chilling injury in eggplant fruit. II. The effects of maturation and harvesting season on pitting injury and browning of seeds and pulp during storage, *J. Jpn. Soc. Hortic. Sci.,* 45, 307, 1976.
90. **Abe, K. and Chachin, K.,** Influence of conditioning on the occurrence of chilling injury and the changes of surface structure of eggplant fruits, *J. Jpn. Soc. Hortic. Sci.,* 54, 247, 1985.
91. **Ito, T. and Nakamura, R.,** The effect of fluctuating temperatures on chilling injury of several kinds of vegetables. *J. Jpn. Soc. Hortic. Sci.,* 53, 202, 1984.
92. **Nakamura, R., Inaba, A., and Ito, A.,** Effect of cultivating condition and postharvest stepwise cooling on the chilling sensitivity of eggplant and cucumber fruits, *Sci. Rep. Fac. Agric. Okayama Univ.,* 66, 19, 1985.
93. **Akamine, E. K.,** Storage of fresh ginger rhizome, *Hawaii Agric. Exp. Stn. Bull.,* No. 130, 1962.
94. **Paull, R. E., Chen, N. J., and Goo, T. T. C.,** Control of weight loss and sprouting of ginger rhizome in storage, *HortScience,* 23, 734, 1988.

95. **Liu, M. S., Chang, T. C., Liao, M. L., and Yang, J. S.,** Inhibition of sprouting and decay of ginger, shallot and garlic during storage, *Food Ind. Res. Dev. Inst. Taiwan Bull.,* No. 258, 1982.

96. **Paull, R. E., Chen, N. J., and Goo, T. T. C.,** Postharvest changes in ginger rhizome during storage, *J. Am. Soc. Hortic. Sci.,* 1988, in press.

97. **Wills, R. B. H., Mulholland, E. E., Brown, B. L., and Scott, K. J.,** Storage of two new cultivars of Guava (*Psidium guajava*) fruit for processing, *Trop. Agric.,* 60, 175, 1983.

98. **Boyle, F. P., Seagrave-Smith, H., Sakata, S., and Sherman, G. D.,** Commercial guava processing in Hawaii, *Univ. Hawaii Agric. Exp. Stn. Bull.,* No. 111, 1957.

99. **Singh, K. K. and Mathur, P. B.,** Cold storage of guavas, *Indian J. Hortic.,* 11, 1, 1954.

100. **Pantastico, E. B., Chattopadhyaya, T. K., and Subramanyam, H.,** Harvesting and handling: storage and commercial storage operations, in *Postharvest Physiology, Handling and Utilization of Tropical and Subtropical Fruits and Vegetables,* Pantastico, E. B., Ed., AVI Publishing, Westport, CT, 1975, 314.

101. **Kripal Singh, K.,** Jackfruit, Minist. Agric. New Delhi Dir. Ext. Farm Inf. Leafl., No. 71, 1972.

102. **San Pedro, A. V.,** Studies on the storage-temperature requirements of lanzones (*Lansium domesticum* Correa), *Philipp. Agric.,* 25, 411, 1936.

103. **Pantastico, E. B., Mendoza, D. B., and Abilay, R. M.,** Some chemical and physiological changes during storage of lanzones (*Lansium domesticum* Correa), *Philipp. Agric.,* 52, 505, 1968.

104. **Srivastava, H. C. and Mathur, P. B.,** Studies in the cold storage of Langsats, *J. Sci. Food Agric.,* 6, 511, 1955.

105. **Aziz, A. B. A., El-Nabawy, S. M., Abdel, Wahab, F. K., and Abdel-Kader, A. S.,** The effect of storage temperature on quality and decay percentage of 'Pair' and 'Taimour' mango fruit, *Sci. Hortic.,* 5, 65, 1976.

106. **Mathur, P. B., Kirpal Singh, K., and Kapur, N. S.,** Cold storage of mangoes, *Indian J. Agric. Sci.,* 23, 65, 1953.

107. **Thomas, P. and Oke, M. S.,** Improvement in quality and storage of mangoes *Mangifera indica* cultivar Alphonso by cold adaptation, *Sci. Hortic.,* 19, 257, 1983.

108. **Lizada, M. C. C., Kosiyachinda, S., and Mendoza, D. B.,** Physiological disorders of mango, in *Mango,* Mendoza, D. B. and Wills, R. B. H., Eds., ASEAN Food Handling Bureau, Kuala Lumpur, Malaysia, 1984, 68.

109. **Veloz, C. S., Torres, F. E., and Lakshminarayana, S.,** Effect of refrigerated temperatures on the incidence of chilling injury and ripening quality of mango fruit, *Proc. Fla. State Hortic. Soc.,* 90, 205, 1977.

110. **Akamine, E. K.,** Haden mango storage, *Hawaii Farm Sci.,* 12(4), 6, 1963.

111. **Raman, K. R., Raman, N. V., and Sadasivam, R.,** A note on storage behaviour of mangosteen, *South Indian Hortic.,* 19, 85, 1971.

112. **Augustin, M. A. and Azudin, M. N.,** Storage of Mangosteen, *ASEAN Food J.,* 2, 78, 1986.

113. **Srivastava, H. C., Singh, K. K., and Mathur, P. B.,** Refrigerated storage of mangosteen, *Food Sci. (Mysore),* 11, 226, 1962.

114. **Lipton, W. J.,** Chilling injury of 'Honey Dew' muskmelons: symptoms and relation to degree of ripeness at harvest, *HortScience,* 13, 45, 1978.

115. **Lipton, W. J. and Mackey, B. E.,** Ethylene and low temperature treatments of Honey Dew melons to facilitate long distance shipment, *U.S. Dep. Agric. Agric. Res. Serv. Rep.,* 10, 28, 1984.

116. **Woodruff, J. G. and Shelor, E.,** Okra for processing, *Ga. Agric. Exp. Stn. Bull.,* 80, 91, 1958.

117. **Ilker, Y. and Morris, L. L.,** Alleviation of chilling injury of Okra, *HortScience,* 10 (Abstr.), 324, 1975.

118. **Ilker, Y.,** Physiological manifestation of chilling injury and its alleviation in Okra fruits (*Abelmoschus esculentus* (L) Moench.), *Diss. Abstr. Int. B,* 38, 13, 1977.

119. **Ogata, K., Yamauchi, N., and Minamide, T.,** Physiological and chemical studies on ascorbic acid of fruits and vegetables. I. Changes in ascorbic acid content during maturation and storage period in Okuras, *J. Jpn. Soc. Hortic. Sci.,* 44, 192, 1975.

120. **Glazener, J. A., Couey, H. M., and Alvarez, A.,** Effect of postharvest treatments on Stemphylium rot of papaya, *Plant Dis.,* 68, 986, 1984.

121. **Jones, W. W.,** Respiration and chemical changes of the papaya fruit in relation to temperature, *Plant Physiol.,* 17, 481, 1942.

122. **Nazeeb, M. and Broughton, W. J.,** Storage conditions and ripening of papaya cultivars 'Bentong' and 'Taiping', *Sci. Hortic.,* 9, 265, 1978.

123. **Wilcox, E. V. and Hunn, C. J.,** Cold storage of tropical fruits, *Hawaii Agric. Exp. Stn. Bull.,* 47, 36, 1914.

124. **Anon.,** Storage of Passion Fruit, 8th Annu. Rep. Commonw. Australia, Counc. Sci. Ind. 1934, 55.

125. **Pruthi, J. S.,** Physiology, chemistry and technology of passion fruit, *Adv. Food Res.,* 12, 203, 1963.

126. **Pruthi, J. S. and Lal, G.,** Refrigerated and common storage of purple passion fruit, *Indian J. Hortic.,* 12, 204, 1955.

127. **Patterson, B. D., Murata, T., and Graham, D.,** Electrolyte leakage induced by chilling in Passiflora species tolerant to different climates, *Aust. J. Plant Physiol.,* 3, 435, 1976.

128. **McColloch, L. P.,** Chilling injury and alternaria rot of bell peppers, *U.S. Dep. Agric. Mark. Res. Rep.,* 536, 16, 1962.

129. **Hughes, P. A., Thompson, A. K., Plumbley, R. A., and Seymour, G. B.,** Storage of capsicums (*Capsicum annuum* (L.) Stendt.) under controlled atmospheres, modified atmosphere and hypobaric conditions, *Hortic. Sci.,* 56, 261, 1981.

130. **Hartman, J. D. and Isenberg, F. M.,** Waxing Vegetables, *N.Y. State Coll. Agric. Cornell Ext. Bull.,* 965, 14, 1956.

131. **Teisson, C.,** Le brunissement interne de låanas. I. Historique, *Fruits,* 34, 245, 1979.

132. **Akamine, E. K., Goo, T., Steepy, T., Greidanus, T., and Iwaoka, N.,** Control of endogenous brown spot of fresh pineapple in postharvest handling, *J. Am. Soc. Hortic. Sci.,* 100, 60, 1975.

133. **Cannon, R. C., Berrill, F. W., and King, K.,** Can blackheart of pineapple be avoided, *Queensl. Agric. J.,* 81, 321, 1955.

134. **Sun, S.,** A study of black heart disease of the pineapple fruits, *Plant Prot. Bull. (Taiwan),* 13, 39, 1971.

135. **Van Lelyveld, L. J. and De Bruyn, J. A.,** Polyphenols, ascorbic acid and related enzyme activities associated with black heart in Cayenne pineapple fruit, *Agrochemophysica,* 9, 1, 1977.

136. **Smith, L. G.,** Cause and development of black heart in pineapple, *Trop. Agric. (Trinidad),* 60, 31, 1983.

137. **Rohrbach, K. G. and Paull, R. E.,** Incidence and severity of chilling induced internal browning of waxed 'Smooth Cayenne' pineapples, *J. Am. Soc. Hortic. Sci.,* 107, 453, 1982.

138. **Mendoza, D. B., Pantastico, E. B., and Javier, F. B.,** Storage and handling of rambutan (*Nephelium lappaceum* L.), *Philipp. Agric.,* 55, 322, 1971.

139. **Paull, R. E. and Chen, N. J.,** Changes in longan and rambutan during postharvest storage, *HortScience,* 22, 1303, 1987.

140. **Broadus, E. W., Collins, W. W., and Pharr, D. M.,** Incidence and severity of hardcore in sweet potatoes as affected by genetic line, curing, and lengths of 1°C and 21°C storage, *Sci. Hortic.,* 13, 105, 1980.

141. **Buescher, R. W.,** Hardcore sweet potato roots as influenced by cultivar, curing, and ethylene, *HortScience,* 12, 326, 1977.

142. **Buescher, R. W., Sistrunk, W. A., and Kasaian, A. E.,** Induction of textural changes in sweet potato roots by chilling, *J. Am. Soc. Hortic. Sci.,* 101, 516, 1976.

143. **Buescher, R. W. and Balmoori, M. R.,** Mechanism of hardcore formation in chill-injured sweet potato (*Ipomoea batatas*) roots, *J. Food Biochem.,* 6, 1, 1982.

144. **Lieberman, M., Craft, C. C., and Wilcox, M. S.,** Effect of chilling on the chlorogenic acid and ascorbic acid content of Porto Rico sweetpotatoes, *Proc. Am. Soc. Hortic. Sci.,* 74, 642, 1959.

145. **Picha, D. H.,** Carbohydrate changes in sweet potatoes during curing and storage, *J. Am. Soc. Hortic. Sci.,* 111, 89, 1986.

146. **Picha, D. H.,** Chilling injury, respiration, and sugar changes in sweet potatoes stored at low temperature, *J. Am. Soc. Hortic. Sci.,* 112, 497, 1987.

147. **Daines, R. H., Hammond, D. F., Haard, N. F., and Ceponis, M. J.,** Hardcore development in sweet-potatoes, a response to chilling and its remission as influenced by cultivar, curing temperatures, and time and duration of chilling, *Phytopathology,* 66, 582, 1976.

148. **Kushman, L. J. and Wright, F. S.,** Sweetpotato storage, *U.S. Dep. Agric. Handb.,* No. 358, 1969.

149. **Broughton, W. J. and Wong, H. C.,** Storage conditions and ripening of chiku fruits, *Achras sapota* L., *Sci. Hortic.,* 10, 377, 1979.

150. **Hobson, G. E.,** Low-temperature injury and the storage of ripening tomatoes, *J. Hortic. Sci.,* 62, 55, 1987.

151. **McColloch, L. P. and Worthington, J. T.,** Low temperature as a factor in the susceptibility of mature-green tomatoes to Alternaria rot, *Phytopathology,* 42, 425, 1952.

152. **McColloch, L. P., Yeatman, J. N., and Loyd, P.,** Color changes and chilling injury of pink tomatoes held at various temperatures, *U.S. Dep. Agric. Mark. Res. Rep.,* 735, 1966.

153. **Salveit, M. E. and Cabrera, R. M.,** Tomato fruit temperature before chilling influences ripening after chilling, *HortScience,* 22, 452, 1987.

154. **Hashad, M. N., Stino, K. R., and El Hinnawy, S. I.,** Transformation and translocation of carbohydrates in taro plants during storage, *Ann. Agric. Sci.,* 1, 269, 1956.

155. **Hirata, K., Chachin, K., and Iwata, T.,** Chilling injury of water convolvulus (*Ipomoea aquatica* Forsk) with reference to its phenolic metabolism, in *Food Science and Nutrition,* Vol. 1, McLoughlin, J. V. and McKenna, B. M., Eds., Boole Press, Dublin, 1983, 11.

156. **Lyrene, P. M.,** The jujube tree (*Zizyphus jujuba* Lam.), *Fruit Var. J.,* 33, 100, 1979.

157. **De Candolle, A.,** *Origin of Cultivated Plants,* D. Appleton, New York, 1908.

158. **Kader, A. A., Li, Y., and Chordas, A.,** Postharvest respiration, ethylene production and compositional changes of Chinese jujube fruits, *Zizyphus jujuba, HortScience,* 17, 678, 1982.

159. **Jawanda, J. S., Bal, J. S., Josan, J. S., and Mann, S. S.,** Studies on the storage of ber fruit. II. Cool temperatures, *Punjab Hortic. J.,* 20, 171, 1980.

160. **Eaks, I. L.,** Rind disorders of oranges and lemons in California, *Proc. 1st Int. Citrus Symp.,* 3, 1343, 1969.

161. **Pantastico, E. B., Mattoo, A. K., Murata, T., and Ogata, K.,** Chilling injury, in *Postharvest Physiology, Handling and Utilization of Tropical and Subtropical Fruits and Vegetables,* Pantastico, E. B., Ed., AVI Publishing, Westport, CT, 1975, 339.

162. **Eaks, I. L.,** Effect of temperature and holding period on some physical and chemical characteristics of lemon fruits, *J. Food Sci.,* 26, 593, 1961.

163. **Eaks, I. L.,** Effects of chilling injury on the respiration of oranges and lemons, *Proc. Am. Soc. Hortic. Sci.,* 87, 181, 1965.

164. **Smoot, J. J., Houck, L. G., and Johnson, H. B.,** Market diseases of citrus and other subtropical fruits, *U.S. Dep. Agric. Agric. Handb.,* No. 398, 1971.

165. **Chalutz, E., Waks, J., and Schiffmann-Nadel, M.,** A comparison of the response of different citrus fruit cultivars to storage temperature, *Sci. Hortic.,* 25, 271, 1985.

166. **Eaks, I. L.,** The physiological breakdown of the rind of lime fruits after harvest, *Proc. Am. Soc. Hortic. Sci.,* 66, 141, 1955.

167. **Wardowski, W. F., Grierson, W., and Edwards, G. J.,** Chilling injury of stored limes and grapefruit as affected by differentially permeable packing films, *HortScience,* 8, 173, 1973.

168. **Pantastico, E. B., Soule, J., and Grierson, W.,** Chilling injury in tropical and subtropical fruits. II. Limes and grapefruit, *Proc. Trop. Reg. Am. Soc. Hortic. Sci.,* 12, 171, 1968.

169. **Rodriguez, J., Slutzky, B., Ruiz, A., and Ancheta, O.,** Morphological study of the preservation of the lime *Citrus aurantifolia* cultivar Persa at different temperatures, *Rev. Jard. Bot. Nac.,* 2, 163, 1981.

170. **Chace, W. G.,** Controlled atmosphere storage of Florida citrus fruits, *Proc. 1st Int. Citrus Symp.,* 3, 1365, 1969.

171. **Davis, P. L. and Hofmann, R. C.,** Reduction of chilling injury of citrus fruits in cold storage by intermittent warming, *J. Food Sci.,* 38, 871, 1973.

172. **Eaks, I. L.,** Effect of chilling on respiration and volatile of California lemon fruit, *J. Am. Soc. Hortic. Sci.,* 105, 865, 1980.

173. **McDonald, R. E., Hatton, T. T., and Cubbedge, R. H.,** Chilling injury and decay of lemons as affected by ethylene, low temperature and optimum storage, *HortScience,* 20, 92, 1985.

174. **McDonald, R. E.,** Effects of vegetable oils, CO_2 and film wrapping on chilling injury and decay of lemons, *HortScience,* 21, 476, 1986.

175. **Wild, B. L., McGlasson, W. B., and Lee, T. H.,** Effect of reduced ethylene levels in storage atmosphere on lemon keeping quality, *HortScience,* 11, 114, 1976.

176. **Wild, B. L., McGlasson, W. B., and Lee, T. H.,** Long term storage of lemon fruit, *Food Technol. Aust.,* 29, 351, 1977.

177. **Khalifah, R. A. and Kuykendall, J. R.,** Effect of Maturity, storage temperature, and prestorage treatment on storage quality of valencia oranges, *Proc. Am. Soc. Hortic. Sci.,* 86, 228, 1965.

178. **Trout, S. A., Tindale, G. B., and Huelin, F. E.,** The storage of oranges with special reference to locality, maturity, respiration and chemical composition, Australian Council of Science and Industry Research Pamphlet, 80, 1938.

179. **Van der Plank, J. E. and Davies, R.,** Temperature-cold injury of fruit, *J. Pomol. Hortic. Sci.,* 15, 226, 1937.

180. **Ceponis, M. J. and Cappellini, R. A.,** Wholesale and retail losses in grapefruit marketed in Metropolitan New York, *HortScience,* 20, 93, 1985.

181. **Miller, E. V.,** The physiology of citrus fruit in storage. II, *Bot. Rev.,* 24, 43, 1958.

182. **Ismail, M. A. and Grierson, W.,** Seasonal susceptibility of grapefruit to chilling injury as modified by certain growth regulators, *HortScience,* 12, 118, 1977.

183. **Purvis, A. C.,** Relationship between chilling injury of grapefruit and moisture loss during storage: amelioration by polyethylene shrink film, *J. Am. Soc. Hortic. Sci.,* 110, 385, 1985.

184. **Wardowski, W. F., Albrago, L. G., Grierson, W., Barmore, C. R., and Wheaton, T. A.,** Chilling injury and decay of grapefruit as affected by thiabendazole, benomyl, and CO_2, *HortScience,* 10, 381, 1975.

185. **Purvis, A. C.,** Influence of canopy depth on susceptibility of 'Marsh' grapefruit to chilling injury, *HortScience,* 15, 731, 1980.

186. **Hatton, T. T. and Cubbedge, R. H.,** Conditioning Florida grapefruit to reduce chilling injury during low-temperature storage, *J. Am. Soc. Hortic. Sci.,* 107, 57, 1982.

187. **Hatton, T. T. and Cubbedge, R. H.,** Preferred temperature for prestorage conditioning of 'Marsh' grapefruit to prevent chilling injury at low temperatures, *HortScience,* 18, 721, 1983.

188. **Martin, F. W. and Cooper, W. C.,** Cultivation of neglected tropical fruits with promise. III. The pummelo, *U.S. Dep. Agric. Agric. Res. Serv. Bull.,* ARS-S-157, 1977.

189. **Reinking, O. A. and Groff, W. W.,** The kao pan seedless Siamese pummelo and its culture, *Philipp. J. Sci.,* 19, 389, 1921.

190. **Kitagawa, H., Kawada, K., and Tarutani, T.,** Effects of temperature, packaging and curing on the storage of Sudachi (*Citrus sudachi* hort. ex. Shirai), *J. Jpn. Soc. Hortic. Sci.,* 51, 350, 1982.

191. **Fidler, J. C. and Coursey, D. G.,** Low temperature injury in tropical fruit, Proc. Conf. Tropical and Subtropical Fruit, London, 1969, 103.

192. **Akamine, E. K.,** Preventing the darkening of fresh lychees prepare for export, *Hawaii Agric. Exp. Stn. Tech. Prog. Rep.,* 127, 1960.

193. **Hatton, T. T., Reeder, W. F., and Kaufman, J.,** Maintaining market quality of fresh lychees during storage and transit, *U.S. Dep. Agric. Mark. Res. Rep.,* 770, 1966.

194. **Scott, K. J., Brown, B. I., Chaplin, G. R., Wilcox, M. E., and McBain, J. M.,** The control of rotting and browning of litchi fruit by hot benomyl and plastic film, *Sci. Hortic.,* 16, 253, 1982.

195. **Swarts, D. H. and Anderson, T.,** Chemical control of mold growth on litchis during storage and sea shipment, *Subtropica,* 1(10), 13, 1980.

196. **Tongdee, S. C., Scott, K. J., and McGlasson, W. B.,** Packaging and cool storage of litchi fruit, *CSIRO Food Res. Q.,* 42, 25, 1982.

197. **Paull, R. E. and Chen, N. J.,** Effect of storage temperature and wrapping on quality characteristics of lichi fruit, *Sci. Hortic.,* 33, 223, 1987.

198. **Smock, R. M.,** Morphology of the flower and fruit of the loquat, *Hilgardia,* 10, 615, 1937.

199. **Mukerjee, P. K.,** Storage of Loquat (*Eriobotrya japonica* Lindl.), *Hortic. Adv.,* 2, 64, 1958.

200. **Ogata, Y.,** Physiological study on the fruit of loquat during storage, *Tech. Bull. Kagawa Agric. Coll.,* 1, 42, 1950.

201. **Guelfat-Reich, S.,** Conservation de la Nefle du japon, *Fruits,* 25, 169, 1970.

202. **Maxie, E. C.,** Storing olives under controlled temperatures and atmospheres, *Calif. Olive Assoc. Annu. Tech. Rep.,* 42, 34, 1963.

203. **Maxie, E. C.,** Experiments on cold storage and controlled atmosphere, *Calif. Olive Assoc. Annu. Tech. Rep.,* 43, 12, 1964.

204. **Woskow, M. and Maxie, E. C.,** Cold storage studies with olives, *Calif. Olive Assoc. Annu. Tech. Rep.,* 44, 6, 1965.

205. **Hartman, H. T. and Opitz, K. W.,** Olive production in California, *Univ. Calif. Leafl.,* 2474, 47, 1977.

206. **Kato, K.,** Astringency removal and ripening as related to temperature during the de-astringency by ethanol in Persimmon fruits, *J. Jpn. Soc. Hortic. Sci.,* 55, 498, 1987.

207. **Taira, S., Kubo, Y., Sugiura, A., and Tomana, T.,** Comparative studies of postharvest fruit quality and storage quality in Japanese persimmon (*Diospyros kaki* L. cv. Hiratananashi) in relation to different methods for removal of astringency, *J. Jpn. Soc. Hortic. Sci.,* 56, 215, 1987.

208. **Kitagawa, H., Sugiura, A., and Sugiyama, M.,** Effects of gibberellin sprays on storage quality of kaki, *HortScience,* 1, 59, 1960.

209. **Elyatem, S. M. and Kader, A. A.,** Post-harvest physiology and storage behaviour of pomegranate fruits, *Sci. Hortic.,* 24, 287, 1984.

210. **Mukerjee, P. K.,** Storage of pomegranates (*Punica granatum* L.), *Sci. Cult.,* 24, 94, 1958.

211. **Ewing, E. E., Senesac, A. H., and Sieczka, J. B.,** Effects of short periods of chilling and warming on potato sugar content and chipping quality, *Am. Potato J.,* 58, 633, 1981.

212. **Hruschka, H. W., Smith, W. L., and Baker, J. E.,** Reducing chilling injury of potatoes with intermittent warming, *Am. Potato J.,* 46, 38, 1969.

213. **Hruschka, H. W., Smith, W. L., and Baker, J. E.,** Chilling injury syndrome in potato tubers, *Plant Dis. Rep.,* 51, 1014, 1967.

214. **Folsom, D.,** Inheritance of predisposition of potato varieties to internal mahogany browning of the tubers, *Am. Potato J.,* 24, 294, 1947.

215. **Cunningham, H. H., Zaehringer, M. V., and Sparks, W. C.,** Storage temperature for maintenance of internal quality of Idaho Russet Burbank potatoes, *Am. Potato J.,* 48, 320, 1971.

216. **Strachan, G.,** Tamarillo storage and processing, *Food Technol. N.Z.,* 5, 304, 1970.

217. **Picha, D. H.,** Postharvest fruit conditioning reduces chilling injury in watermelons, *HortScience,* 21, 1407, 1986.

218. **Showalter, R. K.,** Watermelon color as affected by maturity and storage, *Proc. Fla. State Hortic. Soc.,* 73, 289, 1960.

219. **Hall, E. G. and Scott, K. J.,** Storage and market diseases of fruit, *CSIRO Food Res. Lab. Invest. Rep.,* 1977.

220. **Cheema, G. S., Karmakar, D. V., and Joshi, B. M.,** Investigations on the cold storage of mangoes, *Indian J. Agric. Sci.,* 20, 259, 1950.

221. **Moline, H. E.,** Ultrastructural changes associated with chilling of tomato fruit, *Phytopathology,* 66, 617, 1976.

222. **Cook, H. T., Parsons, C. S., and McColloch, L. P.,** Methods to extend storage of fresh vegetables aboard ships for the U.S. Navy, *Food Technol. (Chicago),* 12, 548, 1958.

Chapter 3

CHILLING INJURY OF CROPS OF TEMPERATE ORIGIN

William J. Bramlage and Shimon Meir

TABLE OF CONTENTS

I. INTRODUCTION

Chilling injury in plants is recognized as abnormal development after exposure for more or less prolonged periods to temperatures below about 10 to 13°C. Plants of tropical or subtropical origin are generally assumed to be susceptible to chilling, and abnormal development following a chilling exposure might occur on any part of such a plant and be expressed in a wide array of deleterious symptoms.

In a temperate climate, periods when the temperature is below 10 to 13°C are common and may even occur at the peak of the growing season. Thus, plants evolving in a temperate climate can be expected to have tolerance to chilling temperatures: they are survivors in a chilling environment. However, this tolerance is not necessarily an immunity to chilling injury. It may represent a resistance to mild chilling temperatures (e.g., down to 3 to 5°C), fairly short exposures to chilling temperatures (e.g., 1 week or more), or a combination of the two conditions. It may also represent a strong capacity to reverse chilling damage, since in nature chilling exposures are often transient. This tolerance may be accomplished by a variety of physiological mechanisms, since the capacity to survive chilling experiences may have evolved differently in different species and in different environments. Yet, it may only mask an underlying susceptibility to cellular changes at low but nonfreezing temperatures, changes that can have adverse effects on plants under certain cellular and environmental conditions.

Interest in and research on chilling injury have focused on plants of tropical and subtropical origin, given their rapid and clear responses to chilling stress and the economic significance of damage to these plants. The possible effects of chilling on temperate-origin plants is seldom given serious consideration, and these plants are used as "controls" in physiological studies of chilling injury. Yet, many unique physiological events do occur in temperate-origin plants when they are kept for prolonged periods of time at low but nonfreezing temperatures, and these can be defined as chilling responses. Not all of them may be injurious to plants, and, indeed, some may be beneficial. However, in this review we shall examine adverse responses to chilling, and we shall argue that all plants probably are susceptible to chilling damage under certain conditions.

Since temperate-origin crops are survivors of chilling environments, it can be expected that chilling injury in them (as compared to tropical or subtropical crops) may require significantly lower temperatures, longer continuous exposure periods, or both, and that physiological consequences may be much more subtle. Experimentally, these environmental conditions are most often employed in cold-storage studies, and environmentally these conditions most commonly occur in postharvest handling of the plants; so this review will emphasize postharvest effects of chilling on crops of temperate origin. However, analogous preharvest responses likely occur when the tissues experience prolonged, continuous exposure to low temperature.

II. SYMPTOMS OF CHILLING INJURY

Symptoms of chilling injury to plants are diverse. Morris[1] catalogued them as follows: surface lesions, water soaking of tissues, internal discoloration (browning), breakdown of tissues, failure of fruits to ripen in the expected pattern, accelerated rate of senescence, increased susceptibility to decay, shortened storage or shelf life, compositional changes related to consumer acceptance, and loss of normal growth capacity. Most of these symptoms are not unique to chilling injury, making it often difficult to diagnose the cause of commercial losses of these crops. This problem is confounded further when we examine possible symptoms of chilling injury in temperate-origin plants, given the more demanding conditions for symptom occurrences, and there may be disagreement as to whether or not a given disorder is truly a symptom of chilling injury. For example, "superficial scald" of apples and pears is considered by some,[2] but not others,[3] to be a chilling disorder. We consider it to be a symptom of chilling injury.

Chilling injuries to temperate-origin crops include most of the symptoms described by Morris.[1] The following examples are taken from Hardenburg et al.[4] unless otherwise noted.

A. SURFACE SYMPTOMS

Apples can develop "soft scald" and "superficial scald", irregularly shaped necrotic lesions on fruit surfaces. Soft scald lesions tend to be more sunken and to extend more deeply into the flesh than those of superficial scald and to have different etiology.[5] Pears also develop superficial scald. Apricots may develop pitting or browning of the peel,[6] and Chinese cabbage may develop browning of the midrib.[7] In addition, the surfaces of cranberries and peaches can lose their sheen, and asparagus can become discolored and lose turgor at the tips of spears.[8]

B. INTERNAL DISCOLORATION AND TISSUE BREAKDOWN

Certain cultivars of apple can develop "low-temperature breakdown", "brown core" ("core flush"), and "soggy breakdown", each disorder possessing distinctive symptoms and etiology.[9] Cranberries, peaches, and nectarines can develop red discoloration of the flesh, and cranberries can develop a rubbery texture. Apricots, peaches, nectarines, and plums can develop severe flesh browning, especially near the pit. Certain potato cultivars can develop "mahogany browning" of their flesh.

C. COMPOSITIONAL CHANGES

Peaches and nectarines can rapidly lose their fruity flavor at chilling temperatures, and plums can develop abnormal flavors, presumably due to altered production of volatiles. Underground storage organs of many plants sweeten due to starch or inulin hydrolysis at temperatures below 4 to 5°C.[10] Beets may become abnormally hard and develop "earthy" flavors.[11] Peaches lose some of their ability to solubilize pectins, leading to the textural condition called "woolliness" after 3 to 4 weeks at near 0°C.[12] Eldorado pears lose most of their ability to solubilize pectins after 9 months at 0°C.[13]

D. ABNORMAL RIPENING

Some cultivars of pears require a period of chilling to induce normal ripening, although they will not ripen to acceptable quality while at low temperature; e.g., Bosc pears were induced to autocatalytic ethylene production by 7 d at 10°C or less, with an optimum at 5°C.[14] Bartlett pears may be induced to premature ripening by frequent cool night temperatures (8°C) in the orchard.[15] However, after prolonged storage at 2.5 to 10°C, Bartlett ripen to a mealy texture and inferior flavor,[16] somewhat similar to "woolliness" of peaches and nectarines. Pears that lose their capacity for normal ripening do not synthesize protein normally after transfer to a ripening temperature.[17] For Eldorado pears, ethylene synthesis at 0°C and ripening at 20°C were induced by 4 weeks at 0°C.[13] Their failure to soften normally after 9 months at 0°C, referred to above, occurred despite development of other aspects of ripening at 20°C.[13] Pears that do not lose their ability to ripen normally during cold storage often develop superficial scald, similar to that on apples, while ones that lose their ripening ability correspondingly develop unique symptoms referred to as "senescent scald", severe necrotic lesions on the fruit surface.[18] Clearly, the pear fruit ripening mechanism has complex responses to chilling temperature.

E. LOSS OF GROWTH CAPACITY

Cut gladiolus and chrysanthemum buds may fail to open properly after storage beyond

7 or 21 d, respectively, at near 0°C. Flowers with a tendency for petal abscission may "shatter" quickly after low-temperature storage.

F. HIGH PERISHABILITY

In addition to all of the above symptoms recognized as consequences of storage at chilling temperatures,[1] it is strange that many crops possess potentials for only brief periods of storage at near 0°C.[4] In most cases, the limiting factor in low-temperature storage is said to be "high perishability". However, when specific causes are stated, they often mimic chilling damage: decay (raspberries), flavor loss (blueberries, cherries, strawberries), internal collapse (gooseberries), color loss (brussels sprouts, cherries, strawberries), and surface lesions (brussels sprouts, cabbage). Many flowers are not recommended for storage below 4°C, without a stated reason.[4] Broccoli "stores best" at 0°C,[4] yet off odors under restricted ventilation were more pronounced at 0°C than at 2.5 or 5°C.[19]

The point here is not to challenge storage recommendations, but to point out the possibility that chilling injury is much more common among crops of temperate origin than is generally recognized. No sharp distinction can be made at the physiological level between any chilling-sensitive and any chilling-resistant plant species.[20] Perhaps a recognition that all crops of tropical or subtropical origin probably are to some extent chilling sensitive will make researchers more alert to symptoms of damage on them.

III. FACTORS AFFECTING CHILLING INJURY

Susceptibility of temperate-origin plants to chilling injuries is by no means constant for a species. Many factors influence form and intensity of injury symptoms as well as the susceptibility to these symptoms.

A. CULTIVAR DIFFERENCES

Genetic diversity of chilling susceptibility within a species is observed clearly in apples, which are commonly stored for many months and thus exposed to continuous chilling temperatures for long durations of time. Cultivars vary greatly[3,9,18] in their susceptibility to the diverse symptoms described above. For example, storage at near 0°C results in "brown core" in McIntosh, "low-temperature breakdown" in Cox's Orange Pippin, "internal breakdown" in Yellow Newtown, "soft scald" in Jonathan, and "soggy breakdown" in Grimes Golden. Other cultivars, e.g., Delicious, can be kept at 0°C for many months without the appearance of any of these disorders. However, a 4-year study concluded that Delicious is subject to a low-temperature disorder described simply as "flesh browning".[21] One low-temperature disorder that most apple cultivars are susceptible to is "superficial scald", but degree of susceptibility varies enormously. Often, two or more disorders occur simultaneously, such as core flush and superficial scald on Granny Smith apples, and the disorders are considered to be separate entities, but low-temperature breakdown and soggy breakdown often develop concurrently on the same fruit and may be different manifestations of the same disorder.

Pears are another species often subjected to long-term storage at low temperature, and, like apples, there is large variability in responses among cultivars.[3,18] Examples of specific low-temperature disorders in certain European cultivars of pears were described by Fidler et al.[3] Probably of greater significance are the general effects of low temperature on pear ripening, described in Section II.D. Pears are harvested and stored in an unripe condition, and different cultivars require different storage conditions for satisfactory quality after ripening at warm temperatures.

Cultivar differences in response to chilling are not as clearly defined in other temperate-origin species, perhaps because most of them are seldom exposed to such rigorous conditions, and responses are not as dramatic. Yet, examples of genetic diversity are found. For example, in potatoes, only certain cultivars such as Chippewa and Sebago develop "mahogany browning", even though all potatoes sweeten when kept at near 0°C for several months.[22] Brecht, et al.[23] describe a nectarine breeding line that possesses some resistance to low-temperature breakdown. In plums, ripening of Early Italian prunes can be enhanced by prior exposure to several days at 1°C,[24] although most plums are adversely affected by low-temperature storage.

B. ENVIRONMENTAL AND NUTRITIONAL CONDITIONS BEFORE CHILLING

Susceptibility to chilling injuries in crops of temperate origin can be markedly altered by environmental conditions preceding chilling. This has been well documented for apples. Cool, wet growing conditions increase the occurrence of low-temperature breakdown and core flush, and possibly of soft scald.[3] In contrast, warm, dry growing seasons are associated with increased susceptibility to superficial scald.[25] These relationships are strongest when conditions existing near harvest are examined. For example, both solar radiation and temperature over the last 6 weeks of the growing season were negatively correlated with brown core development in McIntosh apples.[26] Similarly, superficial scald of Stayman apples was markedly reduced following consistently cool temperatures during the week or two immediately preceding harvest.[27]

In the case of superficial scald, this might be interpreted as a hardening response, but for core flush, brown core, and low-temperature breakdown, the relationship is inverted. However, interpretations are made difficult because these are field conditions and other environmental factors are confounded with temperature. Cool weather often is also cloudy, damp weather, and warm weather often is also clear, dry weather; so solar intensity and water relations are interrelated with temperature. In fact, shading increases brown core in McIntosh apples,[28] so at least part of the effect of cool weather may be due to lower light intensity. Similarly, Fidler's study[25] with superficial scald indicated that water stress may be more important than temperatures as a predisposing factor, and he subsequently found that irrigation reduced superficial scald incidence.[3] The best that can be concluded safely is that certain environmental conditions can predispose apples to develop specific symptoms during low-temperature storage. Furthermore, growing conditions influence the critical temperature for disorders, Cox's Orange Pippin apples can not be stored below 4°C in England, but can be stored at 2.5 to 3°C in New Zealand and at 1.5°C in mainland Australia.[3] Whether or not this represents a hardening response, a well-documented aspect of chilling,[29] cannot be determined from available information.

Development of symptoms in apples is influenced by mineral composition of the fruit. High nitrogen levels can increase the incidences of superficial scald, soft scald, and low-temperature breakdown, and reduce the incidence of core flush.[3] Potassium applications to trees generally reduce incidence of low-temperature breakdown.[3] Low levels of phosphorus in fruit have been related to increased low-temperature breakdown.[3] Calcium is often deficient in apple fruit; foliar sprays of calcium salts can reduce superficial scald,[30] and postharvest calcium applications can reduce low-temperature breakdown.[31] However, annual variation in susceptibility of apples to core flush or low-temperature breakdown could not be predicted from mineral analyses of apples at harvest,[32] suggesting that mineral composition is associated only loosely with symptom development from chilling injury.

C. STAGE OF DEVELOPMENT

Developing plant organs undergo constant chemical, physical, and physiological changes as long as they are attached. It is to be expected, therefore, that stage of development would have some impact on development of chilling injury symptoms, and perhaps upon occurrence of the injury itself. Preclimacteric apples are usually more susceptible to brown core,[7] core flush,[3] and superficial scald[3] than are climacteric fruit, but the reverse is true for soggy breakdown and internal browning.[7] Perhaps the most precise evaluation of the relationship of developmental stage to chilling injury was by Kidd and West,[33] using Bramley's Seedling apples, who found that the fruit developed more symptoms when chilled at the climacteric rise than either before or after this point.

Developmental changes can be rather intricately woven into expression of chilling injury. Although susceptibility of apples to superficial scald usually decreases with maturity, ethylene treatment and delayed cooling after harvest — both of which would hasten ripening — can increase susceptibility to the disorder.[3] Likewise, "ethylene scrubbing" during storage, which delays the onset of ripening, can greatly suppress scald development.[34] These findings suggest that the maturation process can suppress, but the ripening process can enhance, susceptibility to the disorder even though the two processes can occur simultaneously if ripening begins before fruit harvest. Developmental processes may also interact with environmental conditions during chilling. A low-temperature disorder of McIntosh apples called "corky flesh browning" occurs during long-term storage in 1.0 to 1.2% oxygen, and is much more severe in fruit harvested at advanced maturity.[35]

Given the complexity of these interactions and the massive chemical changes associated with maturation and ripening of fruits, these effects seem more likely to be associated with symptom expression than with induction of chilling damage per se.

D. ENVIRONMENTAL CONDITIONS DURING CHILLING

Environmental conditions other than temperature during the chilling period can greatly affect chilling damage of temperate-origin plant structures.

Very low oxygen concentrations (1 to 3%) can reduce superficial scald development on apples[36] and pears.[37] In general, the lower the oxygen concentration, the less likely that the disorder will occur.[36] However, low-oxygen stressing can increase core flush in Granny Smith apples,[36] and low-temperature breakdown of Cox's Orange Pippin apples.[38] High levels of carbon dioxide (3 to 6%) in the atmosphere can reduce superficial scald on apples, but can increase the incidence of core flush, brown core, and low-temperature breakdown.[3,36]

Controlled atmosphere storage of harvested plant materials usually combines conditions of low oxygen, high carbon dioxide, and low temperature and often delays deterioration more effectively than does low temperature alone. For example, peaches can be kept at chilling temperatures longer in controlled atmosphere than in air.[39] Yet, there is no evidence that the effects of atmosphere composition go beyond influences over symptom development, i.e., there is no evidence that atmospheric conditions during chilling can prevent the primary events in chilling injury.

Relative humidity also can influence development of chilling damage. High relative humidity can increase the occurrence of superficial scald and low-temperature breakdown of apples, but core flush is increased by low relative humidity.[3] The effects of high humidity have been ascribed to suppressed evolution of toxic volatiles[40] or to retention of excessive cellular turgor,[41] each of which would influence symptom development.

An important characteristic of chilling injury in plants, regardless of origin, is its reversibility by periodic transfer to nonchilling temperatures.[42] Intermittent warming can avoid development of symptoms on apples, cranberries, peaches, plums, nectarines, and potatoes[42] stored at a chilling temperature. Although the mechanism of reversal is unknown, it probably involves symptom development rather than a primary effect, since warming is effective even after prolonged periods of chilling as long as symptoms have not yet occurred.

E. TIME/TEMPERATURE RELATIONSHIPS

Lyons and Breidenbach[43] summarized approximate time/temperature conditions for development of chilling injury symptoms in a number of vegetable crops. All but asparagus

TABLE 1
Time/Temperature Relationships for Development of Chilling Injury Symptoms for Some Crops of Temperate Origin Stored in Air

Commodity	Approximate time/temperature conditions for symptoms	Normal storage conditions
Apple		
McIntosh	3 months at 0°C	6—8 months at 3°C
Cox's Orange Pippin	3 months at 0°C	6—7 months at 3°C
Jonathan	4 months at 0°C	6 months at 2°C
Delicious	8—11 months at 0°C	8—11 months at 0°C
Apricot	2—3 weeks at 0°C	2—3 weeks at 0°C
Asparagus	10 d at 0°C	3 weeks at 1.5—2.5°C
Chinese cabbage	3—4 months at 0°C	2—3 months at 0°C
Chrysanthemum	3 weeks at 0°C	2—3 weeks at 0°C
Cranberry	2—3 weeks at 0°C	6 weeks at 4°C
Gladiolus	8 d at 2°C	5—8 d at 2—'5°C
Peach and nectarine	3—4 weeks at 0°C	2—3 weeks at 0°C
Pear		
Bartlett	3 months at 0°C	2—3 months at 0°C
Comice	4.5 months at 0°C	4—4.5 months at 0°C
Anjou	7 months at 0°C	6—7 months at 0°C
Plum	5 weeks at 0°C	4—5 weeks at 0°C

From Hardenburg, R. E., Watada, A. E., and Wang, C. Y., Agricultural Handb. No. 66, U.S. Department of Agriculture, Washington, DC, 1986.

and potatoes were of tropical or subtropical origin, and except for potatoes, they developed symptoms after 3 to 15 d at 0 to 10°C.

In Table 1 are summarized commercial recommendations[4] for maximum storage conditions for most of the crops cited in our review. There are two prominent differences between the information in this table and that presented by Lyons and Breidenbach.[43] First, lower temperatures are involved. "Normal storage conditions" for temperate-origin crops do not exceed 5°C, whereas recommendations for tropical or subtropical crops usually do. Second, longer times of chilling are involved before symptoms usually develop. For tropical or subtropical crops, Lyons and Breidenbach show a maximum exposure period of 15 d (except for potatoes), whereas for temperate crops we found reports only for asparagus and gladiolus where symptoms developed after that short an episode. In fact, apples, Chinese cabbage, and pears typically can endure up to 3 months of continual exposure to 0°C before symptoms occur.

It should be recognized, however, that few systematic evaluations of these time/temperature relationships have been conducted on these crops except for apples and pears, perhaps because so many factors influence symptom development. Fidler et al.[3] constructed a graph (Figure 1) to represent the time/temperature relationships of three types of apple cultivars: one purportedly not chilling sensitive (critical temperature of 0°C), one moderately sensitive (critical temperature of 3°C), and one highly sensitive (critical temperature of 4.5°C). This figure in fact may depict the physiological framework of chilling susceptibility of temperate-origin plants, i.e., depict the time for response at a given temperature of plant materials with different resistances to low-temperature stress.

Relating Table 1 to Figure 1, several patterns appear. One is an apparent lack of sophistication in coping with chilling damage in these crops. In most cases, we simply recognize chilling symptoms as the limitation to long-term, low-temperature exposure. For example, the commercial solution to chilling damage to peaches is to keep them no more than 3 weeks at 0°C, even though temperature cycling can avoid chilling damage.[44] In

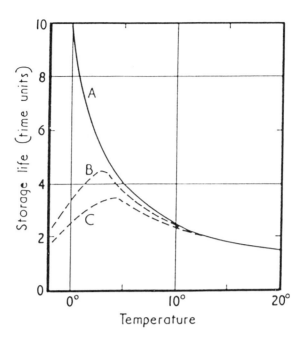

FIGURE 1.   The relationship between temperature and storage life
of apples. (A) Apples purportedly not subject to low-temperature
injury; (B and C) apples subject to low-temperature injury at 3 and
4.5°C, respectively. (From Fidler, J. C., Wilkenson, B. G., Edney,
K. L., and Sharples, R. O., *The Biology of Apple and Pear Storage,*
Commonwealth Agricultural Bureaux, East Malling, Kent, England, 1973. With permission from CAB International, Wallingford,
Oxfordshire, England.)

principle, it should be possible to develop better commercial strategies for avoiding chilling
damage.

Another striking feature is the likely genetic potential for selecting more chilling-tolerant
cultivars of a species. Except for apples and pears, recommendations of optimum conditions
are not tailored to cultivar tolerance, but rather to maximum sensitivity. Careful examination
of cultivar differences, and selection of more tolerant lines, may lead to different recommendations for different cultivars of a species, as is now done for apples and pears.

A third pattern is the commonality of the critical temperature (generally 0 to 3°C), but
the diversity of the critical time. This suggests a fundamental site for injury, but highly
variable biochemical mechanisms for translating stress into visible cellular damage.

IV. POSSIBLE MECHANISMS OF CHILLING RESPONSES

A. PRIMARY EFFECTS

Many differing views on the mechanisms for chilling injury in plants have evolved since
the unifying hypothesis of Lyons.[2] Most of these concede that cell membranes are likely
sites of primary effects of chilling, but opinions vary widely on whether the effects center
on membrane lipids, proteins, or lipid-protein interactions.[45]

Extending Lyons'[2] hypothesis, McGlasson and Raison[46] compared Arrhenius plots for
mitochondrial oxidative activity of six cultivars of apples that varied in susceptibility to low-
temperature breakdown during storage. While all of the plots showed abrupt discontinuities
at low temperature, the temperature at which they occurred did not correlate with susceptibility to the disorder. Similarly, when phase transition temperatures were compared for

chilling-sensitive Starking Delicious and chilling-resistant Ralls Janet apples at harvest, no differences were found.[47] However, after 6 months of storage, the phase transition temperature was lower than at harvest for both cultivars and was lower in the cold-tolerant than in the cold-sensitive cultivar. This corresponded with a higher proportion of unsaturated fatty acids and a higher phosphatidylcholine/phosphatidylethanolamine ratio in Ralls Janet than in Starking Delicious mitochondrial lipids.[48] Although interpretations of Arrhenius plots and bulk lipid analyses are suspect,[45] these results and numerous others at least suggest that compositions and phase transitions of lipids may be important in initiating chilling effects in temperate-origin plants.

Alternatively, a direct effect of low temperature on proteins in the membranes could trigger a chilling effect. Adenosine triphosphatase activity of cauliflower plasma membranes exhibited increased activation energy at low temperature, independent of any phase transition in bulk membrane lipids, which were 75% unsaturated.[49]

Attempting to define the primary event in these tissues is made difficult by the long periods of continued exposure that typically are required for injury to occur. Many of these tissues are very metabolically active at these temperatures, and turnover of both lipids[50] and proteins[51] occurs. During the chilling episode, "hardening" responses can occur, such as increased unsaturation of lipid components, which could account for lowered phase transition temperature following storage of apples.[47] This could make these tissues "moving targets" for cold damage.

Whatever the primary event is in chilling of temperate-origin plants, it likely differs in some fundamental way from that of tropical or subtropical plants. The literature abounds with comparisons of chilling-sensitive and chilling-resistant plants, using short intervals of chilling. This difference well may be quantitative, rather than qualitative, as indicated, for example, by the many reports of greater unsaturation of lipids in resistant plants. If only a relatively small proportion of total lipids or of total proteins in membranes of temperate origin plants was altered at low temperature, relative to proportions altered in tropical-origin plants, then the change would be hard to detect experimentally. More importantly, adverse consequences to cell functioning could be greatly attenuated over time. This also would increase the significance of small differences in secondary factors, such as metabolic differences between cultivars or compositional differences due to environmental conditions during growth or changes during maturation and senescence.

B. SECONDARY EFFECTS

The diversity of visual symptoms of chilling injury indicates many avenues for translating the primary effects of temperature into cellular damage and collapse. In no system is this more clear than in the apple, where a single species can develop at least five unique symptoms, depending on cultivar, maturity, growing conditions, and postharvest environmental conditions. Wade's[52] statement 10 years ago that, "In no case does it seem possible to set out in detail a sequence of events starting with a change in membrane structure and ending in cell death," is equally applicable today. Yet, some conjecture may stimulate research toward that end.

General stress to cell physiology will result from disruption of normal functioning of organelles such as chloroplasts and mitochondria. There is little doubt that such disruption occurs over time at chilling temperatures. For example, in apples stored at 0°C, CO_2 production increased substantially prior to occurrence of low-temperature breakdown; at 2 or 4°C, respiration did not increase and the disorder did not occur.[53] However, consequences of such dysfunctions may be general, leading to faster senescence and loss of disease resistance, rather than leading to unique expressions of cellular damage. Accumulation of fermentation products is often cited and attributed causal significance (see Reference 2), but may be a symptom of morbid tissues rather than a causal factor in cell damage.[52]

Many metabolic changes have been recorded during prolonged chilling of plant tissues.[42] Prominent among these are changes involving oxidative reactions. Recent studies of chilling-enhanced photooxidations suggest that free radical generation is promoted by chilling, leading to lipid peroxidation, and that chilling-resistant tissue has the ability to remove the toxic oxygen species before lipid peroxidation occurs.[54,55]

Effects of long-term chilling on temperate-origin crops generally occur during storage of plant tissues in the dark, but circumstantial evidence that oxidative degradation is involved in the cellular damage is abundant. For example, superficial scald of apples is reduced when they are maintained at 1.5% O_2, and reduced even more if O_2 concentration is maintained at 0.5% during the first 9 d of storage.[56] Pretreatment of cucumbers and sweet peppers with the free radical scavengers ethoxyquin or sodium benzoate reduced subsequent chilling injury.[57] Treatment of apples with the antioxidants ethoxyquin, diphenylamine, and butylated hydroxyanisole reduced soft scald development on apples during storage and correspondingly increased the amount of unsaturated fatty acids in the surface lipids and reduced accumulation of hexanol in the fruit flesh.[58] Ascorbic acid decreased during chilling,[59,60] but increased when tissue was transferred to a nonchilling temperature.[60]

Superficial scald of apples is perhaps the best illustration of how adverse oxidative reactions, if uncontrolled, might lead to symptoms of chilling injury.[61] Susceptible fruit slowly accumulate the terpenoid α-farnesene (usually absent at time of harvest) during the first few months at low temperature. After 2 to 3 months at low temperature, the α-farnesene begins to be oxidized to conjugated trienes which then accumulate for as long as the fruit are at low temperature. The disorder can be controlled by application of the antioxidants diphenylamine or ethoxyquin before storage, which suppress the oxidation of α-farnesene to conjugated trienes.[62] Conjugated trienes are presumed to lead to symptom development, but high-temperature pretreatment before storage (38°C for 3 or 4 d) reduced superficial scald of Granny Smith apples without substantially lowering conjugated triene accumulation at 0°C.[65] Susceptibility of fruit to superficial scald varies greatly with cultivar, maturity, and growing conditions and has been correlated with level of activity of natural antioxidants in the fruit surface at harvest.[61,63] Thus, whether or not this disorder develops during chilling may depend on the ability of the tissue to control oxidative reactions that can produce toxic metabolites.

We suggest that near-freezing temperatures impose stresses on all plant materials by fundamentally altering cellular metabolism in response to primary effects on membrane functioning. Plants of temperate origin, being survivors of a potentially chilling environment, are ones with substantial resistance to low-temperature stress, perhaps via both a smaller primary response and a greater ability to control secondary responses. This attenuated susceptibility would be expected to vary greatly among genetic strains, growing conditions, and stages of plant development, as we have shown here, and to be responsive to a variety of experimental techniques, as described elsewhere.[42,64] However, these plants should be considered to be *resistant but not immune* to chilling damage, with an inherent potential to exhibit symptoms of chilling damage under the appropriate set of conditions.

REFERENCES

1. **Morris, L. L.,** Chilling injury of horticultural crops: an overview, *HortScience,* 17, 161, 1982.
2. **Lyons, J. M.,** Chilling injury in plants, *Annu. Rev. Plant Physiol.,* 24, 445, 1973.

3. **Fidler, J. C., Wilkinson, B. G., Edney, K. L., and Sharples, R. O.,** *The Biology of Apple and Pear Storage,* Commonwealth Agricultural Bureaux, East Malling, Kent, England, 1973.

4. **Hardenburg, R. E., Watada, A. E., and Wang, C. Y.,** The Commercial Storage of Fruits, Vegetables, and Florist and Nursery Stocks, Agricultural Handb. No. 66, U.S. Department of Agriculture, Washington, DC, 1986.

5. **Trout, S. A., Tindale, G. B., and Huelin, F. E.,** Investigations on the storage of Jonathan apples grown in Victoria, CSIRO Bull. No. 135, Melbourne, Australia, 1940.

6. **Iwata, T. and Kinoshita, M.,** Studies of storage and chilling injury of Japanese apricot fruits. II. Chilling injury in relation to storage temperature, cultivar, maturity, and polyethylene packaging, *J. Jpn. Soc. Hortic. Sci.,* 47, 97, 1978.

7. **Apeland, J.,** Chilling injury in Chinese cabbage, *Brassica campestris* pekinensis (Lour) Olsson, *Acta Hortic.,* 157, 261, 1985.

8. **Ryall, A. L. and Lipton, W. J.,** *Handling, Transportation, and Storage of Fruits and Vegetables,* AVI Publishing, Westport, CT, 1972, 344.

9. **Smock, R. M.,** Nomenclature for internal storage disorders of apples, *HortScience,* 12, 306, 1977.

10. **Ap Rees, T., Dixon, W. L., Pollock, C. J., and Franks, F.,** Low temperature sweetening of higher plants, in *Recent Advances in the Biochemistry of Fruits and Vegetables,* Friend, J. and Rhodes, M. J. C., Eds., Academic Press, London, 1981, chap. 2.

11. **Stoll, K. and Weichmann, J.,** Root vegetables, in *Postharvest Physiology of Vegetables,* Weichmann, J., Ed., Marcel Dekker, New York, 1987, chap. 30.

12. **Ben-Arie, R. and Lavee, S.,** Pectic changes occurring in Elberta peaches suffering from woolly breakdown, *Phytochemistry,* 10, 531, 1971.

13. **Wang, C. Y., Sams, C. E., and Gross, K. C.,** Ethylene, ACC, soluble polyuronides, and cell wall noncellulosic neutral sugar content in 'Eldorado' pears during cold storage and ripening, *J. Am. Soc. Hortic. Sci.,* 110, 687, 1985.

14. **Sfakiotakis, E. M. and Dilley, D. R.,** Induction of ethylene production in 'Bosc' pears by postharvest cold stress, *HortScience,* 9, 336, 1974.

15. **Wang, C. Y., Mellenthin, W. M., and Hansen, E.,** Effect of temperature on development of premature ripening in 'Bartlett' pears, *J. Am. Soc. Hortic. Sci.,* 96, 122, 1971.

16. **Porritt, S. W.,** The effect of temperature on postharvest physiology and storage life of pears, *Can. J. Plant Sci.,* 45, 90, 1964.

17. **Li, P. H. and Hansen, E.,** Effect of modified atmosphere storage on organic acid and protein metabolism of pears, *Proc. Am. Soc. Hortic. Sci.,* 85, 100, 1964.

18. **Porritt, S. W., Meheriuk, M., and Lidster, P. D.,** Postharvest disorders of apples and pears, Publ. 1737/E, Communications Branch, Agriculture Canada, Ottawa, Ontario, 1982.

19. **Kasmire, R. F., Kader, A. A., and Klaustermeyer, J. A.,** Influence of aeration rate and atmospheric composition during simulated transit on visual quality and off-odor production by broccoli, *HortScience,* 9, 228, 1974.

20. **Graham, D. and Patterson, B. D.,** Responses of plants to low, nonfreezing temperatures: proteins, metabolism, and acclimation, *Annu. Rev. Plant Physiol.,* 33, 347, 1982.

21. **Meheriuk, M., Lau, O. L., and Hall, J. W.,** Effects of some postharvest and storage treatments on the incidence of flesh browning in controlled-atmosphere-stored 'Delicious' apples, *J. Am. Soc. Hortic. Sci.,* 109, 290, 1984.

22. **Smith, W. L., Jr. and Wilson, J. B.,** Market diseases of potatoes, Agricultural Handb. No. 479, U.S. Department of Agriculture, Washington, DC, 1978.

23. **Brecht, J. K., Kader, A. A., and Ramming, D. W.,** Description and postharvest physiology of some slow-ripening nectarine genotypes, *J. Am. Soc. Hortic. Sci.,* 109, 596, 1984.

24. **Proebsting, E. L., Jr., Carter, G. H., and Mills, H. H.,** Interaction of low temperature storage and maturity on quality of 'Early Italian' prunes, *J. Am. Soc. Hortic. Sci.,* 99, 117, 1974.

25. **Fidler, J. C.,** Scald and weather, *Food Sci. Abstr.,* 28, 545, 1956.

26. **Smock, R. M.,** Some effects of climate during the growing season on keeping quality of apples, *Proc. Am. Soc. Hortic. Sci.,* 62, 272, 1953.

27. **Merritt, R. H., Stiles, W. C., Havens, A. V., and Mitterling, L. A.,** Effects of pre-harvest air temperatures on storage scald of Stayman apples, *Proc. Am. Soc. Hortic. Sci.,* 78, 24, 1961.

28. **Smock, R. M.,** Some factors affecting the brown core disease of McIntosh apples, *Proc. Am. Soc. Hortic. Sci.,* 47, 67, 1946.

29. **Lyons, J. M., Raison, J. K., and Steponkus, P. L.,** The plant membrane in response to low temperature: an overview, in *Low Temperature Stress in Crop Plants: The Role of the Membrane,* Lyons, J. M., Graham, D., and Raison, J. K., Eds., Academic Press, New York, 1979, 1.

30. **Drake, M., Bramlage, W. J., and Baker, J. H.,** Effects of foliar calcium on McIntosh apple storage disorders, *Commun. Soil Sci. Plant Anal.,* 10, 303, 1979.

31. **Scott, K. J. and Wills, R. B. H.,** Postharvest application of calcium as a control for storage breakdown of apples, *HortScience,* 10, 75, 1975.

32. **Autio, W. R., Bramlage, W. J., and Weis, S. A.,** Predicting poststorage disorders of 'Cox's Orange Pippin' and 'Bramley's Seedling' apples by regression equations, *J. Am. Soc. Hortic. Sci.,* 111, 738, 1986.

33. **Kidd, F. and West, C.,** The cause of low-temperature breakdown in apples, Report of the Food Investigation Board for 1933, 57, 1933.

34. **Knee, M. and Hatfield, S. G. S.,** Benefits of ethylene removal during apple storage, *Ann. Appl. Biol.,* 98, 157, 1981.

35. **Lau, O. L., Yastremski, R., and Meheriuk, M.,** Influence of maturity, storage procedure, temperature, and oxygen concentration on quality and disorders of 'McIntosh' apples, *J. Am. Soc. Hortic. Sci.,* 112, 93, 1987.

36. **Little, C. R. and Peggie, I. D.,** Storage injury of pome fruit caused by stress levels of oxygen, carbon dioxide, temperature, and ethylene, *HortScience,* 22, 783, 1987.

37. **Mellenthin, W. M., Chen, P. M., and Kelly, S. B.,** Low oxygen effects on dessert quality, scald prevention, and nitrogen metabolism of 'd'Anjou' pear fruit during long-term storage, *J. Am. Soc. Hortic. Sci.,* 105, 522, 1980.

38. **Sharples, R. O.,** Effect of ultra-low oxygen conditions on the storage quality of English Cox's Orange Pippin apples, in *Controlled Atmospheres for Storage and Transportation of Perishable Agricultural Commodities,* Richardson, D. G. and Meheriuk, M., Eds., Timber Press, Beaverton, Oregon, 1982, 131.

39. **Wang, C. Y. and Anderson, R. E.,** Progress on controlled atmosphere storage and intermittent warming of peaches and nectarines, in *Controlled Atmospheres for Storage and Transportation of Perishable Agricultural Commodities,* Richardson, D. G. and Meheriuk, M., Eds., Timber Press, Beaverton, Oregon, 1982, 221.

40. **Wills, R. B. H.,** Influence of water loss on the loss of volatiles by apples, *J. Sci. Food Agric.,* 19, 354, 1982.

41. **Simon, E. W.,** The symptoms of calcium deficiency in plants, *New Phytol.,* 80, 1, 1978.

42. **Wang, C. Y.,** Physiological and biochemical responses of plants to chilling stress, *HortScience,* 17, 173, 1982.

43. **Lyons, J. M. and Breidenbach, R. W.,** Chilling injury, in *Postharvest Physiology of Vegetables,* Weichmann, J., Ed., Marcel Dekker, New York, 1987, chap. 15.

44. **Anderson, R. E.,** Long-term storage of peaches and nectarines intermittently warmed during storage in CA, *J. Am. Soc. Hortic. Sci.,* 107, 214, 1982.

45. **Lyons, J. M., Graham, D., and Raison, J. K.,** Epilogue, in *Low Temperature Stress in Crop Plants,* Lyons, J. M., Graham, D., and Raison, J. K., Eds., Academic Press, New York, 1979, 543.

46. **McGlasson, W. B. and Raison, J. K.,** Occurrence of a temperature-induced phase transition in mitochondria isolated from apple fruit, *Plant Physiol.,* 52, 390, 1973.

47. **Kimura, S., Yamashita, S., and Okamoto, T.,** Differential scanning calorimetry of lipids of stored apples, *Agric. Biol. Chem.,* 50, 707, 1986.

48. **Kimura, S., Kanno, M., Yamada, Y., Takahashi, K., Murashige, H., and Okamoto, T.,** The contents of conjugated lipids and fatty acids and the cold tolerance in the mitochondria of Starking Delicious and Ralls Janet apples, *Agric. Biol. Chem.,* 46, 2895, 1982.

49. **Wright, L. C., McMurchie, E. J., Pomeroy, M. K., and Raison, J. K.,** Thermal behavior and lipid composition of cauliflower plasma membranes in relation to ATPase activity and chilling sensitivity, *Plant Physiol.,* 69, 1356, 1982.

50. **Thayer, S. S., Choe, H. T., Tang, A., and Huffaker, R. C.,** Protein turnover during senescence, in *Plant Senescence: Its Biochemistry and Physiology,* Thomson, W. W., Nothnagel, E. A., and Huffaker, R. C., Eds., American Society of Plant Physiologists, Rockville, MD, 1987, 71.

51. **Mazliak, P.,** Plant membrane lipids: changes and alterations during aging and senescence, in *Post-Harvest Physiology and Crop Preservation,* Lieberman, M., Ed., Plenum Press, New York, 1983, 123.

52. **Wade, N. L.,** Physiology of cool-storage disorders of fruit and vegetables, in *Low-Temperature Stress in Crop Plants,* Lyons, J. M., Graham, D., and Raison, J. K., Eds., Academic Press, New York, 1979, 81.

53. **Johnson, D. and Ertan, U.,** Interaction of temperatures and oxygen level on the respiration rate and storage quality of Idared apples, *J. Hortic. Sci.,* 58, 527, 1983.

54. **Wise, R. R. and Naylor, A. W.,** Chilling-enhanced photooxidation. The peroxidative destruction of lipids during chilling injury to photosynthesis and ultrastructure, *Plant Physiol.,* 83, 272, 1987.

55. **Wise, R. R. and Naylor, A. W.,** Chilling-enhanced photooxidation. Evidence for the role of singlet oxygen and superoxide in the breakdown of pigments and endogenous antioxidants, *Plant Physiol.,* 83, 272, 1987.

56. **Little, C. R., Faragher, J. D., and Taylor, H. J.,** Effects of initial oxygen stress treatments in low oxygen modified atmosphere storage of 'Granny Smith' apples, *J. Am. Soc. Hortic. Sci.,* 107, 320, 1982.

57. **Wang, C. Y. and Baker, J. E.,** Effects of two free radical scavengers and intermittent warming on chilling injury and polar lipid composition of cucumber and sweet pepper fruits, *Plant Cell Physiol.,* 20, 243, 1979.

58. **Wills, R. B. H., Hopkirk, G., and Scott, K. J.,** Reduction of soft scald in apples with antioxidants, *J. Am. Soc. Hortic. Sci.,* 106, 569, 1981.

59. **Yamauchi, N., Minamide, T., and Ogata, K.,** Physiological and chemical studies on ascorbic acid of fruits and vegetables. II. Changes of ascorbic acid content during development of chilling injury, *J. Jpn. Soc. Hortic. Sci.,* 44, 303, 1975.

60. **Lieberman, M., Craft, C. C., and Wilcox, M. S.,** Effect of chilling on the chlorogenic acid and ascorbic acid content of Porto Rico sweetpotatoes, *Proc. Am. Soc. Hortic. Sci.,* 74, 642, 1959.

61. **Meir, S. and Bramlage, W. J.,** Antioxidant activity in 'Cortland' apple peel and susceptibility to superficial scald after storage, *J. Am. Soc. Hortic. Sci.,* 113, 412, 1988.

62. **Huelin, F. E. and Coggiola, I. M.,** Superficial scald, a functional disorder of stored apples. IV. Effect of variety, maturity, oiled wraps, and diphenylamine on the concentration of alpha-farnesene in the fruit, *J. Sci. Food Agric.,* 19, 297, 1968.

63. **Anet, E. F. L. J.,** Superficial scald, a functional disorder of stored apples. XI. Apple antioxidants, *J. Sci. Food Agric.,* 25, 299, 1974.

64. **Bramlage, W. J.,** Chilling injury of crops of temperate origin, *HortScience,* 17, 165, 1982.

65. **Klein, J.,** Volcani Center, Bet Dagan, Israel, personal communication.

Chapter 4

THE EFFECTS OF CHILLING TEMPERATURES DURING SEED GERMINATION AND EARLY SEEDLING GROWTH

Robert C. Herner

TABLE OF CONTENTS

I. INTRODUCTION

Plants can be divided into chilling-tolerant or chilling-sensitive species, depending on their susceptibility to chilling injury when exposed to temperatures between 0°C and 10 to 15°C. Susceptibility is, in turn, related to the place of origin of the crops, with those arising from tropical or subtropical areas being chilling sensitive and those arising in temperate areas being chilling resistant.[1]

Chilling injury is a physiological disorder which affects plants at all stages of development, so it is not surprising to find that seeds and seedlings of crops of tropical or subtropical origin are severely affected by exposure to temperatures between 0°C and 10 to 15°C. On the other hand, seeds and seedlings of crops of temperate origin are not injured by these temperatures.[2,3] The following discussion presents the current thoughts concerning the injury which occurs to seeds and seedlings of chilling-sensitive crops at low but non-freezing temperatures.

II. IMBIBITIONAL CHILLING INJURY

Chilling-resistant crops produce seeds that are able to germinate at temperatures less than 10°C.[4] Germination rate may be very slow at low temperatures, and, as a result, emergence percentage may be decreased because of factors such as soil crusting and seed decay. In the absence of pathogens, seeds are not injured by low-temperature exposure for long periods and the seeds germinate when temperatures rise. Even after the radicle has emerged from the seed coat, chilling-resistant seeds are not injured by exposure to temperatures in the chilling range.

Chilling-sensitive plants produce seeds that do not germinate below 10 to 15°C and can be subdivided into two general categories based on when injury occurs in relation to the initiation of germination.[2]

The first group of plants producing seeds that are sensitive to low temperatures are those injured during the initial stages of imbibition. Seeds of the warm-season legumes, such as snap beans,[5] lima beans,[6,7] soybeans,[8,9] chick-peas,[10] sweet corn,[11] and cotton[12,13] have been shown to be particularly sensitive during the imbibition period. Symptoms of imbibitional-chilling injury include the production of low-vigor plants, decreased germination, increased decay, production of abnormal plants, and seed death.

The second group of plants produce seeds which are not injured by imbibition at low temperatures. Seeds of this type of plant remain fully imbibed at low temperatures for fairly long periods of time, but do not germinate unless temperatures rise; germination then proceeds normally with little indication of injury. However, once radicle growth has been initiated, injury occurs if the seed is exposed to chilling temperatures. Typical symptoms of injury include necrosis of the tissue just behind the radicle tip[14] or damage to the root cortex.[15] Degree of injury is dependent on how low the temperature is and the length of exposure.[16,17] Tomato,[18] pepper,[16,17] and watermelon[19] are examples of plants with this type of seed. A response of this type is probably true of other warm-season solanaceous and cucurbitaceous plants.

A. FACTORS AFFECTING IMBIBITIONAL CHILLING INJURY

Much of the research done on the effect of low temperatures on germination has been on crops subject to injury which occurs during the initial stages of imbibition. This injury has been related to (1) temperature and relation between the timing of low-temperature exposure and stage of germination, (2) initial moisture content of the seed at the start of imbibition, (3) rate of imbibition and seed coat characteristics or integrity, and (4) seed vigor and cultivar or species.

1. Temperature and Timing

More injury occurs in chilling-sensitive seeds if imbibition begins under low temperatures than if seeds are imbibed in moderate temperatures.[7,12,13] However, little or no injury results if imbibition begins under warm conditions followed by low-temperature exposure. The low-temperature-sensitive stage is the initial stage of germination when imbibition is occurring rapidly. An exposure of a few minutes or hours is sufficient to cause injury during the initiation of germination.

2. Initial Moisture Content

Increasing the initial seed moisture content of seeds by placing them in a high-relative humidity environment prior to imbibition also protects the seeds from injury when subsequently imbibed at low temperatures.[5,6,11,20,21] The combination of low initial moisture content and imbibition at low temperature is very damaging.

3. Rate of Imbibition and Seed Coat Characteristics

Rapid imbibition is reportedly related to poor germination at low temperatures. The imbibition rate is related to seed coat characteristics and integrity. Simon and Mills[22] reported that when *Pisum sativum* embryos without seed coats were imbibed, water uptake was more rapid, more potassium leaked from the seed, and more damage occurred than intact seeds with good seed coat integrity. Seeds with damaged testas behaved more like naked embryos than intact seeds. Tully et al.[23] also indicated that the seed coat controlled water uptake, and this, in turn, determined whether injury occurred during imbibition at low temperatures.

Snap beans with pigmented testas absorbed water more slowly and germinated better than those with white seed coats.[24-27] Taylor and Dickson[28] showed that snap beans which contain the semihard seed coat character have decreased imbibition rates at low seed moisture levels, and imbibitional chilling injury is reduced. A semihard seed is one which does not imbibe water when seed moisture is less than 5%, takes up water very slowly when seed moisture is between 5 and 10%, and when seed moisture is raised above 10%, normal imbibition occurs.[29] Seeds that imbibe water rapidly tend to have ''loose'' adhering seed coats, and those which imbibe water slowly tend to have ''tight'' adherence of the seed coat to the cotyledons.[24]

Thus, one of the functions of the intact seed coat may be to slow the rate of imbibition. If the seed coat is damaged during seed development or during harvesting and handling, imbibition rate of the seed may be increased resulting in damage and poor seed germination.

Two areas of concern must be recognized concerning the studies correlating imbibition rate and poor germination. One is that measurements of the rate of water uptake by intact seeds may not reflect the rate of hydration of the cotyledonary and axis tissue, but may instead reflect hydration of the seed coat and/or filling of air cavities within the testa and between the cotyledons. There is also a considerable amount of evidence that the absolute rate of water imbibition is not correlated with injury, a topic discussed in more detail later.

4. Vigor and Cultivar Differences

Low-vigor seeds seem to be particularly injured by imbibition at low temperatures.[30-33] The environment during seed development can have a large effect on seed vigor and the capacity to germinate at low temperatures.[9,30,34-37] Soybean and snap bean seeds which matured under cooler temperatures germinated better at low temperatures,[9,35,36] perhaps due to effects on seed coat integrity, the composition of membranes in the cells, or to some other unexplained factors. There is also a great deal of evidence for cultivar differences in susceptibility to low temperatures during germination.

B. CAUSES OF IMBIBITIONAL CHILLING INJURY

Several ideas have been proposed to explain the injury to seeds of warm-season crops that occurs during imbibition at low temperatures. These include (1) membrane compositional differences, (2) pathogen attack, (3) rate of water uptake and leakage, (4) seed coat characteristics, (5) membrane disruption, (6) membrane reorganization, and (7) tissue wetting phenomena.

1. Membrane Compositional Differences

Since some seeds of crops which are chilling sensitive are injured during imbibition under cold temperature conditions, and seeds of chilling-resistant crops are not, it has been suggested that imbibitional chilling injury was the result of membrane compositional differences. These compositional differences could result in a phase change of discrete portions of the membrane when exposed to low temperature, resulting in a series of events that lead to chilling injury of the chilling-sensitive species.[1]

Several attempts have been made to correlate membrane composition of seeds to susceptibility to chilling injury during germination. Some reports have shown a correlation between germination at cold temperatures and a higher unsaturated to saturated fatty acid ratio in seed lipids of chilling resistant species,[38,39] and one report suggests that there is a difference in the type of phospholipid synthesized in cotyledonary and axis tissue during germination of chilling-sensitive and -resistant species.[40] The activity of membrane-bound enzymes is adversely affected by exposure to low temperatures during germination and early seedling growth of chilling-sensitive species,[41,42] and it has been suggested that a phase change of the membranes does occur at low temperatures in these species.[41-43] An adverse effect on respiration has also been demonstrated in germinating seeds which may be a reflection of a phase change occurring.[41-45]

In contrast, other researchers have not been able to detect differences in saturation of membrane lipids between chilling-sensitive and chilling-resistant cultivars of the same species.[46-48] Others could not demonstrate any detectable phase transition of soybean phospholipids or mitochondrial lipids in the temperature range where chilling injury occurs.[49] Cohn and Obendorf[50] showed that in corn there was no difference in oxygen uptake of whole kernels after imbibitional chilling, nor were there differences in ATP levels as a result of low-temperature imbibition. This indicated that a disruption of energy metabolism was not the primary cause of imbibitional chilling injury. Finally, Simon et al.[51] could not demonstrate any indication of a discontinuity of the Arrhenius plot (which may be indicative of a membrane phase change) in the respiration rate of cucumber or mung bean seeds during germination compared over a range of temperatures.

The role, if any, that differences in membrane composition may have in differential sensitivity to cold temperatures during imbibition of dry seeds is still not clear. It is possible that once the seed is fully hydrated and radicle protrusion has occurred, differences in membrane composition could account for differences in sensitivity, although this has not been proven.

2. Pathogen Attack

One of the most frequently observed effects of imbibing seeds in a low-temperature medium is increased or sustained leakage of substances from the seed.[52] Substances leaking out of seeds during imbibition include amino acids, sugars, organic acids, gibberellic acid, phosphates, phenolics,[53] succinate,[54] and enzymes.[55] Associated with this leakage is increased decay, probably promoted by the loss of sugars and other substances into the germinating medium.[55-58] Leach[59] has suggested that soil organisms have a lower optimum growth temperature compared to that for seed germination and seedling growth, which allows soil pathogens to grow, multiply, and attack the seed. Pathogen attack is a very important problem in the cold soil environment, and some protection against pathogens can be provided by chemical seed treatments. However, it is probable that decay is a secondary factor in decreased seed germination of chilling-sensitive species and that some primary lesion causes the damage that allows the pathogens to attack the seed.

3. Leakage of Substances from the Seed

Dry seeds placed in a moist medium not only imbibe water rapidly, but there is also a large amount of materials leaked from the tissue into the medium. The amount of material is initially very high and rapidly decreases as imbibition proceeds.[52] Generally, the greater the leakage or the longer the period that a high leakage rate is sustained, the greater is the damage to the tissue.[52] Several studies have shown that the amount of leakage is negatively correlated to vigor and field emergence.[9,60] It has also been shown that when chilling-sensitive seeds are imbibed at low temperatures, the leakage of material is increased compared to higher, nonchilling temperatures.[8,61] Leakage from the seeds can be decreased by the presence of an intact seed coat or by raising the initial moisture content of the seed.[62] Although leakage from the seed can undoubtedly result in damage, the loss of materials from the seed is not the primary cause of injury, but is simply a reflection of changes in the membrane physiology. There have been many studies trying to explain the changes in permeability of membranes during imbibition.

4. Membrane Reorganization

Simon[52] has suggested that membranes in dry seeds are not in a bilayer configuration typical of hydrated plant membranes, but are arranged in ''hexagonal arrays'' so that the phospholipids are oriented around water molecules. The membrane changes from a semipermeable membrane to a rather porous, disorganized state as water is lost from the seed during maturation.

Simon further suggested that as water enters dry seeds, the membranes become reorganized into a typical bilayer and leakage is decreased as the membranes regain their semipermeability. This is supported by the fact that dry seeds placed in a moist environment have a high rate of electrolyte leakage which rapidly decreases with time as the seed hydrates.[52,63] This hypothesis is also supported by the fact that pea seeds harvested when still succulent and placed in water leak very little material, but the leakage rate is initially very high if they are dried.[63]

Too rapid movement of water into dry seed would presumably interfere with membrane reorganization, and, therefore, the high rate of leakage would be sustained and injury would result.[22] Low temperatures during imbibition also would be expected to prevent or delay membrane reorganization. If the seeds imbibe slowly, membrane reorganization could occur more readily, thus decreasing leakage and injury.[63] If the seed moisture content was raised prior to being exposed to free water, membrane reorganization would already have occurred and one would expect less leakage and injury when high-moisture seeds are imbibed.[64]

The high initial rate of leakage of materials could be due to washing off of surface deposits, but Simon and Wiebe[64] have stated that the amounts leaked are too large to account for as surface deposits, and the material leaked suggested cytoplasmic origin. Repeated wetting and drying cycles result in large initial leakage rates when dried seeds are wetted. One would expect a complete washing off of surface deposits after two or three rinses if they were responsible for the large leakage rates.[64] A word of caution has been raised by Duke et al.[54] concerning the measurement of seed leachates and the interpretation of the state of membranes. They state that electrolyte measurements reflect only passive diffusion through the membrane, UV measurements reflect both passive diffusion and cellular or membrane rupture, and enzymatic activity measurements measure only cellular or membrane rupture. To accurately measure the state of the membranes, they suggest that the latter measurement technique is the most useful.

Investigations concerning ultrastructural changes of seed membranes during hydration have indicated that membranes of dry seeds do show some irregularities and disorganization which disappear upon hydration.[65,66] On the other hand, several researchers have stated that there is no evidence in plants that membranes exhibit a hexagonal array configuration, and, in fact, most reports indicate that the membranes are in a typical bilayer form at very low moistures levels.[67-70]

5. Membrane Rupture and Rate of Water Uptake

Another hypothesis concerning imbibitional-chilling injury was proposed by Larson[71] when he suggested that rapid intake of free water during imbibition causes a disruption of cellular membranes, increasing leakage from the tissue and causing cell death.

Powell and Matthews[52,72-74] showed that rapid imbibition of seeds that were then stained with 2,3,5-triphenyltetrazolium chloride (TTC) had a layer of cells on the adaxial surface (curved outer surface) of the cotyledons that did not stain red, indicating damaged cells. Seeds with high rates of water uptake had cracks in the seed coats, a high proportion of cells not stained by TTC, low vigor, high amounts of electrolyte leakage, and exhibited poor field emergence.[73] At first, the nonstaining cells were described as "dead", but later work indicated that the proportion of nonstaining cells exposed to TTC could be decreased if succinate was present during imbibition or subsequent staining.[53] This finding indicated that, during imbibition, succinate was one of the constituents being leached from the cotyledons.

Reducing the rate of water uptake, either by imbibing with intact and undamaged seed coats or imbibing in polyethylene glycol (PEG), decreased the amount of leakage and the amount of injury.[72] Wolk[75] investigated the effects of four methods of slowing imbibition rate on the percent germination of intact snap bean seeds. He found that when (1) incremental water additions to a petri dish vs. all 5 ml of water at once were added, (2) the hilum was placed up vs. down and in contact with moisture, (3) the number of layers of filter paper were varied using the same volume of water, and (4) wax was applied to the hilum region, the rate of imbibition was slowed and percent germination was increased after a 24-h period of imbibition at 5°C. All of the above support the hypothesis that the rapid movement of water into dry seeds causes damage, but it does not explain why severity of injury is increased at low temperatures.

There are several reports that indicate that the absolute rate of water uptake is not the sole factor that controls imbibitional-chilling injury. For example, in peanuts the testa is very thin and membranous. The presence or absence of the peanut testa has no effect on the rate of imbibition by the embryo, but the removal of the testa increased leakage of material into the surrounding medium and decreased vigor of the embryo, thus illustrating that in this case there is no correlation between imbibition rate and injury.[76]

As reported earlier, raising the initial moisture content of chilling-sensitive seeds decreases injury that occurs during imbibition at low temperatures.[5,6,21] However, raising the initial moisture content of the seed actually increases the imbibition rate and results in decreased rates of leakage from the seed;[8,75] these results are opposite from those expected if the rate of water uptake was solely responsible for damage that occurs during low temperature imbibition. Chen et al.[10] imposed slow and fast imbibition rates on chick-pea seed at 2 and 20°C. The slow 20°C imbibition rate and fast 2°C imbibition rate were identical, but germination was very good at 20°C and very poor at 2°C. Again, the absolute rate of imbibition is not related to injury.

Finally, imbibition rates at low temperatures are slower than at high temperatures, but damage to sensitive seeds imbibed at low temperatures is greater than those imbibed at high temperatures.[5,6,77-79] Leopold showed that as temperatures decrease below 10 to 15°C, imbibition of soybeans decreases, but leakage rates actually increase.[61] Imbibition rate is only indicative of injury when comparisons are made under identical conditions of temperature, initial seed moisture content, plant species, and other factors.[75]

6. Seed Coat Characteristics

Several reports have shown that the seed coat can have a beneficial or adverse effect on the development of imbibitional chilling injury depending upon the characteristics of the seed coat or its integrity. Tully et al.[23] showed that intact pea seeds imbibed water slowly in cold water with little or no loss in vigor, while soybeans imbibed water more rapidly and were injured. If the pea seed coat was nicked, the peas imbibed water more rapidly and were then susceptible to injury. The susceptibility of intact soybean seeds to imbibitional injury was reduced by imbibition in a PEG solution. Similar protective effects of the seed coat have been observed by others for peas,[72-74] beans,[27,80] and soybeans.[81]

It has long been known that transverse cotyledon cracking occurs during imbibition of low moisture seeds and is more severe when imbibed at low temperatures.[82] Removing the seedcoat prior to imbibition increases cracking, and raising the initial seed moisture decreases cracking, responses which are similar to imbibitional chilling injury. It has also been observed that snap beans with white seed coats are much more susceptible to transverse cotyledon cracking than those with pigmented seed coats.[83,84]

It has been a frequent observation that snap beans with pigmented seed coats germinate better when imbibed at low temperatures than those with white seed coats.[24-27,80] This has been associated with phenolics in the seedcoat,[26] rate of imbibition,[24] rate of leakage loss from the seed,[27,80] the adherance of the seedcoat to the cotyledons,[24,75] and the semihard seedcoat character.[28,29,85] Seed coat pigmentation was not related to cold temperature germination in lima beans.[86]

Dickson and co-workers[28,29,85] have shown that seeds with semihard character and tight adherance of the seedcoat to the cotyledon do not suffer damage during imbibition at low temperatures. They attribute this response to slow initial water uptake regulated by the seed coat.

Wolk[75] has studied extensively the response of the chilling-sensitive Tendercrop and the chilling-tolerant Kinghorn Wax snap bean cultivars. Tendercrop seeds have pigmented and very loosely attached seed coats, while Kinghorn Wax seeds have white and very tightly attached seed coats and possess the semihard seed character.

Wolk found that Tendercrop seeds take up water faster than Kinghorn Wax seeds at all temperatures and at all initial moisture contents tested. Because of the loosely attached seed coats of the Tendercrop cultivar, there was a larger amount of water taken up by intact seeds, but not all of the water was imbibed, thus resulting in a larger amount of free water presumably existing between the seed coat and the cotyledons, between the cotyledons, and in other air spaces within the seed. This was not the case in Kinghorn Wax seeds.

Crawford[87] suggested that all germinating seeds undergo a period of anoxia after imbibition and before rupture of the testa. Seeds resistant to injury regulate glycolysis so that a minimal production of ethanol occurs, while sensitive species have increased alcohol dehydrogenase activity during imbibition which results in higher amounts of ethanol accumulation. Orphanos and Heydecker[88] showed that if soaked seeds were treated after soaking by removal of the testa, complete or partial drying, or by adding hydrogen peroxide, they did not show any injury. Low-vigor seeds were more susceptible to such soaking injury, and slowing imbibition rates or increasing the initial moisture levels could inhibit or prevent injury.[31-33] The implication of these studies is that free water trapped within the seed but not absorbed by the embryo could cause anaerobic conditions resulting in injury due to the build up of toxins such as acetaldehyde and ethanol. It is possible that the injury to seeds with "loosely" adhering seed coats, such as the snapbean cultivar Tendercrop, is due to the free water inside the seed.

There may be different modes of entry of free water depending upon the species of plant and the structure of the seed.[89,90] Wolk[75] and Holubowicz et al.[85] have suggested that the entry point of water into the seed may be of importance in causing or preventing injury, and Wolk[75] has suggested that the sequence of tissues being wetted may also play a role in injury that occurs during imbibition.

Based on the fact that detached axis tissue is injured less when chilled during imbibition than if attached to the cotyledons, either the axis tissue is less sensitive to imbibitional chilling injury than cotyledonary tissue or the injured cotyledons may have a detrimental effect on the axis tissue.[75] Also, undamaged cotyledonary tissue is often found immediately surrounding the axis connection to the cotyledon after chilling, which suggests that the axis tissue provides some protection to the cotyledonary tissue.[75] Much more work needs to be done to clarify the relationship of the axis and cotyledonary tissue and which tissue is most sensitive to injury.

7. Initial Wetting Reaction

Vertucci and Leopold[91,92] suggested that the initial wetting reaction is important in imbibitional chilling injury. They also suggested that water binding in dry seed could be very important in determining whether injury occurs or not. In seeds with less than 8% moisture content, water is very strongly bound, respiration is not measurable, and imbibition damage is severe. Between 8 and 24% moisture content, binding forces are moderate, oxygen uptake is small, and imbibitional damage is moderate. Above 24% moisture, there is little imbibitional injury, respiration increases rapidly, and water is weakly bound. It has also been suggested that expansion of cells during initial wetting could cause stresses that result in injury, particularly at low temperatures.[93,94] Wolk[75] also suggests that inbibitional injury is related to bound water transitions in seed tissues. There is often a sharp front separating the fully hydrated and dry portions of seeds.[95] Some seeds appear to have a very uniform wetting of the tissues during imbibition, while others, such as soybeans and snap beans which are injured by imbibition in cold temperatures, have a nonuniform wetting of tissues.[81] This aspect of imbibitional chilling injury also needs further attention.

C. ALLEVIATION OF IMBIBITIONAL CHILLING INJURY

Several strategies have been suggested to overcome low-temperature injury during imbibition. Various pregermination treatments have been somewhat successful in decreasing injury and increasing germination of seeds of chilling-sensitive plants.

1. Hardening, Osmoconditioning, Pregermination

''Hardening'' treatments in which seeds are imbibed for relatively short periods at a nonchilling temperature, dried, and then planted give some protection against imbibitional chilling injury in cotton.[13] Adding small concentrations of phosphorous to the seed increased seedling growth of corn planted in cold soils,[96] and gibberellin application has been shown to increase germination in some cases.[97,98]

Although osmoconditioning and pregermination of small seeds of chilling-sensitive species has been successful in increasing germination at low temperatures, large seeds (such as legumes) present problems because the seed coats slough off, the cotyledons can be easily broken from the embryonic axis,[99] and these seeds have a propensity for being susceptible to soaking injury.[87,88]

2. Raising the Initial Moisture Content

Raising the initial moisture of seeds prior to planting in cold soils has increased germination percentage, germination rate, plant stand, and prevented imbibitional chilling injury in corn,[11,50,100-102] snap beans,[5] lima bean,[6,103] cotton,[97] and soybeans.[19,31] Seeds that have increased seed moisture should be planted very soon after raising the moisture level in order to avoid decay and loss of vigor and to avoid readjustments in moisture content if the seed is placed in a dry environment.

3. Seed Coating

Priestley and Leopoid[104] showed that soybean and cotton seeds coated with lanolin were protected from imbibitional chilling injury. Lanolin is very sticky and probably could not be used on a practical basis. Makoni[78] showed that coating seeds with the anti-transpiration waxes Vapor Gard® and Wiltpruf and postharvest fruit wax, Pacrite 383, in which water had been removed from the commercial formulations, also decreased imbibitional chilling injury of snap beans. These materials dried without forming a sticky residue and could be use during normal planting procedures. All of these materials decreased water uptake during imbibition. It may be possible to raise the initial moisture levels of seeds, coat with a wax to prevent the loss of moisture from the seeds, and gain an additional advantage of planting high-moisture seeds and decreasing initial rates of water uptake. In addition, the waxes could be impregnated with materials which could stimulate growth and control pathogens.

4. Genetic Changes

Cultivar differences in the ability to germinate at low temperatures and overcoming imbibitional chilling injury have been shown for soybean,[9,20] corn,[11,34,105] snap bean,[5,27,80,106] and cotton,[107,108] suggesting that further improvements can be made through plant breeding. It is interesting to note that radicles of cold-tolerant snap bean lines did not emerge from the seed coats until temperatures rose to 10 to 12°C, while radicles of chilling-sensitive lines emerged at 8°C, but rotted in the soil.[106] Selection for seeds with semihard seeds[28,29,85] and tight adherance of the seed coat[75] would also be useful in new cultivars.

Finally, a better understanding of the role of environmental conditions during seed development on seed coat characteristics and seed germination under cold temperature conditions could be very useful in developing strategies for improving the ability to germinate at low temperatures.

III. NONIMBIBITIONAL CHILLING INJURY DURING GERMINATION

As pointed out earlier, seeds of the Solanaceae and Cucurbitaceae families are not subject to injury when the seeds are imbibed at low temperature. Fully imbibed seeds of these groups will not germinate at temperatures below about 10°C,[4,109] but will germinate without apparent injury when temperatures rise.[4,19] However, if seeds of these groups are germinated at higher temperatures to the point where radicle growth has occurred, injury will occur if the time is too long and the temperature is in the chilling range.[14,16,17,110] Most of the research done on this type of seed has been done on increasing germination at low temperatures, but very little work has been done on understanding the causes of injury which occurs after radicle growth has occurred or in alleviation of this injury.

A. FACTORS AFFECTING LOW-TEMPERATURE GERMINATION

Several reports have shown differences in ability to germinate at low temperatures in tomato,[111-116] tomato-related species,[117] melons,[118] and cucumbers,[119] which suggests that improvements can be made genetically in the ability to germinate at low temperatures.

Maluf and Tigchelaar[115] suggested that differences in germination at low temperatures between tomato lines is due to differences in the ability to initiate growth at low temperatures which, in turn, may be related to the fatty acid composition of the seeds.[120] Abdul-Baki and Stoner[121] showed that leachates from a cold-germinating line of tomato promoted seed germination at cold temperatures and leachates from a non-cold-germinating line inhibited germination. This suggests a role of substances in the seed coat in the germination at low temperatures. The promoting substances from tomato were ineffective in promoting germination at cold temperatures of pepper, eggplant, cucumber, or okra.

Removal of the seed coat of tomato[110] or the endosperm immediately in front of the radicle of pepper seed[122] results in a promotion of germination at low temperatures. This suggests that physical restraint of the seed coat or endosperm is important in allowing germination to proceed in these crops.

Treatment of the seed with gibberellin has resulted in better germination of pepper,[122,123] cucumber, and muskmelon,[124] but not watermelon.[124] Osmoconditioning with salt[125] or PEG[126,127] results in better germination at low temperatures, as does pregermination.[18] The rate of germination is increased, and the spread in time of germination of the majority of seeds is decreased by these treatments; however, the germinated seed or young seedlings can be injured if temperatures are too low or the length of exposure time is too long.[16,17]

B. CAUSES AND ALLEVIATION OF INJURY TO THE HYDRATED, GROWING SEED

Once radicle growth has begun, injury occurs to the radicle; usually the tissue just behind the growing point collapses.[14] This potentially causes problems because pregerminated seeds cannot be held at too low a temperature or for too long a time before chilling injury occurs and germination is decreased significantly.[16,17,128] Such injury could also occur in the soil if radicle growth had begun and soil temperatures were to decrease, or if pregerminated seeds were planted into very cold soils. Very little work has been done to try to determine the cause of this type of injury or to develop methods to alleviate the injury. It is probable that the chilling injury which occurs after radicle growth has occurred is similar to that of other fully hydrated plant tissues.

IV. CHILLING INJURY OF SEEDLINGS

A. SYMPTOMS OF CHILLING INJURY OF SEEDLINGS
1. Wilting

Seedlings of plants of tropical or subtropical origin are also damaged by exposure to low, nonfreezing temperatures. The most common visible symptom of chilling injury in intact seedlings is water loss, which results in severe wilting during the chilling exposure.[129-131] If the chilling temperature has not been too low or the duration too long, seedlings will recover turgor, and no further symptoms develop when seedlings are subsequently exposed to a nonchilling temperature.[132] If severe damage has been sustained, dessication and necrosis of leaves or death of the whole seedling occurs.[133,134] Seedlings which have been chilled, but not killed, may be stunted in growth temporarily or permanently, depending on the plant species.[96,133-135]

Wilting that occurs during chilling has been attributed to changes in membrane permeability, presumably the plasmalemma, which allows water and soluble material to leak out into the intercellular space where the water is lost through evaporation.[130,131] If excised leaves, leaf disks, or other plant parts of chilled tissue are placed in water, there is an increased leakage of electrolytes and other materials into the water, indicating a more freely permeable membrane as a result of chilling exposure.[129,130,136,137] It also has been suggested that chill-induced wilting is caused by stomatal opening during the transfer to low temperatures, thus allowing water loss through the stomates.[138] Water uptake by roots and transport to the shoots also may decrease at low temperatures. There are conflicting reports concerning the extent of injury when the shoots only, the roots only, or the whole plant is chilled.[130,138] This conflict may be due to species differences in response to chilling.

2. Respiration

Another symptom of chilling exposure is a greatly decreased respiration rate of intact seedlings and the decreased oxidation of certain substrates by membrane-bound enzymes of isolated mitochondria.[42,139] Stewart and Guinn[139] have shown that certain TCA cycle enzymes appear to be more sensitive to low temperatures than others. For example, they found that malate oxidation was more affected by low temperature than oxidation of succinate or α-ketoglutarate. The respiration rate of intact castor bean seedlings and succinate oxidation by isolated mitochondria showed discontinuities in Arrhenius plots in the range of 9 to 13°C.[42,43] This condition suggested changes in the physical state of the mitochondrial membranes at these temperatures. In contrast, when glyoxysomes were isolated and the activity of several soluble enzymes was assayed at various temperatures, there were no discontinuities in the Arrhenius plots, which indicates that there was no functional relationship between the activity of these enzymes and the glyoxysomal membrane.[43] These enzymes included isocitrate lyase, malate synthase, citrate synthase, malate dehydrogenase, glycolate oxidase and β-oxidation enzymes.

The respiration rate of leaves may increase during prolonged exposure to low temperatures,[140] and it often increases dramatically when transferred to a warmer environment.[141] The respiration rate and size of increase have been shown to be dependent upon chilling temperature and growth conditions prior to chilling.[140] Respiration of chilled tissue is not as sensitive to 2,4-dinitrophenol (DNP) as unchilled tissue, indicating some uncoupling of oxidative phosphorylation as a result of chilling.[141] One would expect lower adenosine triphosphate (ATP) levels as a result of reduced respiration and partial uncoupling at chilling temperatures which, indeed, has been demonstrated.[141,142] Cotton seedlings chilled 1 d lost ATP, but were shown to restore the initial ATP levels when returned to a nonchilling temperature. ATP levels remained depressed, however, when seedlings were chilled 2 d before being returned to a nonchilling temperature.[142] Respiration of unchilled tissue is insensitive to salicylhydroxamic acid (SHAM).[141] The respiratory burst following chilling is sensitive to treatment with SHAM, but the tissue loses sensitivity after 24 h. This suggests that the alternative respiratory pathway becomes operative during the time that seedlings are recovering from chilling.[141]

3. Photosynthesis

Photosynthesis of chilling-sensitive plants has also been shown to decrease after exposure to low temperatures. Specifically, the membrane-localized Hill reaction activity of chloroplasts declines in chilled leaves of chilling-sensitive, but not chilling-resistant, plants.[143-145]

B. FACTORS AFFECTING AND ALLEVIATION OF SEEDLING CHILLING INJURY

1. Diurnal Variation in Sensitivity

Tomato seedlings have been shown to be much more sensitive to chilling injury if exposed to chilling temperatures just at the end of the dark period of the diurnal light/dark cycle.[146-148] Exposure to 10 min of light allowed plants to regain tolerance to chilling, and seedlings held at low night temperatures prior to being exposed to chilling stress were less injured.[146] The rhythm of sensitivity seemed to be controlled by the diurnal changes in light rather than temperature.[147] Mung bean, pepper, cosmos, corn, snap bean, and eggplant all behaved similarly.[147] Shoots with roots excised retained their response to light, and supplying the cut shoots with sucrose, glucose, or fructose reduced chilling sensitivity. The endogenous carbohydrate levels in the shoots were correlated with chilling sensitivity, leading King et al.[146] to suggest that chilling tolerance following light or dark exposure was related to carbohydrate depletion or accumulation.

2. Hardening

Chilling tolerance can be increased in many plants by a process known as "hardening". Hardening may be induced by low temperature,[138,149,150] withholding water,[138] or by growing plants in concentrated salt solutions.[134] Not all plants, however, may be hardened by the above treatments.[137,150,151]

Seedlings may be conditioned or hardened by exposure to temperatures slightly above the chilling range.[149] Plants conditioned in this way are more tolerant to subsequent chilling than seedlings transferred directly to chilling temperatures.[149] In readily hardened, chilling sensitive plants, the degree of fatty acid unsaturation associated with the phospholipid fraction increases during low-temperature conditioning.[150] Hardening does not cause an increase in total leaf fatty acids or in the degree of unsaturation of glycolipids. It also was shown that in leaves of plants grown at 25°C, the degree of unsaturation and the weight of phospholipids decreases with age. This was associated with an increase in sensitivity of older leaves to chilling injury.[150] Some plants, such as *Episcia reptans,* do not harden readily. When these plants are grown at typical conditioning or hardening temperatures, there is no increase in unsaturation of fatty acids of the phospholipid fraction and the plants are not protected against injury when chilled.[140,150] Hardening, which protects some plants against chilling injury, could be due to increases in carbohydrate content.

The use of low temperatures to harden plants has also been shown to prevent the loss of ATP, which normally occurs during chilling.[140,142]

Plants also may be hardened by withholding water. Drought hardening beans for 4 d at 25°C and 40% RH is as effective as chill hardening for 4 d at 12°C and 85% RH in preventing chilling injury.[138] In contrast to chill hardening, drought hardening does not increase the fatty acid unsaturation of phospholipids, which suggests that either lipid composition is not directly involved in chilling injury or that drought hardening induces some other means of protection against chilling injury.[138]

3. High Humidity during Chilling Exposure

If plants are placed in chilling temperatures at 100% RH, water loss, and thus wilting, are prevented and other symptoms of chilling injury do not occur, even upon transfer to warmer temperatures.[131] Leaves from plants chilled at 100% RH lose fewer electroylytes when placed in water than do leaves chilled at 85% RH.[131] Plants which do not harden are not protected from chilling injury when placed at 100% RH.[138] Furthermore, bean plants placed at 12°C for 4 d, a normal hardening temperature, but at 100% RH, do not harden and are injured to the same extent as nonhardened plants during subsequent chilling.[138] It would appear that water loss prior to chilling may be necessary in the hardening process, while water loss during chilling may be essential for chilling symptoms to be expressed in seedlings. King and Reid[152] have suggested that the diurnal differences in sensitivity to chilling injury is not related to differences in water loss or stomatal conductance.

4. Chemical Alteration of Chilling Sensitivity

Modification of the fatty acid composition of polar lipids or of the type of phospholipid produced may alter chilling sensitivity. For example, as the temperature at which cotton seeds germinate is lowered from 30 to 15°C, the linolenic acid content of polar lipids increases. Treatment with a pyridazinone derivative reduces the increase in linolenic acid at low temperatures and reduces the ability of the seedlings to withstand chilling injury.[153] Waring et al.[154] showed that treatment of tomato seedlings with ethanolamine could alter the composition of the phospholipids and the degree of saturation of the acyl chains. However, effects of this alteration on chilling sensitivity were not reported.

There have been reports of modifying chilling sensitivity of seedlings by applications of growth regulators or plant hormones. Cucumber plants grown at 16 to 21°C and treated with *N*-dimethylamine succinamic acid (B-Nine) expressed the same degree of chilling tolerance to 0°C as plants hardened by exposure to 10 to 16°C.[155] Untreated plants transferred directly from 16 to 21°C to 0°C were killed.

Abscisic acid (ABA) has been shown to increase as a result of drought-hardening treatments.[134] Increased ABA levels, induced by water stress or exogenous ABA pretreatments, provide some protection against chilling injury.[141,156] By inducing stomatal closure and increasing root permeability to water, ABA may act to prevent the excessive water loss often observed in chilled seedlings.[156] Ethephon treatment of tomato seedlings has been shown to improve frost tolerance,[157] and perhaps this treatment has the same effect on chilling injury. The use of antitranspirants reduced water loss and chilling injury of cotton seedlings.[158]

It appears that there may be promise in using certain growth regulators or hormones to provide at least partial protection against chilling injury. Those that appear most promising alter growth rate or provide some means of improving plant water balance, especially during the chilling exposure. These materials may alter the carbohydrate levels within the plant also.

5. Genetic Improvement

In the long term, it would appear that genetic manipulation of plant material using a combination of available wild species with cold tolerance,[159-163] cell culture, and molecular biological techniques may provide the best answer to solving the problem of chilling sensitivity.[164-166]

A word of caution concerning the use of those techniques is in order. Although chilling sensitivity of whole plants of *Lycopersicon hirsutum* is less than those of *L. esculentum*,[160] there is some evidence that on the cellular level there is no difference in the base temperature for growth of these species.[168] Dix[169] also showed that whole plants regenerated from tobacco cells selected for chilling tolerance were not improved in chilling tolerance from plants regenerated from unselected cells. Chilling resistance at the cellular level may not be reflected by chilling resistance at the tissue, organ, or whole plant level and vice versa. In addition, selection for the ability to withstand chilling temperatures at one stage of growth and development is no guarantee that all other stages will also show the same degree of resistance.[170,171]

Obviously much more work must be done to understand the basic problem of chilling injury in order to develop strategies to overcome it.

REFERENCES

1. **Lyons, J. M.,** Chilling injury in plants, *Annu. Rev. Plant Physiol.,* 24, 445, 1973.
2. **Herner, R. C.,** Germination under cold soil conditions, *HortScience,* 21, 1118, 1986.
3. **Wolk, W. D. and Herner, R. C.,** Chilling injury of germinating seeds and seedlings, *HortScience,* 17, 169, 1982.
4. **Kotowski, F.,** Temperature relations to germination of vegetable seed, *Proc. Am. Soc. Hortic. Sci.,* 23, 176, 1926.
5. **Pollock, B. M., Roos, E. E., and Manalo, J. R.,** Vigor of garden bean seeds and seedlings influenced by initial seed moisture, substrate oxygen, and imbibition temperature, *J. Am. Soc. Hortic. Sci.,* 94, 577, 1969.
6. **Pollock, B. M.** Imbibition temperature sensitivity of Lima bean seeds controlled by initial seed moisture, *Plant Physiol.,* 44, 907, 1969.
7. **Pollock, B. M. and Toole, V. K.,** Imbibition period as the critical temperature sensitive stage in germination of Lima bean seeds, *Plant Physiol.,* 41, 221, 1966.
8. **Bramlage, W. J., Leopold, A. C., and Parrish, D. J.,** Chilling stress to soybeans during imbibition, *Plant Physiol.,* 61, 525, 1978.
9. **Bramlage, W. J., Leopold, A. C., and Specht, J. E.,** Imbibitional chilling sensitivity among soybean cultivars, *Crop Sci.,* 19, 811, 1979.

10. **Chen, T. H. H., Yamamoto, S. D. K., Gusta, L. V., and Slinkard, A. E.,** Imbibitional chilling injury during chickpea germination, *J. Am. Soc. Hortic. Sci.,* 108, 944, 1983.
11. **Cal, J. P. and Obendorf, R. L.,** Imbibitional chilling injury in *Zea mays* L. altered by initial kernel moisture and maternal parent, *Crop Sci.,* 12, 369, 1972.
12. **Christiansen, M. N.,** Periods of sensitivity to chilling in germinating cotton, *Plant Physiol.,* 42, 431, 1967.
13. **Christiansen, M. N.,** Induction and prevention of chilling injury to radicle tips of imbibing cottonseed, *Plant Physiol.,* 43, 743, 1968.
14. **Harrington, J. F. and Kihara, G. M.,** Chilling injury of germinating muskmelon and pepper seed, *Proc. Am. Soc. Hortic. Sci.,* 75, 485, 1959.
15. **Christiansen, M. N.,** Influence of chilling upon seedling development of cotton, *Plant Physiol.,* 38, 520, 1963.
16. **Irwin, C. C. and Price, H. C.,** Sensitivity of pregerminated pepper seed to low temperatures, *J. Am. Soc. Hortic. Sci.,* 106, 187, 1981.
17. **Irwin, C. C. and Price, H. C.,** The relationship of radicle length to chilling sensitivity of pregerminated pepper seed, *J. Am. Soc. Hortic. Sci.,* 108, 484, 1983.
18. **Bussell, W. T. and Gray, D.,** Effects of pre-sowing seed treatments and temperatures on tomato seed germination and seedling emergence, *Sci. Hortic.,* 5, 101, 1976.
19. **Sachs, M.,** Priming of watermelon seeds for low-temperature germination, *J. Am. Soc. Hortic. Sci.,* 102, 175, 1977.
20. **Hobbs, P. R. and Obendorf, R. L.** Interaction of initial seed moisture and imbibitional temperature on germination and productivity of soybean, *Crop Sci.,* 12, 664, 1972.
21. **Obendorf, R. L. and Hobbs, P. R.,** Effect of seed moisture on temperature sensitivity during imbibition of soybean, *Crop Sci.,* 10, 563, 1970.
22. **Simon, E. W. and Mills, L. K.,** Imbibition, leakage and membranes, *Recent Adv. Phytochem.,* 17, 9, 1983.
23. **Tully, R. E., Musgrave, M. E., and Leopold, A. C.,** The seed coat as a control of imbibitional chilling injury, *Crop Sci.,* 21, 312, 1981.
24. **Wyatt, J. E.,** Seed coat and water absorption properties of seed of near-isogenic snap bean lines differing in seed coat color, *J. Am. Soc. Hortic. Sci.,* 102, 478, 1977.
25. **Dickson, M. H.,** Breeding beans, *Phaseolus vulgaris* L., for improved germination under unfavorable low temperature conditions, *Crop Sci.,* 11, 848, 1971.
26. **Dickson, M. H. and Petzoldt, R.,** Deleterious effects of white seed due to p gene in beans, *J. Am. Soc. Hortic. Sci.,* 113, 111, 1988.
27. **Powell, A. A., Oliveira, M. D., and Matthews, S.,** The role of imbibition damage in determining the vigour of white and colloured seed lots of dwarf French beans *(Phaseolus vulgaris),* J. Exp. Bot., 37, 716, 1986.
28. **Taylor, A. G. and Dickson, M. H.,** Seed coat permeability in semi-hard snap bean seeds: its influence on imbibitional chilling injury, *J. Hortic. Sci.,* 62, 183, 1987.
29. **Dickson, M. H. and Boettger, M. A.,** Heritability of semi-hard seed induced by low seed moisture in beans *(Phaseolus vulgaris* L.) *J. Am. Soc. Hortic. Sci.,* 107, 69, 1982.
30. **Styer, R. C. and Cantliffe, D. J.,** Relationship between environment during seed development and seed vigor of two endosperm mutants of corn, *J. Am. Soc. Hortic. Sci.,* 108, 717, 1983.
31. **Woodstock, L. W. and Taylorson, R. B.,** Ethanol and acetaldehyde in imbibing soybean seeds in relation to deterioration, *Plant Physiol.,* 67, 424, 1981.
32. **Woodstock, L. W. and Taylorson, R. B.,** Soaking injury and its reversal with polyethylene glycol in relation to respiratory metabolism in high and low vigor soybean seeds, *Physiol. Plant.,* 53, 263, 1981.
33. **Woodstock, L. W. and Tao, K. J.,** Prevention of imbibitional injury in low vigor soybean embryonic axes by osmotic control of water uptake, *Physiol. Plant.,* 51, 133, 1981.
34. **Styer, R. C. and Cantliffe, D. J.,** Changes in seed structure and composition during development and their effects on leakage in two endosperm mutants of sweet corn, *J. Am. Soc. Hortic. Sci.,* 108, 721, 1983.
35. **Siddique, M. A. and Goodwin, P. B.,** Maturation temperature influences on seed quality and resistance to mechanical injury of some snap bean genotypes, *J. Am. Soc. Hortic. Sci.,* 105, 235, 1980.
36. **Siddique, M. A. and Goodwin, P. B.,** Seed vigour in bean *(Phaseolus vulgaris* L. cv. Apollo) as influenced by temperature and water regime during development and maturation, *J. Exp. Bot.,* 31, 313, 1980.
37. **Ashworth, E. N. and Obendorf, R. L.,** Imbitional chilling injury in soybean axes: relationship to stelar lesions and seasonal environments, *Agron. J.,* 72, 923, 1980.
38. **Bartkowski, E. J., Buston, D. R., Katterman, F. R. H., and Kircher, H. W.,** Dry seed fatty acid composition and seedling emergence of Pima cotton at low soil temperatures, *Agron. J.,* 69, 37, 1977.
39. **Clay, W. F., Bartkowski, E. J., and Katterman, F. R. H.,** Nuclear deoxyribonucleic acid metabolism and membrane fatty acid content related to chilling resistance in germinating cotton *(Gossypium barbadense),* Physiol. Plant., 88, 171, 1976.

40. **Dogras, C. C., Dilley, D. R., and Herner, R. C.,** Phospholipid biosynthesis and fatty acid content in relation to chilling injury during germination of seeds, *Plant Physiol.,* 60, 897, 1977.

41. **Raison, J. K. and Chapman, E. A.,** Membrane phase changes in chilling-sensitive *Vigna radiata* and their significance to growth, *Aust. J. Plant Physiol.,* 3, 291, 1976.

42. **Wade, N. L., Breidenbach, R. W., Lyons, J. M., and Keith, A. D.,** Temperature-induced phase changes in the membranes of glyoxysomes, mitochondria, and proplastids from germinating castor bean endosperm, *Plant Physoil.,* 54, 320, 1974.

43. **Breidenbach, R. W., Wade, N. L., and Lyons, J. M.,** Effect of chilling temperatures on the activities of glyoxysomal and mitochondrial enzymes from castor bean seedlings, *Plant Physiol.,* 54, 324, 1974.

44. **Hoekstra, F. A.,** Imbibitional chilling injury in pollen. Involvement of the respiratory chain, *Plant Physiol.,* 74, 815, 1984.

45. **Leopold, A. C. and Musgrave, M. E.,** Respiratory changes with chilling injury of soybeans, *Plant Physiol.,* 64, 702, 1979.

46. **Priestley, D. A. and Leopold, A. C.,** The relevance of seed membrane lipids to imbibitional chilling effects, *Physiol. Plant.,* 49, 198, 1980.

47. **Stewart, R. R. C. and Bewley, J. D.,** Protein synthesis and phospholipids in soybean axes in response to imbibitional chilling, *Plant Physiol.,* 68, 516, 1981.

48. **Wolk, W. D.,** Chilling Injury and Membrane Fatty Acid Saturation in Imbibing and Germinating Seeds of *Phaseolus vulgaris* L., M.S. thesis, Michigan State University, East Lansing, 1980.

49. **O'Neil, S. D. and Leopold, A. C.,** An assessment of phase transitions in soybean membranes, *Plant Physiol.,* 70, 1405, 1982.

50. **Cohn, M. A. and Obendorf, R. L.,** Independence of imbibitional chilling injury and energy metabolism in corn, *Crop Sci.,* 16, 449, 1976.

51. **Simon, E. W., Minchin, A., McMenamin, M. M., and Smith, J. M.,** The low temperature limit for seed germination, *New Phytol.,* 77, 301, 1976.

52. **Simon, E. W.,** Phospholipids and plant membrane permeability, *New Phytol.,* 73, 377, 1974.

53. **Powell, A. A. and Matthews, S.,** A physical explanation for solute leakage from dry pea embryos during imbibition, *J. Exp. Bot.,* 32, 1045, 1981.

54. **Duke, S. H., Kakefuda, G., and Harvey, T. M.,** Differential leakage of intracellular substances from imbibing soybean seeds, *Plant Physiol.,* 72, 919, 1983.

55. **Flentje, N. T. and Saksena, H. K.,** Pre-emergence rotting of peas in South Australia. III. Host-pathogen interaction, *Aust. J. Biol. Sci.,* 17, 665, 1964.

56. **Hayman, D. S.,** The influence of temperature on the exudation of nutrients from cotton seeds and on pre-emergence damping-off by *Rhizoctonia solani, Can. J. Bot.,* 47, 1663, 1969.

57. **Matthews, S. and Bradnock, W. R.,** Relationship between seed exudation and field emergence in peas and French beans, *Hortic. Res.,* 8, 83, 1968.

58. **Schroth, M. N. and Cook, R. J.,** Seed exudation and its influence on pre-emergence damping-off of bean, *Phytopathology,* 54, 670, 1964.

59. **Leach, L. D.,** Growth rates of host and pathogen as factors determining the severity of pre-emergence damping-off *J. Agric. Res.,* 75, 161, 1947.

60. **Yaklich, R. W., Kulik, M. M., and Anderson, J. D.,** Evaluation of vigor tests in soybean seeds. Relationship of ATP, conductivity, and radioactive tracer multiple criteria laboratory tests to field performance, *Crop Sci.,* 19, 806, 1979.

61. **Leopold, A. C.,** Temperature effects on soybean imbibition and leakage, *Plant Physiol.,* 65, 1096, 1980.

62. **Duke, S. H., Kakefuda, G., Henson, C. A., Loeffler, N. L., and Van Hulle, N. M.,** Role of the testa epidermis in the leakage of intracellular substances from imbibing soybean seeds and its implications for seedling survival, *Physiol. Plant.,* 68, 625, 1986.

63. **Simon, E. W. and Raja Harun, R. M.,** Leakage during seed imbibition, *J. Exp. Bot.,* 23, 1076, 1972.

64. **Simon, E. W. and Wiebe, H. H.,** Leakage during imbibition, resistance to damage at low termperature and the water content of peas, *New Phytol.,* 74, 407, 1975.

65. **Chabot, J. F. and Leopold, A. C.,** Ultrastructural changes of membranes with hydration in soybean seeds, *Am. J. Bot.,* 69, 623, 1982.

66. **Webster, B. D. and Leopold, A. C.,** The ultrastructure of dry and imbibed cotyledons of soybean, *Am. J. Bot.,* 64, 1286, 1977.

67. **McKersie, B. D. and Senaratna, T.,** Membrane structure in germinating seeds, *Recent Adv. Phytochem.,* 17, 29, 1983.

68. **McKersie, B. D. and Stinson, R. H.,** Effect of dehydration on leakage and membrane structure in *Lotus corniculatus* L. seeds, *Plant Physiol.,* 66, 316, 1980.

69. **Priestley, D. A. and de Kruijff, B.,** Phospholipid motional characteristics in a dry biological system. A P-nuclear magnetic resonance study of hydrating *Typha latifolia* pollen, *Plant Physiol.,* 70, 1075, 1982.

70. **Seewaldt, V., Priestley, D. A., Leopold, A. C., Feigenson, G. W., and Goodsaid-Zalduondo, F.,** Membrane organization in soybean seeds during hydration, *Planta,* 152, 19, 1981.

71. **Larson, L. A.,** The effect soaking pea seeds with or without seedcoats has on seedling growth, *Plant Physiol,* 43, 255, 1968.
72. **Powell, A. A. and Matthews, S.,** The damaging effect of water on dry pea embryos during imbibition, *J. Exp. Bot.,* 29, 1215, 1978.
73. **Powell, A. A. and Matthews, S.,** The influence of testa condition on the imbibition and vigour of pea seeds, *J. Exp. Bot.,* 30, 193, 1979.
74. **Powell, A. A. and Matthews, S.,** The significance of damage during imbibition to the field emergence of pea (*Pisum sativum* L.) seeds, *J. Agric. Sci.,* 95, 35, 1980.
75. **Wolk, W. D.,** Imbibitional Injury of *Phaseolus vulgaris* L. Seeds, Ph.D. thesis, Michigan State University, East Lansing, 1988.
76. **Abdel Samad, I. M. and Pearce, R. S.,** Leaching of ions, organic molecules, and enzymes from seeds of peanut (*Arachis hypogea* L.) imbibing without testas or with intact testas, *J. Exp. Bot.,* 29, 1471, 1978.
77. **Allerup, S.,** Effect of temperature on uptake of water in seeds, *Physiol. Plant.,* 11, 99, 1958.
78. **Makoni, R. M.,** Alleviation of Imbibitional Chilling Injury in *Phaseolus vulgaris* L. Seeds, M.S. thesis, Michigan State University, East Lansing, 1988.
79. **Shull, C. A.,** Temperature and rate of moisture intake in seeds, *Bot. Gaz.,* 69, 361, 1920.
80. **Powell, A. A., Oliveira, M. D. and Matthews, S.,** Seed vigour in cultivars of dwarf French bean (*Phaseolus vulgaris*) in relation to the colour of the testa, *J. Agric. Sci.,* 106, 419, 1986.
81. **Duke, S. H. and Kakefuda, G.,** Role of the testa in preventing cellular rupture during imbibition of legume seeds, *Plant Physiol.,* 67, 449, 1981.
82. **McCollum, J. P.,** Factors affecting cotyledonal cracking during the germination of beans (*Phaseolus vulgaris*), *Plant Physiol.,* 28, 267, 1953.
83. **Atkin, J. D.,** Relative susceptibility of snap bean varieties to mechanical injury of seed, *Proc. Am. Soc. Hortic. Sci.,* 72, 370, 1958.
84. **Deakin, J. R.,** Association of seed color with emergences and seed yield of snap beans, *J. Am. Soc. Hortic. Sci.,* 99, 110, 1974.
85. **Holubowicz, R., Taylor, A. G., Goffinet, M. C., and Dickson, M. H.,** Nature of the semihard seed characteristic in snap beans, *J. Am. Soc. Hortic. Sci.,* 113, 248, 1988.
86. **Dickson, M. H.,** Cold tolerance in lima beans, *HortScience,* 8, 410, 1973.
87. **Crawford, R. M. M.,** Tolerance of anoxia and ethanol metabolism in germinating seeds, *New Phytol.,* 79, 511, 1977.
88. **Orphanos, P. I. and Heydecker, W.,** On the nature of the soaking injury of *Phaseolus vulgaris* seeds, *J. Exp. Bot.,* 19, 770, 1968.
89. **Korban, S. S., Coyne, D. P. and Weihing, J. L.,** Rate of water uptake and sites of water entry in seeds of different cultivars of dry bean, *HortScience,* 16, 545, 1981.
90. **Kyle, J. H. and Randall, T. E.,** A new concept of the hard seed character in *Phaseolus vulgaris* L. and its use in breeding and inheritance studies, *Proc. Am. Soc. Hortic. Sci.,* 83, 461, 1963.
91. **Vertucci, C. W. and Leopold, A. C.,** Dynamics of imbibition by soybean embryos, *Plant Physiol.,* 72, 190, 1983.
92. **Vertucci, C. W. and Leopold, A. C.,** Bound water in soybean seed and its relation to respiration and imbibitional damage, *Plant Physiol.,* 75, 114, 1984.
93. **Leopold, A. C.,** Volumetric components of seed imbibition, *Plant Physiol.,* 73, 677, 1983.
94. **Willing, R. P. and Leopold, A. C.,** Cellular expansion at low temperature as a cause of membrane lesions, *Plant Physiol.,* 71, 118, 1983.
95. **Waggoner, P. E. and Parlange, Y.,** Water uptake and water diffusivity of seeds, *Plant Physiol.,* 57, 153, 1976.
96. **Gubbels, G. H.,** Emergence, seedling vigor and seed yields of corn after pre-sowing treatments and the addition of phosphorus with the seed, *Can. J. Plant. Sci.,* 56, 749, 1976.
97. **Cole, D. F. and Wheeler, J. E.,** Effect of pregermination treatments on germination and growth of cottonseed at suboptimal temperatures, *Crop Sci.,* 14, 451, 1974.
98. **Wittwer, S. H. and Bukovac, M. J.,** Gibberellin and higher plants. VIII. Seed treatments for beans, peas, and sweet corn, *Mich. Agric. Exp. Stn. Q. Bull.,* 40, 215, 1957.
99. **Bodsworth, S. and Bewley, J. D.,** Osmotic priming of seeds of crop species with polyethylene glycol as a means of enhancing early and synchronous germination at cool temperatures, *Can. J. Bot.,* 59, 672, 1981.
100. **Bennett, M. A. and Water, L., Jr.,** Germination and emergence of high-sugar sweet corn is improved by presowing hydration of seed, *HortScience,* 22, 236, 1987.
101. **Bennett, M. A. and Waters, L., Jr.,** Seed hydration treatments for improved sweet corn germination and stand establishment, *J. Am. Soc. Hortic. Sci.,* 112, 45, 1987.
102. **Bennett, M. A. and Waters, L., Jr., and Curme, J. H.,** Kernel maturity, seed size, and seed hydration effects on the seed quality of a sweet corn inbred, *J. Am. Soc. Hortic. Sci.,* 113, 348, 1988.
103. **Bennett, M. A. and Waters, L., Jr.,** Influence of seed moisture on lima bean stand establishment and growth, *J. Am. Soc. Hortic. Sci.,* 109, 623, 1984.

104. **Priestley, D. A. and Leopold, A. C.**, Alleviation of imbibitional chilling injury by use of lanolin, *Crop Sci.*, 26, 1252, 1986.
105. **Pinnell, E. L.**, Genetic and environmental factors affecting corn seed germination at low temperatures, *Agron. J.*, 49, 562, 1949.
106. **Dickson, M. H. and Boettger, M. A.**, Emergence, growth, and blossoming of bean (*Phaseolus vulgaris*) at suboptimal temperatures, *J. Am. Soc. Hortic. Sci.*, 109, 257, 1984.
107. **Christiansen, M. N. and Lewis, C. F.**, Reciprocal defferences in tolerance to seed-hydration chilling in F_1 progeny of *Gossypium hirsutom* L., *Crop Sci.*, 13, 210, 1973.
108. **Marani, A. and Dag, J.**, Inheritance of the ability of cotton seeds to germinate at low temperature in the first hybrid generation, *Crop Sci.*, 2, 243, 1962.
109. **Smith, P. G. and Millett, A. H.**, Germinating and sprouting responses of the tomato at low temperatures, *Proc. Am. Soc. Hortic. Sci.*, 84, 480, 1964.
110. **Liptay, A. and Schopfer, P.**, Effect of water stress, seed coat restraint, and abscisic acid upon different germination capabilities of two tomato lines at low temperature, *Plant Physiol.*, 73, 935, 1983.
111. **Cannon, O. S., Gatherum, D. M., and Miles, W. G.**, Heritability of low temperature seed germination in tomato, *HortScience*, 8, 404, 1973.
112. **DeVos, D. A., Hill, R. R., Jr., and Hepler, R. W.**, Response to selection for low temperature sprouting ability in tomato populations, *Crop Sci.*, 22, 876, 1982.
113. **DeVos, D. A., Hill, R. R., Jr., Hepler, R. W., and Garwood, D. L.**, Inheritance of low temperature sprouting ability in F_1 tomato crosses, *J. Am. Soc. Hortic. Sci.*, 106, 352, 1981.
114. **El Syed, M. N. and John, C. A.**, Heritability studies of tomato emergence at different temperatures, *J. Am. Soc. Hortic. Sci.*, 98, 440, 1973.
115. **Maluf, W. R. and Tigchelaar, E. C.**, Responses associated with low temperature seed germinating ability to tomato, *J. Am. Soc. Hortic. Sci.*, 105, 280, 1980.
116. **Ng, T. J. and Tigchelaar, E. C.**, Inheritance of low temperatures seed sprouting in tomato, *J. Am. Soc. Hortic. Sci.*, 98, 314, 1973.
117. **Scott, S. J. and Jones, R. A.**, Quantifying seed germination responses to low temperature: variation among *Lycopersicon* spp., *Environ. Exp. Bot.*, 25, 129, 1985.
118. **Nerson, H., Cantliffe, D. J., Paris, H. S., and Karchi, Z.**, Low-temperature germination of birdsnest-type muskmelons, *HortScience* 17, 639, 1982.
119. **Nienjuis, J., Lower. R. L., and Staub, J. E.**, Selection for improved low temperature germination in cucumber, *J. Am. Soc. Hortic. Sci.*, 108, 1040, 1983.
120. **Maluf, W. R. and Tigchelaar, E. C.**, Relationship between fatty acid composition and low-temperature seed germination in tomato, *J. Am. Soc. Hortic. Sci.*, 107, 620, 1982.
121. **Abdul-Baki, A. A. and Stoner, A.**, Germination promoter and inhibitor in leachates from tomato seeds, *J. Am. Soc. Hortic. Sci.*, 103, 684, 1978.
122. **Watkins, J. T. and Cantliffe, D. J.**, Mechanical resistance of the seed coat and endosperm during germination of *Capsicum annuum* at low termperature, *Plant Physiol.*, 72, 146, 1983.
123. **Watkins, J. T., Cantliffe, D. J., and Sachs, M.**, Temperature and gibberellin-induced respiratory changes in *Capsicum annuum* during germination at varying oxygen concentrations, *J. Am. Soc. Hortic. Sci.*, 108, 356, 1983.
124. **Nelson, J. M. and Sharples, G. C.**, Effect of growth regulators on germination of cucumber and other cucurbit seeds at suboptimal temperatures, *HortScience*, 15, 253, 1980.
125. **Ells, J. E.**, The influence of treating tomato seed with nutrient solutions on emergence rate and seedling growth, *Proc. Am. Soc. Hortic. Sci.*, 83, 684, 1963.
126. **Liptay, A. and Tan, C. S.**, Effect of various levels of available water on germination of polyethylene glycol (PEG) pretreated or untreated tomato seeds, *J. Am. Soc. Hortic. Sci.*, 110, 748, 1985.
127. **Rumpel, J. and Szudyga, I.**, The influence of pre-sowing seed treatments on germination and emergence of tomato 'New Yorker' at low temperatures, *Sci. Hortic.* 9, 119, 1978.
128. **Pill, W. G. and Fieldhouse, D. J.**, Emergence of pregerminated tomato seed stored in gels up to twenty days at low temperatures, *J. Am. Soc. Hortic. Sci.*, 107, 722, 1982.
129. **Rikin, A. and Richmond, A. E.**, Amelioration of chilling injuries in cucumber seedlings by abscisic acid, *Physiol. Plant.*, 38, 95, 1976.
130. **Wright, M.**, The effect of chilling on ethylene production, membrane permeability and water loss of leaves of *Phaseolus vulgaris*, *Planta*, 120, 63, 1974.
131. **Wright, M. and Simon, E. W.**, Chilling injury in cucumber leaves, *J. Exp. Bot.*, 24, 400, 1973.
132. **Creencia, R. P. and Bramlage, W. J.**, Reversibility of chilling injury to corn seedlings, *Plant Physiol.*, 47, 389, 1971.
133. **Rikin, A., Atsmon, D., and Gitler, C.**, Chilling injury in cotton (*Gossypium hirsutum* L.): prevention by abscisic acid, *Plant Cell Physiol.*, 20, 1537, 1979.
134. **Rikin, A., Blumenfeld, A., and Richmond, A. E.**, Chilling resistance as affected by stressing environments and abscisic acid, *Bot. Gaz.*, 137, 307, 1976.

135. **Christiansen, M. N. and Thomas, R. O.,** Season-long effects of chilling treatments applied to germinating cotton seeds, *Crop Sci.,* 9, 672, 1969.

136. **Paull, R. E.,** Temperature-induced leakage from chilling-sensitive and chilling-resistant plants, *Plant Physiol.,* 68, 149, 1981.

137. **Rikin, A. and Richmond, A. E.,** Factors affecting leakage from cucumber cotyledons during chilling stress, *Plant Sci. Let.,* 14, 263, 1979.

138. **Wilson, J. M.,** The mechanism of chill- and drought-hardening of *Phaseolus vulgaris* leaves, *New Phytol.,* 76, 257, 1976.

139. **Stewart, J. M. and Guinn, G.,** Response of cotton mitochondria to chilling temperatures, *Crop Sci.,* 11, 908, 1971.

140. **Wilson, J. M.,** Leaf respiration and ATP levels at chilling temperatures, *New Phytol.,* 80, 325, 1978.

141. **Sasson, N. and Bramlage, W. J.,** Effects of chemical protectants against chilling injury of young cucumber seedlings, *J. Am. Soc. Hortic. Sci.,* 106, 282, 1981.

142. **Stewart, J. M. and Guinn, G.,** Chilling injury and changes in adenosine triphosphate of cotton seedlings, *Plant Physiol.,* 44, 605, 1969.

143. **Lasley, S. E., Garber, M. P., and Hodges, C. F.,** After effects of light and chilling temperatures on photosynthesis in excised cucumber cotyledons, *J. Am. Soc. Hortic. Sci.,* 104, 477, 1979.

144. **Smillie, R. M. and Nott, R.,** Assay of chilling injury in wild and domestic tomatoes based on photosystem activity of the chilled leaves, *Plant Physiol.,* 63, 796, 1979.

145. **Yakir, D., Rudich, J., and Bravdo, B. A.,** Adaption to chilling: photosynthetic characteristics of the cultivated tomato and a high altitude wild species, *Plant Cell Environ.,* 9, 477, 1986.

146. **King, A. I., Joyce, D. C., and Reid, M. S.,** Role of carbohydrate in diurnal chilling sensitivity of tomato seedlings, *Plant Physiol.,* 86, 764, 1988.

147. **King, A. I., Reid, M. S., and Patterson, B. D.,** Diurnal changes in the chilling sensitivity of seedlings, *Plant Physiol.,* 70, 211, 1982.

148. **Patterson, B. D., Paul, R., and Graham, D.,** Adaptation to chilling: survival, germination, respiration and protoplasmic dynamics, in *Low Temperature Stress in Crop Plants: The Role of the Membrane,* Lyons, J. M., Graham, D., and Raison, J. K., Eds., Academic Press, New York, 1979, 25.

149. **Wheaton, T. A. and Morris, L. L.,** Modification of chilling sensitivity by temperature conditioning, *Proc. Am. Soc. Hortic. Sci.,* 91, 529, 1967.

150. **Wilson, J. M. and Crawford, R. M. M.,** The acclimatization of plants to chilling temperatures in relation to the fatty-acid composition of leaf polar lipids, *New Phytol.,* 73, 805, 1974.

151. **Wilson, J. M. and Crawford, R. M. M.,** Leaf fatty-acid content in relation to hardening and chilling injury, *J. Exp. Bot.,* 25, 121, 1974.

152. **King, A. E. and Reid, M. S.,** Diurnal chilling sensitivity and desiccation in seedlings of tomato, *J. Am. Soc. Hortic. Sci.,* 112, 821, 1987.

153. **St. John, J. B. and Christiansen, M. N.,** Inhibition of linolenic acid synthesis and modification of chilling resistance in cotton seedlings, *Plant Physiol.,* 57, 257, 1976.

154. **Waring, A. J., Breidenbach, R. W., and Lyons, J. M.,** In vivo modification of plant membrane phospholipid composition, *Biochim. Biophys. Acta,* 443, 157, 1976.

155. **Jaffe, M. J. and Isenberg, F. M.,** Some effects of N-dimethyl amino succinamic acid (B-nine) on the development of various plants, with special reference to the cucumber, *Cucumis sativus., Proc. Am. Soc. Hortic. Sci.,* 87, 420, 1965.

156. **Rikin, A., Atsmon, D., and Gitler, C.,** Chilling injury in cotton (*Gossypium hirsutum* L.): effects of antimicrotubular drugs, *Plant Cell Physiol.,* 21, 829, 1980.

157. **Liptay, A., Phatak, S. C., and Jaworski, C. A.,** Ethephon treatment of tomato transplants improves frost tolerance, *HortScience,* 17, 400, 1982.

158. **Christiansen, M. N. and Ashworth, E. N.,** Prevention of chilling injury to seedling cotton with antitranspirants, *Crop Sci.,* 18, 907, 1978.

159. **Kamps, T. L., Isleib, T. G., Herner, R. C., and Sink, K. C.,** Evaluation of techniques to measure chilling injury in tomato, *HortScience,* 22, 1309, 1987.

160. **Patterson, B. D., Paull, R., and Smillie, R. M.,** Chilling resistance in *Lycopersicon hirsutum* Humb. & Bonpl., a wild tomato with a wide altitudinal distribution, *Aust. J. Plant Physiol.,* 5, 609, 1978.

161. **Patterson, B. D. and Payne, L. A.,** Screening for chilling resistance in tomato seedlings, *HortScience,* 18, 340, 1983.

162. **Vallejos, C. E.,** Genetic diversity of plants for response to low temperatures and its potential use in crop plants, in *Low Temperature Stress in Crop Plants: The Role of the Membrane,* Lyons, J. M., Graham, D., and Raison, J. K., Eds., Academic Press, New York, 1979, 473.

163. **Scott, S. J. and Jones, R. A.,** Cold tolerance in tomato. II. Early seedling growth of *Lycopersicon* spp., *Physiol. Plant.,* 66, 659, 1986.

164. **Breidenbach, R. W. and Waring, A. J.,** Response to chilling of tomato seedlings and cells in suspension cultures, *Plant Physiol.,* 60, 190, 1977.

165. **Dix, P. J. and Street, H. E.,** Selection of plant cell lines with enhanced chilling resistance, *Ann. Bot.,* 40, 903, 1976.

166. **Smillie, R. M., Melchers, G., and von Wettstein, D.,** Chilling resistance of somatic hybrids of tomato and potato, *Carlsberg Res. Commun.,* 44, 127, 1979.

167. **Zamir, D., Tanksley, S. D., and Jones, R. A.,** Haploid selection for low temperature tolerance of tomato pollen, *Genetics,* 101, 129, 1982.

168. **DuPont, F. M., Staroci, L. C., Chou, B., Thomas, B., Williams, B. G., and Mudd, J. S.,** Effect of chilling temperatures on cell cultures of tomato, *Plant Physiol.,* 77, 64, 1984.

169. **Dix, P. J.,** Chilling resistance is not transmitted sexually in plants regenerated from *Nicotiana sylvestris* cell lines, *Z. Pflanzenphysiol.,* 84, 223, 1977.

170. **Dickson, M. H. and Petzholdt, R.,** Inheritance of low temperature tolerance in beans at several growth stages, *HortScience,* 22, 481, 1987.

171. **Kemp, G. A.,** Low-temperature growth responses of the tomato, *Can. J. Plant Sci.,* 48, 281, 1968.

Chapter 5

ULTRASTRUCTURAL CHANGES DURING CHILLING STRESS

Kazuhiro Abe

TABLE OF CONTENTS

I. INTRODUCTION

Many physiological and biochemical studies have been conducted concerning chilling injury of various plants. Some of the studies were on the ultrastructural changes of plant cells exposed to low temperatures for the purpose of elucidating the mechanism of chilling injury. The findings in these papers are as follows: disorganization of matrix and cristae of mitochondria in snap bean hypocotyls exposed to chilling temperature;[1] dilation of endo-plasmic reticulum, the loss of ribosomes, and the clumping of nuclear chromatin in chilled tomato seedling cotyledon cells;[2] and ultrastructural changes in proplastids and rough en-doplasmic reticulum of cultured cells of *Cornus stolonifera* after 12 h at 0°C.[3]

Changes in plastids were also found in several plant species, including swelling of plastid and disorganization of internal lamella in leaf cells of cucumber,[4,5] abnormal development of mesophyll plastids,[6] and ultrastructural damage of mesophyll chloroplasts[7] in some C4-pathway species. In cotton plants, swollen plastids and disorganized chloroplasts were found after the plants had been exposed to 5°C for 4 d,[8] and in cells of *Anacystis nidulans*, irregular photosynthetic lamella was noted after chilling near 0°C for 30 min.[9]

After prolonged chilling, the mitochondria generally appeared less severely altered than the other organelles in grapefruits[10] and tomato seedling cotyledons.[2] However, some workers reported that relatively short chilling exposures (2,4, and 8 h) of the tomato seedling co-tyledons resulted in discontinuities in the mitochondrial envelope, followed by swelling of the cristae and increased opacity of the matrix.[11] These results suggest that the structure of mitochondria may somehow stabilize during the longer chilling exposure, whereas that of other organelles continues to change.

Most of the studies in this field have been conducted with young organs of growing plants for the purpose of convenience. Some workers, however, have been interested in studies on ultrastructural changes in fruits and vegetables during storage at low temperatures. Such studies may be worthwhile to understand the mechanism of chilling injury and also to improve handling of chilling-sensitive commodities. Unfortunately, very few of these cy-tological reports are available, as compared with the great number of reports of physiological and biochemical investigations, and further studies are needed.

This chapter mainly deals with ultrastructural changes that are related to chilling injury of fruits and vegetables during storage. In connection with this topic, histological and cytological studies on the characteristics of the injury as observed by microscopy and scanning electron microscopy are also reviewed.

II. HISTOLOGICAL STUDIES

A. PITTING INJURY

It has been well known that many vegetables and fruits of tropical and subtropical origin are sensitive to chilling and are damaged during storage or in transit at low temperatures. Cucumber is such a fruit and is apt to suffer from chilling injury at low temperatures in a relatively short period of time. The first symptom of chilling injury is the formation of circular or irregular-shaped pits on the surface of the fruit. This sympton, called pitting, results in a whitish leakage oozing from the fruit surface. The pitting is usually followed by the development of infection by pathogenic organisms.

Morris and Platenius[12] reported that the epidermal cells of cucumber fruit held at low temperatures sloughed off entirely, the bordering tissue lost most of its water content, and the final stage of pitting appeared to be a localized desiccation process near the epidermis of the fruit. The low relative humidity of the storage room accelerated pitting injury; however, desiccation is not the initial cause of pitting, and the primary injury to cells may be produced by low temperatures.

73

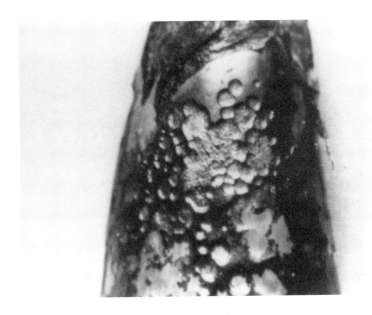

FIGURE 1. Typical pitting injury of eggplant fruits stored at 1°C. (From Abe, K., Iwata, T., and Ogata, K., *J. Jpn. Soc. Hortic. Sci.*, 42, 402, 1974.)

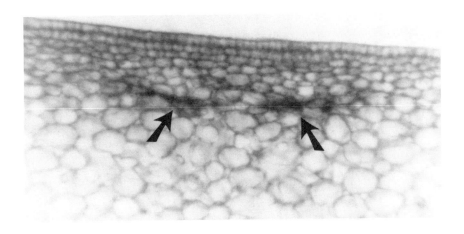

FIGURE 2. A cross-section of eggplant fruit on which slight pitting appeared. Injured cells (arrows) are shown at several layers below its surface. (From Abe, K., Iwata, T., and Ogata, K., *J. Jpn. Soc. Hortic. Sci.*, 42, 402, 1974.)

Since this study, there have been a few histological reports relating to pitting injury of horticultural crops. Eggplant is also known to be very sensitive to low temperatures, with pitting as a typical symptom of chilling injury when eggplant is stored at low, but nonfreezing, temperatures (Figure 1). Using this material, in 1974, Abe et al.[13] microscopically observed that pitting injury in the fruit (cv. Senryo) stored at 1°C for 4 d started with deformation and browning of cells located several layers inside the surface (Figure 2); however, fruit stored at 20°C showed no such cellular changes until decay. Pitting injury was observed even at high humidity, and bursting of epidermal cell walls in the pitted portion did not occur, even when the injury became extremely severe (Figure 3). These observations demonstrate that for the occurrence and development of pitting, water loss from the surface is

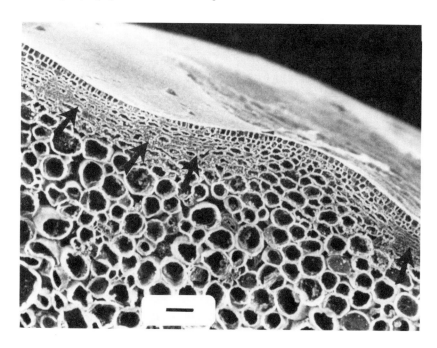

FIGURE 3.   Surface and inner structure of pitted portion in eggplant fruit stored at 1°C for 11 d, observed by scanning electron microscopy. Arrows indicate destructed parenchymal cells. Bar = 75 μm. (From Abe, K. and Chachin K., *J. Jpn. Soc. Hortic. Sc.*, 54, 247, 1985.)

not always necessary. This observation, namely, the pitting injury in eggplant fruit caused by collapse of cells located several layers inside its surface, was confirmed by Rhee and Iwata[14] and Tsay[15] with eggplant fruit, cv. Kokko Shin No. 2 stored at 4°C and cv. Pingtung Chang-Chei stored at 1°C, respectively.

Using cucumber fruit, Rhee and Iwata[16] microscopically observed that before the appearance of pitting, cell flattening following plasmolysis originated from the parenchymal cells adjoining the substomatal chamber. Flattened cells increased in parallel with the epidermis and developed into the inner cell layers. Flattening of epidermal cells also occurred following the flattening of parenchymal cells, which progressed to the depth of eight to ten layers under the epidermis. At this stage, depression occurred around the stoma and pitting was observed externally. At the more advanced stage of pitting, the areas of cell flattening around the respective stoma extended to a depth of 15 to 20 layers under the epidermis. Thus, it became clear that the surface pitting of low-temperature-stored fruits, such as eggplant and cucumber, is caused by collapse of parenchymal cells instead of by the collapse or bursting of epidermal cells. In *Phalaenopsis* orchids, collapse of mesophyll cells contributes to the pitting of the leaves in response to chilling at 2°C.[17]

Rhee and Iwata[16] reported that pitting of cucumber fruit is apt to appear at the lower relative humidity of 80 to 95% in cold storage as compared with higher relative humidity over 95%. Yamawaki et al.[18] also reported that under low relative humidity, the pitting injury in cucumber fruit was more conspicuous than under high humidity. On the contrary, Abe and Ogata[19] found that eggplant fruits held in higher relative humidity in a polyethylene bag showed more severe pitting injury than those held at low relative humidity. Thus, the influence of storage humidity on the development of pitting seems to differ between cucumber and eggplant fruits, and it is obvious that the pitting injury was caused by low temperature.

B. OBSERVATION OF PITTING BY SCANNING ELECTRON MICROSCOPY

Temperature manipulation is often effective in reducing chilling injury, as described in detail in other chapters.

FIGURE 4.   Pitting injury of eggplant fruit. (Left) Type I pitting occurred on fruit stored at 1°C directly after harvest. (Right) Type II pitting occurred on the fruit stored at 1°C after prestorage conditioning at 20°C for 5 d. (From Abe, K. and Chachin, K., *J. Jpn. Soc. Hortic. Sci.*, 54, 247, 1985.)

Abe and Chachin[20] observed that prestorage conditioning of eggplant fruit at 10 or 20°C for 5 to 15 d extended the shelf life of fruit stored at 1°C because of the slowed chilling injury. The occurrence of pitting, the first symptom of chilling injury, was delayed by 2 to 3 d. In fruit stored at 1°C after the conditioning, two types of pitting were recognized: in one, the pits were larger in size (100 to 1500 μm) and smaller in number (referred to as type I pitting), while in the other, the pits were smaller in size (100 to 300 μm) and larger in number (referred to type II pitting) (Figure 4). In fruit stored at low temperature immediately after harvest, only type I pitting was observed. This type of pitting is regarded as ordinary pitting due to chilling injury.

In fruit stored after conditioning, type II pitting appeared more frequently in fruit packaged in perforated polyethylene bags than in those packaged in sealed ones, whereas the occurrence of type I pitting was just the reverse. Pasteurization of the fruit by dipping in ethanol after conditioning reduced the occurrence of type II pitting.

Scanning electron microscopy was used to observe these materials. Circular protuberances were seen on the surface of fruit stored at 20°C. A fungus (*Alternaria* sp.) was frequently found on and near the protuberances, through which mycelium frequently penetrated into the eggplant fruit.

Sectioning of fruit proved that in type I pitting there was no destruction of epidermal cells at an early stage, and that the destruction of parenchymal cells was the main cause of the pits. In type II pitting, there was penetration of fungus mycelium (Figure 5), indicating that destruction of epidermal as well as parenchymal cells was the cause of this type of pitting (Figure 6).

C. HISTOCHEMICAL OBSERVATIONS

Rhee and Iwata[14] observed the tissues of eggplant fruit histochemically and found that chlorogenic acid appeared to be the major phenolic substance in the internal parenchymal tissues, vascular tissues, and seeds. In the epidermal tissues, phenolic substances other than chlorogenic acid were presumed to be present in addition to anthocyan. Polyphenol oxidase was found in the same tissues where the phenolic substances were distributed.

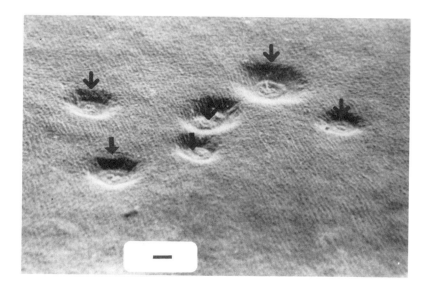

A

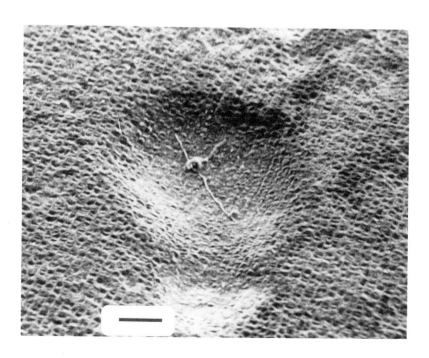

B

FIGURE 5.   Surface structure of type II pitted portion (arrows) on eggplant fruit stored at 1°C for 6 d after prestorage conditioning at 20°C for 5 d, observed by scanning electron microscopy. Bar = 75 μm. (From Abe, K. and Chachin, K., *J. Jpn. Soc. Hortic. Sci.*, 54, 247, 1985.)

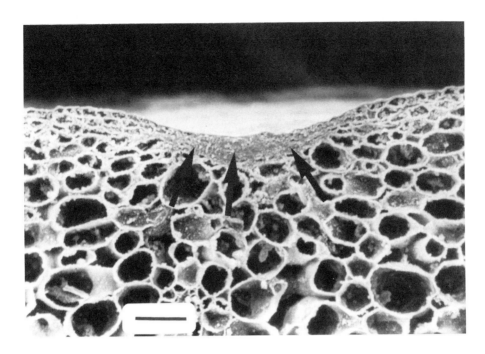

FIGURE 6. Cross-section of type II pitting on eggplant fruit stored at 1°C for 11 d after prestorage conditioning at 20°C for 5 d. Arrows indicate destructed epidermal and parenchymal cells. Bar = 75 μm. (From Abe, K. and Chachin, K., *J. Jpn. Soc. Hortic. Sci.*, 54, 247, 1985.)

It had been known that the chlorogenic acid content in eggplant fruit increases during low-temperature storage before the occurrence of chilling injury.[21] Browning of peel, parenchymal tissues, and seeds is another typical symptom of chilling injury of eggplant fruit, which usually occurs following pitting.[21] The results mentioned above seem to indicate that oxidation of phenolic compounds cause browning in eggplant fruit during storage at low temperatures.

Rhee and Iwata[22] made histological observations on the taro (*Colocasia esculenta* Schott cv. Dotare) tuber during development of internal browning due to chilling injury at 4°C. Before the appearance of internal browning, slight browning occurred in some tannin cells and surrounding parenchymal cells. With the progress of internal browning, the numbers of browned tannin cells with granulation and the degree of browning of surrounding parenchymal cells increased. Furthermore, in parenchymal cells, nuclei swelled, and leukoplasts and amyloplasts assembled around nuclei before the appearance of internal browning. When internal browning progressed, nuclei shrank with plasmolyses and collapsed, accompanied by destruction of leukoplasts and amyloplasts.

Fukushima[23] reported that in the flesh cells prepared from cucumber fruit stored at 20°C for 3 d, heavily stained (indicating SH group) masses of irregular shapes were unevenly scattered; conversely, in the preparation from fruit stored at 0°C, stained precipitation was observed in cell walls. He suggested that in cucumber fruit, chilling may at first break the plasmalemma of sieve tubes, the exudate of which outflows into the cell walls of tissue cells. The subsequent contact between the tube esterase and wall pectin results in pectin demethylation, which makes the elastic cell wall rigid and fragile. This rigidity ultimately leads to a water stress injury.

Murata[24] conducted histological studies with green banana fruit (cv. Sin-zun) and found that brown substances were observed around the vascular tissues of the peel of fruits injured by chilling (6°C for 3 d) and that they increased in number with the progress of chilling injury. The darkened vascular tissues were not found in the tissues of healthy fruit. Taken together, these results suggests that disorganization of sieve tubes is one of the factors of chilling injury.

III. ULTRASTRUCTURAL CHANGES

A few workers have observed the ultrastructural changes associated with chilling injury of fruits and vegetables. Yamaki and Uritani[25] studied the mitochondrial structure prepared from sweet potato roots (cv. Okinawa No. 100) by electron microscopy. Two types of mitochondria were found in the mitochondrial fraction prepared from a healthy sweet potato. One, which had reticulately developed cristae and extremely electron-dense matrix spaces, was called Form A. The other, which had cristae which were not clearly distinguished from the matrix spaces, was called Form B. In the fraction prepared from the 14-d chill-stored sweet potatoes, a third type of mitochondria was found in addition to Forms A and B. This one, which had an extremely swollen form, was called Form C. Form C is thought to occur through the degradation of Form A or B, concomitant with the release of phospholipid from both the inner and outer membranes during chill-storage. It is quite likely that Form C occurs during physiological deterioration in sweet potato root tissues, which proceeds irreversibly during chill storage.

Furthermore, Yamaki and Uritani[26] observed the structure of cells in the intact tissue of sweet potato roots suffering from chilling injury and found that the membranes surrounding starch granules and cytoplasmic membranes appeared to be preserved without damage for the most part; however, the vacuolar membranes were degraded and disappeared in some areas. The cristae of mitochondria in parenchymal cells of chilling-injured tissues developed into a larger reticular form than those from healthy sweet potato root tissue.

Moline[27] studied ultrastructural modification of tomato fruit (cv. Walter) organelles correlated with low-temperature injury. The fruit, chilled for 10 d at 2°C, contained plastids that had begun to swell. They had begun to lose their grana, and the intergranal lamellae were distended. Mitochondria were also swollen, and the cristae had begun to lyse. Numerous small vacuoles had begun to develop in the cytoplasm. Plastids of fruit stored for 15 d at 2°C had obviously degenerated, and many, having lost all grana, had developed large vacuoles. Mitochondria were swollen and cristae appeared to have developed into numerous vesicles. Very few organelles were distinguishable in a tomato fruit stored for 21 d at 2°C. Moline concluded that ultrastructural evidence of damage in chilled tomato fruit is readily apparent and distinct from changes associated with ripening.

Using eggplant fruit, Abe and Ogata[28] reported that the ultrastructural changes and cell wall bursting of epidermal and parenchymal cells did not occur, even after 9 d of storage at 20°C (Figures 7 and 8). On the contrary, eggplant fruit not yet exhibiting pitting injury after 4 d of storage at 1°C showed such ultrastructural changes of the cytoplasm in paren-chymal cells as partial degradation of tonoplast and swelling of mitochondria. The epidermal cells of the pitted portion in the fruit stored for 9 d at 1°C displayed no cell wall bursting; however, nuclear membrane, mitochondria, and ground substance showed marked changes (Figure 9). The parenchymal cells of the pitted portion in the fruit stored for 9 d at 1°C showed cell collapse without cell wall bursting, and ultrastructural changes, such as deg-radation of tonoplast and swelling of mitochondria having irregular distribution of cristae in them, were obvious. In the parenchymal cells adjoining the pitted portion, tonoplast, mitochondria, and ground substance showed slight changes which bore resemblances to the changes in the pitted portion (Figures 10 and 11).

Water convolvulus (*Ipomoea aquatica* Forsk.) is a widely distributed leafy vegetable in Southeast Asia, including the southern area of Japan. Chilling injury is manifested as brown-ing of young leaves and young stems at the top part of each vine when stored at temperatures

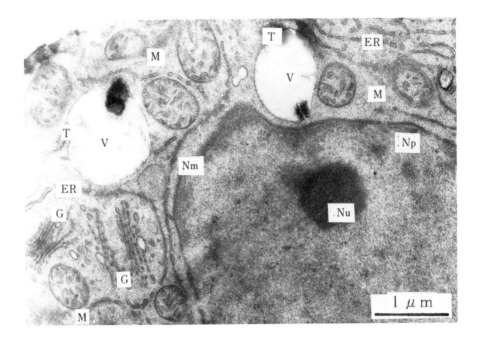

FIGURE 7. Ultrastructure of epidermal cell in noninjured eggplant fruit representing the normal structure. Note mitochondria having tubulous cristae (M), endoplasmic reticulums (ER), tonoplast (T), vacuoles (V), nuclear pore (Np), nuclear membrane (Nm), and nucleolus (Nu). (From Abe, K. and Ogata, K., *J. Jpn. Soc. Hortic. Sci.*, 46, 541, 1978.)

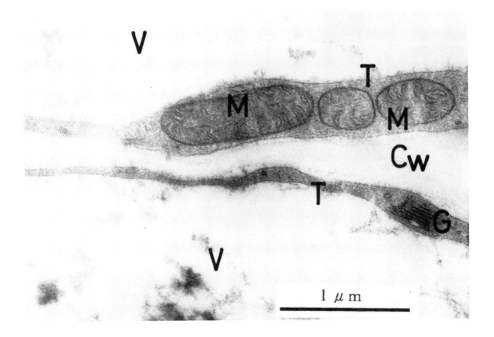

FIGURE 8. Ultrastructure of parenchymal cell in eggplant fruit stored at 20°C for 9 d. Note mitochondria (M), tonoplast (T), and cell wall (Cw) showing normal structure. (From Abe, K. and Ogata, K., *J. Jpn. Soc. Hortic. Sci.*, 46, 541, 1978.)

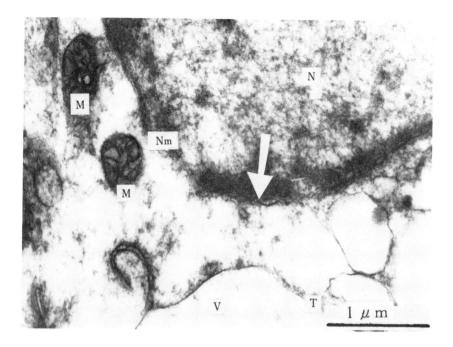

FIGURE 9.   Ultrastructure of epidermal cell of pitted portion in eggplant fruit stored at 1°C for 9 d. Note mitochondria (M) showing cristae which developed into large reticular form, vacuole (V), and injured nuclear membrane indicated by arrows. (From Abe, K. and Ogata, K., *J. Jpn. Soc. Hortic. Sci.*, 46, 541, 1978.)

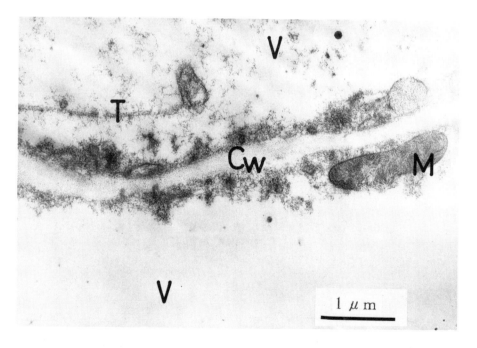

FIGURE 10.   Ultrastructure of parenchymal cell adjoined to the pitted portion in eggplant fruit stored at 1°C for 9 d. Note the partially swollen mitochondria (M) and the detached and partially disappeared tonoplast (T). (From Abe, K. and Ogata, K., *J. Jpn. Soc. Hortic. Sci.*, 46, 541, 1978.)

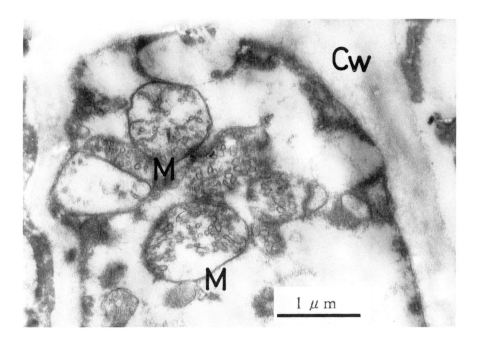

FIGURE 11. Ultrastructure of parenchymal cell adjoined to the pitted portion in eggplant fruit stored at 1°C for 9 d. Note swollen mitochondria having a double membrane and irregular distribution of cristae. (From Abe, K. and Ogata, K., *J. Jpn. Soc. Hortic. Sci.*, 46, 541, 1978.)

below about 9°C.[29] Hirata[30] found that the green part of young leaves of water convolvulus stored at 1°C for 5 d had already shown the ultrastructural deformation preceding the occurrence of visible chilling injury. For example, the chloroplasts were swollen, and the number of grana stacks decreased. The mitochondria cristae had disappeared, and the tonoplasts were invaginated into the vacuoles and had partially disappeared. These subcellular changes were presumed to be associated with chilling injury.

Mitochondria still play an important role after harvest in vegetables and fruits because they are the sites of respiratory metabolism and other physiological processes. According to the electron microscopic observations mentioned above, mitochondria seem to be the most changeable organelles in chilling-sensitive horticultural crops when the crops are exposed to low temperatures. Depression or impairment of oxidative activity of mitochondria by chilling stress has also been reported by plant physiologists.[18,31-35]

On the other hand, ultrastructural changes, such as the partial disappearance of tonoplast in the cells of eggplant fruit, sweet potato roots, and water convolvulus leaves were observed by electron microscopy during low- temperature storage. The tonoplast controls the contact between phenolic substances in vacuoles and polyphenol oxidase in plastids such as chloroplasts. Disappearance of the tonoplast by chilling stress would accelerate the oxidation of phenolic substances, and the oxidation products may impair the function of other organelles or membranes, thus leading to visible symptoms of chilling injury.

IV. CONCLUDING REMARKS

In the fruits and vegetables referred to in this chapter, it is obvious that changes in mitochondria, tonoplasts, and chloroplasts occurred soon after the commodities were exposed to low temperatures, preceding the appearance of visible injury. This fact suggests that chilling stress rapidly affects the ultrastructure of cells, inducing malfunction of organelles and membranes. Figure 12 summarizes the mutual relationship between the ultrastructural

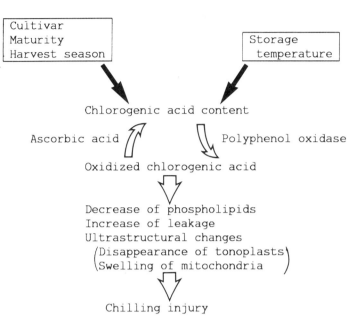

FIGURE 12.   Mechanism of chilling injury in eggplant fruit.

and physiological changes, together with related factors, and proposes a hypothesis of the mechanism of chilling injury by using eggplant fruit as an example.

Chlorogenic acid contents in eggplant fruit vary with cultivars,[36] maturities,[21] and harvest season,[21] and the fruits with high chlorogenic acid content are usually sensitive to chilling injury. Chlorogenic acid is oxidized by polyphenol oxidase, which is activated during storage at low temperatures, then the oxidized products impair the mitochondria and membranes. This idea was supported by a study in which chlorogenic acid was added to sections of eggplant fruit. This addition accelerated swelling of mitochondria, partial disappearance of tonoplast, increase of K^+ leakage, and decrease of phospholipids, etc.[37] Participation of phenol substances in the occurrence of chilling injury was obvious in the histochemical observations of taro and banana, as mentioned above. In any case, it seems that the ultrastructure of the cells of chilling-susceptible plants responds very sensitively to low temperatures. Further detailed studies of ultrastructural changes will become a useful tool in the understanding of chilling injury.

REFERENCES

1. **Jaweed, M. M. and Watada, A. E.,** Electron micrographs of snap bean mitochondria exposed to chilling temperatures *Proc. W. Va. Acad. Sci.,* 41, 150, 1969.
2. **Ilker, R., Waring, A. J., Lyons, J. M., and Breidenbach, R. W.,** The cytological responses of tomato-seedling cotyledons to chilling and the influence of membrane modifications upon these responses, *Protoplasma,* 90, 229, 1976.
3. **Niki, T., Yoshida, S., and Sakai, A.,** Studies on chilling injury in plant cells. I. Ultrastructural changes associated with chilling injury in callus tissues of *Cornus tolinifera, Plant & Cell Physiol.,* 19, 139, 1978.
4. **Kislyuk, I. M.,** Morphological and functional changes of chloroplasts after cooling of leaves of *Cucumis sativus* L., in *The Cell and Environmental Temperature,* Troshin, A. S., Ed., Pergamon Press, Oxford, 1963, 59.

5. **Klein, S.,** The effect of low temperature on the development of the lamellar system in chloroplasts, *J. Biophys. Biochem. Cytol.,* 8, 529, 1960.

6. **Slack, C. R., Roughan, P. G., and Bassett, H. C. M.,** Selective inhibition of mesophyll chloroplast development in some C4-pathway species by low temperature, in *Mechanisms of Regulation of Plant Growth,* Bieleski, R. L., Ferguson, A. R., and Cresswell, M. M., Eds., Royal Society of New Zealand Bull., 1974, 499.

7. **Taylor, A. O. and Craig, A. S.,** Plants under climatic stress. II. Low temperature, high light effects on chloroplast ultrastructure, *Plant Physiol.,* 47, 719, 1971.

8. **Wise, R. R., McWilliam, J. R., and Naylor, A. W.,** The effect of chilling stress on chloroplast ultrastructure and oxygen exchange in cotton and collard, *Plant Physiol.,* 67, 61, 1981.

9. **Brand, J. J., Kirchanski, S. J., and Ramirez-Mitchell, R.,** Chilling-induced morphological alterations in *Anacystis nidulans* as a function of growth temperature, *Planta,* 145, 63, 1979.

10. **Platt-Aloia, K. A. and Thomson, W. W.,** An ultrastructural study of two forms of chilling-induced injury to the rind of grapefruit (*Citrus paradisi*, Macfed), *Cryobiology,* 13, 95, 1976.

11. **Ilker, R., Breidenbach, R. W., and Lyons, J. M.,** Sequence of ultrastructural changes in tomato cotyledons during short periods of chilling, in *Low Temperature Stress in Crop Plants: The Role of the Membrane,* Lyons, J. M., Graham, D., and Raison, J. K., Eds., Academic Press, New York, 1979, 97.

12. **Morris, L. L. and Platenius, H.,** Low temperature injury to certain vegetables after harvest, *Proc. Am. Soc. Hortic. Sci.,* 36, 609, 1938.

13. **Abe, K., Iwata, T., and Ogata, K.,** Chilling injury in eggplant fruits. I. General aspects of the injury and microscopic observation of pitting development, *J. Jpn. Soc. Hortic. Sci.,* 42, 402, 1974.

14. **Rhee, J. and Iwata, M.,** Histological observations on the chilling injury of eggplant fruit during cold storage, *J. Jpn. Soc. Hortic. Sci.,* 51, 237, 1982.

15. **Tsay, L.,** Studies on the chilling injury of eggplant, *J. Jpn. Soc. Cold Preserv. Food,* 12, 119, 1986.

16. **Rhee, J. and Iwata, M.,** Histological observations on the chilling injury of cucumber fruit during cold storage, *J. Jpn. Soc. Hortic. Sci.,* 51, 231, 1982.

17. **McConnell, D. B. and Sheehan, T. J.,** Anatomical aspects of chilling injury to leaves of *Phalaenopsis* BL, *HortScience,* 13, 705, 1978.

18. **Yamawaki, K., Tomiyama, M., Chachin, K., and Iwata, T.,** Relationship between change in mitochondrial function and chilling injury of cucumber fruit, *J. Jpn. Soc. Hortic. Sci.,* 52, 332, 1983.

19. **Abe, K. and Ogata, K.,** Chilling injury in eggplant fruits. III. Effects of temperature and humidity on pitting injury of eggplant fruits during storage, *J. Ins. Cold Chain,* 2, 104, 1976.

20. **Abe, K. and Chachin, K.,** Chilling injury in eggplant fruits. VII. Influence of conditioning on the occurrence of chilling injury and the changes of surface structure of eggplant fruits, *J. Jpn. Soc. Hortic. Sci.,* 54, 247, 1985.

21. **Abe, K., Chachin, K., and Ogata, K.,** Chilling injury in eggplant fruits. II. The effects of maturation and harvesting season on pitting injury and browning of seeds and pulp during storage, *J. Jpn. Soc. Hortic. Sci.,* 45, 307, 1976.

22. **Rhee, J. and Iwata, M.,** Histological observations on the chilling injury of taro tubers during cold storage, *J. Jpn. Soc. Hortic. Sci.,* 51, 362, 1982.

23. **Fukushima, T.,** Chilling-injury in cucumber fruits. VI. The mechanism of pectin de-methylation, *Sci. Hortic.,* 9, 215, 1978.

24. **Murata, T.,** Physiological and biochemical studies of chilling injury in bananas, *Physiol. Plant.,* 22, 401, 1969.

25. **Yamaki, S. and Uritani, I.,** Mechanism of chilling injury in sweet potato. VII. Changes in mitochondrial structure during chilling storage, *Plant & Cell Physiol.,* 13, 795, 1972.

26. **Yamaki, S. and Uritani, I.,** Mechanism of chilling injury in sweet potato. VIII. Morphological changes in chilling injured sweet potato root, *Agric. Biol. Chem.,* 37, 183, 1973.

27. **Moline, H. E.,** Ultrastructural changes associated with chilling of tomato fruit, *Phytopathology,* 66, 617, 1976.

28. **Abe, K. and Ogata, K.,** Chilling injury in eggplant fruits. IV. Ultrastructural changes of cell and organelle membranes associated with chilling injury of eggplant fruits, *J. Jpn. Soc. Hortic. Sci.,* 46, 541, 1978.

29. **Hirata, K., Chachin, K., and Iwata, T.,** Changes of K ion leakage, free amino acid contents and phenylpropanoid metabolism in water convolvulus (*Ipomoea aquatica* Forsk.) with reference to occurrence of chilling injury, *J. Jpn. Soc. Hortic. Sci.,* 55, 516, 1987.

30. **Hirata, K.,** Studies on the Chilling Injury of Water Convolvulus (*Ipomoea aquatica* Forsk.), Ph.D. thesis, University of Osaka Prefecture, Sakai, 1987.

31. **Lyons, J. M. and Raison, J. K.,** Oxidative activity of mitochondria isolated from plant tissues sensitive and resistant to chilling injury, *Plant Physiol.,* 45, 386, 1970.

32. **Lyons, J. M., Wheaton, T. A., and Pratt, H. K.,** Relationship between the physical nature of mitochondrial membranes and chilling sensitivity in plants, *Plant Physiol.,* 39, 262, 1964.

33. **Raison, J. K., Lyons, J. M., Mehlhorn, R. J., and Keith, A. D.,** Temperature-induced phase changes in mitochondrial membranes detected by spin labeling, *J. Biol. Chem.,* 246, 4036, 1971.

34. **Wade, N. L., Breidenbach, R. W., and Lyons, J. M.,** Temperature-induced phase changes in the membranes of glyoxysomes, mitochondria, and proplastids from germinating castor bean endosperm, *Plant Physiol.,* 54, 320, 1974.

35. **Yamawaki, K., Yamauchi, N., Chachin, K., and Iwata, T.,** Relationship of mitochondrial enzyme activity to chilling injury of cucumber fruit, *J. Jpn. Soc. Hortic. Sci.,* 52, 93, 1983.

36. **Abe, K., Chachin, K., and Ogata, K.,** Chilling injury in eggplant fruits. VI. Relationship between storability and contents of phenolic compounds in some eggplant cultivars, *J. Jpn. Soc. Hortic. Sci.,* 49, 269, 1980.

37. **Abe, K. and Ogata, K.,** Chilling injury in eggplant fruits. V. Changes of K ion leakage and contents of phospholipids during storage and effects of phenolic compounds on K ion leakage, phospholipid content and ultrastructural changes of eggplant fruits sections, *J. Jpn. Soc. Hortic. Sci.,* 47, 111, 1978.

Part II. Contributing Factors and Assessment of Chilling Injury

Chapter 6

GENETIC AND ENVIRONMENTAL INFLUENCES ON THE EXPRESSION OF CHILLING INJURY

Brian D. Patterson and Michael S. Reid

TABLE OF CONTENTS

I. INTRODUCTION

Plants respond to chilling according to their genetic makeup and in a way which is greatly influenced by the surrounding conditions. We should therefore be able to minimize the effects of chilling either by modifying the genome, or by manipulating the environment of the plant. This chapter explores these two approaches. While the reader's interest is assumed to be mainly in horticultural crops, the rich source of genetic variation in wild species is mentioned where relevant.

II. GENETIC DIVERSITY IN CHILLING RESISTANCE

A. DIFFERENCES BETWEEN TAXA
1. Chilling-Resistant Species
The most chilling-resistant plants are those that are able to complete their life cycle at 0°C or less. Few land plants appear to be able to do this, but several examples from aquatic habitats are known. One of them, *Pyramimonas gelidicola*, is an alga that completes its life cycle well below 0°C. The Antarctic lakes in which it lives remain unfrozen because of their high salinity.[1] In the polar seas there are several macro- and microalgae (seaweeds, diatoms, and others) that live at temperatures between −2 and 0°C, while melting snowdrifts in seasonal climates are often colored pink by the snow algae which grow in them.[2] The biochemistry of these very chilling-resistant plants is adapted to low temperatures at every stage of development. For instance, *Pyramimonas* is motile at −10°C, and an alga from antarctic soil, *Prasiola*, can even photosynthesize at −20°C.[3]

2. Species that are Predominantly Chilling Resistant
Most land plants are sensitive to chilling at some stage of development. Possible exceptions are the species that grow below snow banks in polar and alpine regions, where even during the growing season they are cooled by meltwater near 0°C. Outside such specialized habitats, temperatures often rise well above 0°C during the growing season, even in arctic and alpine regions.[4] The poor vegetation cover in the arctic tundra is an indirect, rather than direct, effect of low temperature. The short growing season, outside which liquid water is unavailable, the low nutrient status of the soils, and the permafrost in the soil all reduce growth more than does low temperature itself.[5]

Plants from outside the tropics are genetically programmed to succeed in climates where relatively cold winters alternate with warm summers. Despite this, they can suffer from chilling injury (see Chapter 3). For instance, fruit of some apple and peach varieties break down during long storage at 0°C, and this is presumably because their metabolism is ill-adapted to such low temperatures. In the summer growing phase, development can be sensitive to chilling in other ways. For instance, although strawberry plants can be stored at 0°C when they are dormant, the development of their pollen can be disrupted by chilling during the growing period.[6]

3. Species that are Predominantly Chilling Sensitive
Many economically important plants originate from the tropics and subtropics. Unless they are from the highlands (for instance the potato), they are usually sensitive to chilling at several stages of development. The plants may be injured if they are exposed to temperatures below a threshold which is often typical for the species. For instance, many varieties of banana fruit are injured at temperatures below a threshold of 12°C.[7] Figure 1 illustrates this by comparing the response to temperature of the flowers of the tropical plant anthurium[8] with that of brussels sprouts.[9]

If cut flowers of anthurium are exposed to temperatures below a threshold of about

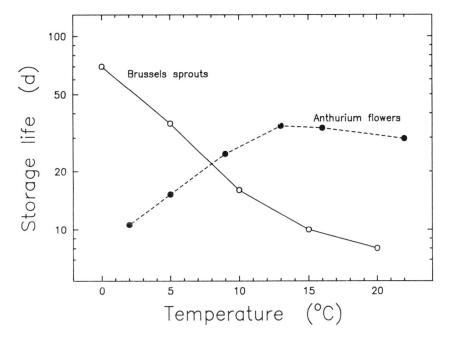

FIGURE 1.   Influence of temperature on the storage life of brussels sprouts[9] and anthurium flowers.[8] Brussels sprouts were maintained at the indicated temperature for the whole of the storage period. Anthurium flowers were exposed to the indicated temperature for 9 d and their subsequent shelf life at 22°C then determined.

13°C, their subsequent shelf life at 22°C is reduced. Below 13°C, the damage becomes progressively more severe as the temperature is lowered.[8] The storage life of anthurium flowers can be seen as being limited by an interaction of senescent changes and damage due to chilling injury. The onset of senescent changes is delayed as the temperature is lowered, while the likelihood of chilling injury is increased. These two opposing effects result in the storage life being longest at about 13°C. In contrast, brussels sprouts are not sensitive to chilling injury, so that only the effect of lowered temperature on reducing senescence is seen.[9] As a result, brussels sprouts are best stored just above their freezing point.

The threshold temperature may vary with tissue or stage of development and is not always a good indicator of overall sensitivity to chilling. For instance, the internal breakdown called blackheart appears in pineapple fruit after they have been stored for a few weeks at temperatures up to 21°C, but not at higher temperatures.[10] Despite the high threshold temperature for this fruit disorder, the pineapple plant is less chilling sensitive than the banana plant, the fruit of which is chilled only below 12°C.

In the permanently humid equatorial lowlands, even extreme minimum temperatures do not fall below 20°C. It is therefore hardly surprising that species from these regions are found to be very sensitive to chilling. *Espiscia reptans* (*Gesneriaceae*), an herb from Central and South America, is more easily damaged by chilling than any other species so far described.[11] Symptoms of chilling that take days to appear in species from cooler climates appear within minutes in leaves of *E. reptans*.[12,13] It is not yet known, however, whether this degree of sensitivity is typical of all species that are confined to the equatorial lowlands.

The response of anthurium flowers is typical of a large group of tropical plants in which the threshold temperature for the onset of a particular kind of chilling injury is well marked, although this temperature may vary appreciably for different physiological stages. Disruption of pollen development, for instance, tends to have a higher threshold than does the development of lesions on leaves. In any one wild species, however, the threshold temperatures for the onset of different chilling injuries may well have tended to converge during evolution. Even when the different chilling injuries are the result of different molecular events, the relevant molecular controls will have evolved in the natural environment under similar constraints of temperature.[14] Because of this, it is difficult to assess whether good coordination between different chilling symptoms results from a causal connection. For instance, the effect of chilling on the kinetics of chlorophyll fluorescence, growth rate, and appearance of necrotic symptoms correlate well.[15] Genetic experiments to indicate whether these forms of chilling injury are inherited together would be a test of causal connection.

There is an important group of plants of tropical affinities in which a threshold temperature for chilling injury is not well defined. An example is skin pitting and/or oleocellosis of citrus fruit.[16,17] Oleocellosis is a scald-like skin blemish that is connected with the rupture of oil glands. Although these injuries are more common if the fruit are stored for some weeks at temperatures much below 10°C, they often do not appear after cool storage and sometimes appear at higher temperatures. Some species, such as limes, seem more susceptible than, for instance, most varieties of orange.[18] Even with the same genotype, however, the incidence of chilling injury can vary strikingly between different seasons, growing areas, and stages of maturity.[19] This behavior contrasts with that of banana fruit[7] which are always injured below at well-marked threshold temperature of about 12°C.

B. DIVERSITY WITHIN TAXA: ALTITUDINAL AND LATITUDINAL VARIATION

Within closely related groups of plants there is often a range of genetic adaptation to chilling temperatures. The extent and nature of this adaptation is likely to be a reflection of the climate in which individual species and varieties have evolved. In searching for genotypes with enhanced chilling resistance, one of the most important aspects to be taken into account is, therefore, the climate of the natural habitat, particularly in regard to mean and minimum temperatures. Among the more important geographical features that influence temperature are latitude, altitude, and degree of continentality. Continentality (the influence of the continental mass in contrast to the maritime influence) has been little investigated with regard to the evolution of chilling resistance. Because diurnal and annual extremes of temperature, both high and low, tend to increase with distance from the sea in a land mass, resistance to chilling might be expected to also increase with distance from the sea.

Such effects are even found in the lowland tropics if they are seasonally dry, and chilling temperatures in such locations are more common than is generally appreciated. The temperatures at one such lowland tropical location, only 13° latitude from the equator and 80 km from the sea, are shown in Figure 2. At this site, night temperatures below 10°C are associated with dry air derived from the continental mass of central Australia. As the protective blanket of water vapor in the air is reduced, cooling by radiation at night becomes more effective. Day maximum temperatures are reduced much less. Plants from such a habitat will frequently experience temperature minima below 10°C in July, when the atmosphere is particularly dry.

Many species which are predominantly tropical either range to cooler latitudes, or extend to high altitudes within the tropics. These outliers are likely sources of genes which confer resistance to chilling. This is illustrated by two plant genera, the orchid genus *Dendrobium* and the tomato genus *Lycopersicon*.

1. Example of the Orchids

In general, species whose natural habitats lie close to the equator are sensitive to chilling, while those from higher latitudes are not. In plant genera for which data are available, it is possible to be somewhat more specific and point to an association between combinations of latitude and altitude of origin and degree of chilling resistance.

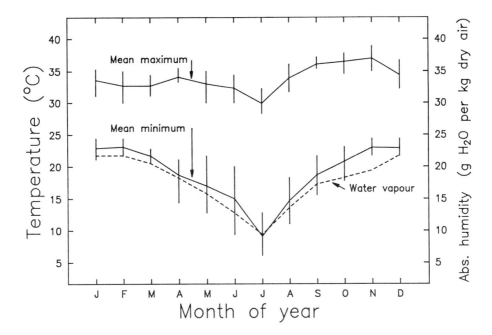

FIGURE 2.   Chilling temperatures in a lowland tropical environment in Australia.[20] Monthly mean maximum and minimum temperatures for Tortilla Flats,(latitude 13° 6'S, longitude 131° 14'E). This is located 80 km from the port of Darwin at an altitude of about 150 m. Temperature is shown by the solid line, with the bars giving the 14 and 86 percentile variation from the mean. About 1 in 7 d will have readings equaling the extremes shown by the bars. The monthly mean content of water vapor in the atmosphere (absolute humidity) is shown by the dashed line.

On the seaboard of eastern Australia, mean minimum temperatures in midwinter decrease by about 8°C for each 10° increase in latitude. Superimposed on this trend is the effect of altitude. Within the tropics, mean temperatures decrease by about 5 to 6°C for each 1000 m of altitude, while the effect on temperature minima is somewhat greater.[21] Both these trends are reflected in the temperature adaptation of plant species. Table 1 shows this for Australian species of the orchid genus *Dendrobium*. In eastern Australia, these species grow naturally as epiphytes in different latitudes,[23] from its northern tip (11° S), where temperatures at sea level rarely fall below 20°C, to Tasmania (41° S), where winter minimum temperatures are near 0°C.[24] The table also indicates those species which will succeed as epiphytes under outdoor garden conditions as in southern Victoria (latitude 38° S).[22] In this region, mean minimum temperatures are below 10°C for about 5 months of the year.[24] Epiphytic orchids grown outdoors on trees, where the root system is exposed to the prevailing air temperature, seem to be particularly good indicators of chilling sensitivity. This may be because their roots are less buffered against extremes of temperature than are those of terrestrial species. Table 1 shows that every species that is resistant to chilling either ranges at least as far south as 33° latitude in the wild, or attains altitudes of at least 1000 m above sea level. In contrast, none of the seven species which require winter minimum temperatures above 10°C in cultivation have a natural habitat which extends south of 20° latitude, or has been recorded at altitudes as high as 1000 m.

Many other orchid genera have species which vary in chilling resistance in an analogous way. Examples are *Cattleya* and *Laelia*, from different altitudes in South and Central America, and *Paphiopedilum* and *Vanda*, which extend over a wide range of altitude and latitude in Asia.[25]

2. Example of the Tomatoes

The *Solanaceae* includes genera such as *Solanum* and *Lycopersicon* which show notable

TABLE 1
Chilling Resistance of Australian Dendrobium Species in Relation to Latitude and Altitude of Origin

Species	Latitude range (degrees south)	Max altitude (m)	Chilling resistance
cucumerimum	28—34	—	High
linguiforme	20—36	—	High
pugioniforme	28—36	1000	High
striolatum	33—41	900	High
tenuissimum	28—33	1000	High
beckleri	28—33	—	High
teretifolium	15—36	—	High
kingianum	22—33	—	High
ruppianum	11—20	1000	High
speciosum	15—39	1000	High
gracilicaule	17—35	—	High
bairdianum	16—20	1100	High
falcorostrum	27—32	1000	High
aemulum	18—36	—	High
tetragonum	16—36	—	High
monophyllum	21—30	—	Medium
schneiderae	21—30	750	Medium
adae	16—20	750	Medium
fleckeri	16—20	900	Medium
toressae	16—17	1200	Medium
rigidum	11—17	0—100	Low
caniculatum	16—23	—	Low
smilliae	15—20	—	Low
bigibbum	11—16	—	Low
discolor	11—24	—	Low
johannis	11—16	—	Low
baileyi	16—20	—	Low

Note: The degree of chilling resistance is termed "high" if the species can grow and flower outdoors at latitude 38° S in coastal southern Victoria,[22] or is known to tolerate temperatures below 10°C. It is termed "low" if the species needs winter minimum temperatures above 10°C. Species termed "medium" have intermediate requirements. Distribution data from Nichols.[23]

variations in chilling resistance and which contain pools of genes for chilling resistance.[26] In the species *Lycopersicon hirsutum* (a wild tomato), chilling sensitivity varies among different varieties. This species grows naturally in Ecuador and Peru between the equator and 13° S.[27] Within this limited range of latitude it grows from sea level to an altitude of 3000 m, where night temperatures are below 10°C throughout the year.[21] A related species, *L. chilense*, ranges further south into Chile. Some of the kinds of cold resistance present in these species are detailed below.

Seed germination — The minimum temperature of germination of most cultivars of domestic tomato is about 10°C. Some accessions of *L. hirsutum*[28] and *L. chilense*[29] will germinate 3 to 4°C below this. Some of this germination at low temperatures may be a function of faster germination at all temperatures between the minimum for germination and 20°C.[28]

Survival after a chilling stress — Figure 3 shows the ability of seedlings of *L. hirsutum* from different altitudes to survive chilling at 0°C.[28] Accessions from the higher altitudes tend to survive chilling exposure better than those from low altitudes, even in the one species.

Growth and development — At high altitudes in the tropics, temperatures fall below the chilling threshold at night and rise above 10°C within an hour or two of dawn.[30] It is

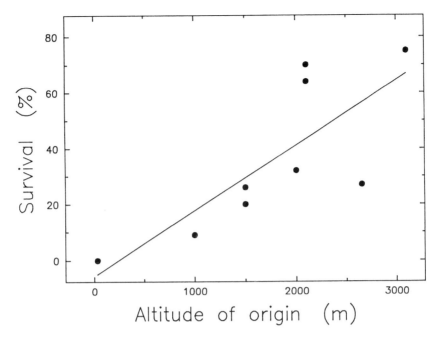

FIGURE 3. Survival of altitudinal forms of *Lycopersicon hirsutum* after chilling at 0°C. Seedlings were chilled for 7 d from midway through the light period after the appearance of the first true leaf. Survival was assessed after a further week at 22°/18°C. (Modified from Patterson, B. D., Paull, R., and Smillie, R. M., *Aust. J. Plant Physiol.*, 5, 609, 1978. With permission.)

likely that genetic adaptation of species from these locations will have been towards resisting low night rather than low day temperatures. When young seedlings of *L. hirsutum* (originally from 3000 m) are kept from the time of radical extrusion under 16-h nights at 0°C, alternating with 8-h days at 20°C, they develop true leaves. In contrast, seedlings of *L. esculentum* develop poorly under these conditions.[31] On more mature plants, the plastochron index has been used as a criterion of development.[32] This technique has indicated that *L. hirsutum*, from around 3000 m,[33] and an accession of *L. chilense* grow well at low temperatures.[34]

Chlorophyll synthesis — Etiolated seedlings of *L. hirsutum* were able to green at lower temperatures than were seedlings of *L. esculentum* when they were exposed to light.[28]

Pollen development — The development of pollen of a wide range of species is particularly sensitive to disruption by low temperatures. Some examples of this are shown in Table 2. This kind of cold sensitivity has been examined in more detail in the tomato. If immature flowers of *L. esculentum* are exposed to repeated night temperatures below 10°C, they produce pollen grains that contain no chromatin and are, therefore, infertile.[40,46] Two periods of pollen development, at about 11 and 6 d before anthesis, are particularly sensitive.[30] Accessions of *L. hirsutum* originating from high altitude[47,30] can produce pollen of normal appearance and liberate their pollen freely when grown under these conditions.

Pollen germination and fertilizing ability — The pollen of some varieties of *L. hirsutum* from high altitudes is able to germinate *in vitro* at 6°C.[48] This better growth at a low temperature may be, at least in part, the reason for the greater ability of the wild species to fertilize at low temperatures. In mixtures of the pollen of a wild tomato (*L. hirsutum*) from high altitudes and that of the domestic tomato, the "wild" pollen is more effective in fertilizing at low temperature.[49] How much of this fertilizing advantage at low temperature is due to the ability to germinate quickly and how much to the growth rate of pollen tubes or other factors is not known. Interestingly, within the altitudinal forms of *L. hirsutum*, the ability of pollen to germinate at low temperatures does not correlate well with seed ger-

TABLE 2
Sensitivity of Pollen Development to Low Temperatures

Species	Temperature causing pollen abnormality (°C)	Ref.
Capsicum annuum	15	35
Carica papaya	12—16	7
Cucumis melo	<11	6
Fragaria virginiana × *chiloensis*	<11	6
Glycine soja	15	38
Gossypium spp.		39
Lycopersicum esculentum	<10	40
Oryza sativa	12	41
Pennisetum typhoides	13—16	42
Saccharum officinarum	21	43
Solanum melongena	<11	6
Sorghum bicolor	10	44
Triticum aestivum	0—2	45

mination at low temperatures.[48] Moreover, within cultivars of *L. esculentum*, it does not correlate with plant growth.[50] This is presumably because the adaptations are under the control of combinations of a number of genes that are present to different extents in the varieties. However, the fertilizing ability of *L. hirsutum* pollen at low temperature is correlated genetically with the capacity for root growth at low temperatures.[51]

Protoplasmic streaming — The effect of temperature on protoplasmic streaming in the stem and leaf trichome cells of tomatoes can be followed under the microscope. The effect of decreasing the temperature to 5°C on the rate of protoplasmic streaming is less in *L. hirsutum* from high altitudes than it is in *L. esculentum* or low-altitude forms of *L. hirsutum*. However, the rate of streaming is decreased by low temperatures to a greater extent than it is in trichome cells of a winter-growing annual, *Veronica persica*.[52] In other words, while the high-altitude accession of this wild tomato is more resistant to chilling by this criterion than is the domestic tomato, it is by no means as resistant as the *Veronica* species.

Uptake of amino acids — Low temperature reduces amino acid uptake into isolated leaf tissue of tomato. The Q_{10} for this effect is initially similar in leaf tissues of a variety of *L. hirsutum* from 3000 m in the Andes and in those of the domestic tomato *L. esculentum*. However, during exposure for 2 to 6 h to temperatures below 10°C, the rate of uptake declines to a greater extent in the domestic tomato than in the wild tomato.[53]

Photosynthetic capacity — Photosynthetic capacity can be estimated from the fluorescence kinetics of chlorophyll in intact leaves (see Wilson and Greaves, Chapter 8). This technique has been used to show that photosynthetic capacity of *L. hirsutum* from higher altitudes is more stable during storage at low temperatures. Tuberous *Solanum* species (wild potatoes), which also occur over a wide range of altitudes show a similar relationship between altitude of origin and stability of photosynthetic capacity at low temperatures.[54] Photosynthetic capacity which has been lost during chilling can be recovered on transfer to higher temperatures. This occurs more rapidly in the wild tomatoes from high altitude than in the domestic tomato.[34,55]

It can be seen from these results that the wild tomatoes from the andean regions of South America, particularly *L. hirsutum* and *L. chilense*, have much to offer as experimental material in the study of chilling injury. Presumably, other plant groups will have evolved along similar lines. In the next section, ways to exploit this genetic variation will be considered.

III. GENETIC MANIPULATION OF CHILLING RESISTANCE

A. BREEDING FOR RESISTANCE

Breeding for chilling resistance has important practical requirements. The species or varieties representing the gene pools to be drawn on must be identified. In addition, the number and nature of the genes responsible for the chilling resistance influences the ease with which they can be transferred to horticulturally important plants. If conventional plant breeding is to be used, it is important to know whether the genes for chilling resistance are scattered over the genome, or whether they are in linked groups. Transfer of genes using more sophisticated techniques requires that the genes be identified and then isolated. In addition, rapid and simple selection techniques for genotypes containing the desired genes must be available.

1. Gene Pools for Chilling Resistance

Gene pools in species which are sexually compatible with horticultural crops and which give fertile offspring are known for a number of species. Besides the orchids and tomatoes noted above, they include some particularly chilling-sensitive crops. The pineapple (*Ananas comosus*) is compatible with other *Ananas* species, one of which, *A. bracteatus*, (syn. *Pseudananas bracteatus*) reaches as far south as Argentina and would be expected to be more cold tolerant than existing pineapple cultivars.[56] Maize is not known as a wild plant, but the land race Confite Puneno is grown in Peru and Bolivia at altitudes between 3600 and 4000 m and so is a likely source for genes for various kinds of chilling resistance.[57] The most chilling-resistant avocadoes are those which derive from the highlands of Mexico, and genes from these are probably responsible for the relative chilling resistance of the modern varieties of avocado, such as Fuerte, Hass, and Sharwil, that can be grown outside the tropics.[58] Similarly, the common varieties of custard apple derive their cold resistance from the highland species *Annona cherimolia*.

As gene technology progresses, so the gene pools from which chilling resistance can be obtained will expand beyond sexually compatible plants. However, genes have evolved in particular genetic backgrounds, where they are adapted to work in concert with the other genes of the organism. For this reason, genes from related plants may perform their function better than ones derived from other organisms.

2. Inheritance of Chilling Resistance

Few detailed studies have been made of the way in which chilling resistance is inherited. There are, however, data concerning the expression of certain kinds of chilling resistance in primary crosses between chilling-resistant and chilling-sensitive parents. Particularly clear examples of the way in which chilling resistance is inherited in primary crosses (F$_1$ hybrids) between chilling-resistant and chilling-sensitive species has been provided by breeders of ornamental orchids. Table 3 shows data for such primary hybrids of species from a number of plant families.

Among the orchids, chilling resistance is particularly well expressed in the F$_1$ generation. As mentioned in Section II.B.1., the roots of epiphytic orchids are not buffered against chilling temperatures, and only those species which are relatively resistant to chilling can be cultivated under conditions where minimum temperatures fall below 10°C.

When a chilling-resistant orchid is crossed with one which is more chilling sensitive, the F$_1$ hybrid has the properties of the more resistant parent. For example, the species *Sarcochilus hartmannii* is native to the mountains of subtropical Australia and is able to grow outdoors in southern Victoria (latitude 38° S) where winter minimum temperatures approach 0°C.[62] In contrast, the tropical species *Phalaenopsis amabilis* does not thrive in cultivation if it is exposed to night temperatures below 10°C. Their bigeneric hybrid (named

TABLE 3
Temperature Responses of F₁ Hybrids Between Chilling-Resistant and Chilling-Sensitive Parents

Parent species		Physiological response to low temperatures	Hybrid response	Ref.
Sensitive	Resistant			
Orchidaceae				
Laelia purpurata	*Cattleya loddigesii*	Survival	Dominant	25
Phalaenopsis amabilis	*Sarcochilus hartmannii*	Survival	Dominant	22
Annonaceae				
Annona squamosa	*A. cherimolia*	Survival	Dominant	59
Solanaceae				
Lycopersicon esculentum	*L. hirsutum*	Pollen production	Intermediate	30
L. esculentum	*Solanum tuberosum*	Chloroplast function	Intermediate	60
L. esculentum	*S. lycopersicoides*	Chloroplast function	Dominant	61
Passifloraceae				
Passiflora edulis flavicarpa	*P. cincinnata*	Ion leakage	Intermediate	62
P. edulis flavicarpa	*P. caerulea*	Ion leakage	Intermediate	62

"*Sarconopsis*") is close to the *Sarcochilus* parent in cold resistance, however, and can be grown outdoors in southern Victoria.[22] This behavior is well known among orchid breeders, but is not peculiar to the *Orchidaceae*. Table 3 shows analogous examples from the families *Passifloraceae (Passiflora), Solanaceae (Lycopersicon),* and *Annonaceae (Annona)* in which F₁ hybrids between chilling-resistant and chilling-sensitive parents often follow the resistant parent.

Some partial exceptions to the dominance of chilling resistance in the F₁ generation have been noted. *Passiflora edulis flavicarpa* (the yellow passionfruit) and the ornamental species *P. cincinnata* differ only moderately in their degree of chilling resistance. The chilling sensitivity of their F₁ hybrid, measured using the technique of ion leakage at 0°C, appears to be intermediate between the parents. However, the behavior of the hybrid between the much more resistant species *P. caerulea* and *P. edulis flavicarpa* follows that of the more resistant parent.[61] Another kind of chilling resistance, the ability to produce pollen under cold nights, is also intermediate in the F₁ hybrid of two tomato species, *L. esculentum* and *L. hirsutum.*[30] Somatic hybrids do not necessarily contain equal amounts of the two parental genomes. This may explain why in somatic hybrids of potato and tomato chilling resistance (measured by the time taken for photosynthetic capacity to decrease in the dark at 0°C) varies between individual hybrids. It was, however, generally intermediate between the parents.[59]

The custard apple grown commercially in eastern Australia is an F₁ hybrid of two species, *Annona cherimolia* (cherimoya) and *A. squamosa* (sweetsop). While the cherimoya derives from the highlands of Peru and Ecuador, the sweetsop is native to the tropical lowlands.[63] The cherimoya can be fruited in the cool conditions of New Zealand (latitude >34° S), while the sweetsop is generally confined to the lowland tropics. The two species and their F₁ hybrid (often called atemoya) are grown together at the Tropical Fruit Station, Alstonville (latitude 29° S in eastern Australia). Despite its name, this research station is outside the tropics and night minimum temperatures below 10°C are frequent there during the winter.

This seriously damages the leaves of the lowland tropical parent, *A. squamosa*. In contrast, both *A. cherimolia* and its F_1 hybrid with *A. squamosa* remain healthy.[64] This would suggest that, just as in many other F_1 hybrids, the chilling resistance of one parent has been dominant.

The bulk of the evidence suggests that the strength with which parental chilling resistance is expressed in its F_1 hybrids depends on the specific kind of chilling resistance involved and the degree of parental resistance. At least among orchids, dominance seen in the F_1 breaks down in subsequent generations. This is consistent with there being a number of genes interacting to give the chilling resistance, with them being active in the heterozygous state, and then segregating in subsequent generations.

Another factor may influence the way in which chilling resistance is inherited. Studies with tomatoes have shown that the frequency of crossing over increases at low temperature.[65] In some procedures, low temperatures are applied as a selection at the time of meiosis, as at the stage of pollination.[66,67] As well as exerting a selective effect on the survival of particular genotypes, low temperatures could influence the extent of recombination of genes derived from the parent species and present on homologous chromosomes. Much remains to be done in elucidating the ways in which the different kinds of chilling resistance are inherited. However, enough is already known about the frequently strong expression of chilling resistance in F_1 crosses for this property to be exploited by plant breeders.

3. Selection Techniques

Effective screening methods are important requirements, so that individual genotypes containing the genes for chilling resistance can be selected from a heterogenous population. The correct conditions for screening must be chosen, because chilling resistance can easily be confused with resistance to other stresses. For instance, chilling injury is nearly always made worse by water loss,[11] so that at low humidities, genotypes with a greater resistance to water loss but equal resistance to chilling may appear to be more chilling resistant. For this reason, selective treatments should in general be made under humid conditions. In addition, diurnal changes in chilling sensitivity and stomatal behavior[48,68] can influence survival more than the genetic differences sought among hybrid progeny.

The best selective environments are likely to be those which most resemble the natural temperature conditions which the improved cultivar must resist.

Selection during pollination — An ovule is successfully fertilized by only one pollen grain out of (potentially) many thousands. If fertilization is performed at a sufficiently low temperature, the growth of chilling-resistant genotypes of pollen will be favored over others. These will reach the ovule first so that their genes will appear in the resulting seed. At no other stage of development can selection be made on such large numbers of genotypes. Evidence that selection at this stage is effective is given by experiments with pollen mixtures of a high-altitude wild tomato (*Lycopersicon hirsutum*) and the domestic tomato.[49] The frequency with which *L. hirsutum* pollen contributes to hybrid zygote formation more than doubles on reducing the temperatures at pollination from 24/19°C day/night to 12/6°C. Using pollen from the F_1 hybrid between these species, low temperatures favor fertilization by pollen genotypes containing particular segments of the *L. hirsutum* chromosomes.[66] This method appears to select for some, but not all kinds of chilling resistance at other stages of development. For instance, root growth at low temperature is selected, but shoot growth is not.[51]

Selection for seed germination — Genotypes whose seeds germinate at low temperatures are easily selected for. Large numbers of seeds are incubated at a temperature at which only the more resistant genotypes germinate at an appreciable rate. Again, the genes that govern low temperature germination may be different to those governing other kinds of chilling resistance.

Seedling selection — Selection of young seedlings using their survival at low temperature, or their degree of development, also allows for large numbers to be screened. It is reasonable to assume that much of the ability of the adult plant to survive after a chilling stress will be expressed in the seedling. Chilling tends to make seedlings more easily invaded by microorganisms, introducing the risk that selection will be for resistance to a microorganism rather than to the chilling stress. This problem can be eliminated by germinating the seed in sterile culture before the chill selection.[31] Under these conditions, chilling tomato seedlings using cold (0°C) nights alternating with warm days over a period of several weeks differentiates chilling-resistant and chilling-sensitive genotypes better than constant chilling.[31] It is conceivable that while a seedling can easily recover from a single night of chilling, repairing the damage caused by the stress repeated every night will consume a significant part of the energy received during the day. This may be the reason why a more resistant genotype will cumulatively be better able to grow and develop under these conditions, despite wide differences in size of seed reserves.

Selection on mature plants — Selection for some kinds of chilling resistance, for instance, those forms of resistance involved with flowering and the setting of fruit, must be at a mature stage of development. An example is selection for pollen development at low temperatures. The effect of low night temperatures on pollen production in a number of species is illustrated by Table 2. In the tomato, good pollen development at low temperatures is not selected when seedlings are screened for good development under low night temperatures.[31] Different genes presumably control the two forms of resistance. Space is likely to be too restricted in artificial environments to accommodate more than a few plants at the flowering stage. For this reason, populations may have to be screened in natural outdoor environments. Although conditions in higher latitudes are often too variable to do this reproducibly, the variability decreases as one approaches the equator. Temperatures do not change with the seasons, and the frontal systems which give much of the day-to-day temperature variation at higher latitudes weaken as they approach the tropics.

The Hawaiian islands offer a natural environment that approaches the ideal for the selection of genotypes adapted to different temperatures.[69] Figure 4 shows the temperatures at a location on the island of Maui, Hawaii, which was selected for its suitability to screen for pollen production at low temperatures.[30] Plants of *L. esculentum* (domestic tomato), a high-altitude form of *L. hirsutum* and their F_1 and F_3 hybrids were raised outdoors in containers at low altitude, where night temperatures remain above 15°C. At the flowering stage, the plants were transferred to an altitude of 2000 m. Figure 4 shows the temperatures recorded at the top of the plant canopy over the following 30 d at this altitude. Mean minimum night temperatures were about 7.5°C. Figure 5 shows that this environment seriously interfered with the development of the *L. esculentum* pollen, but had no effect on the microscopic appearance of the *L. hirsutum* pollen. Among an F_3 population of their hybrid, a proportion of plants produced good pollen, which was used to produce a backcross to the *L. esculentum* parent. In this experiment, 131 plants were exposed to the high-altitude environment, and this number could easily be increased.

Selection on isolated tissues — With isolated tissues, the risk of destroying valuable genotypes is minimized because the bulk of the plant is not subjected to the chilling stress. The stress of chilling provokes either a response, such as ethylene evolution and an increase in respiration (see Chapters 14 and 15), or a decline in function, shown by leakage of electrolytes (see Chapter 12), slowing of protoplasmic streaming,[52,70] and decline in photosynthetic ability. Much interest has been shown in the analysis of chlorophyll fluorescence (see Chapter 8) for the latter stress response. The use of isolated tissues does present problems, however. Sample to sample variation is often high, requiring each genotype to be tested many times. Significant amounts of tissue are therefore consumed in such tests.

Because chilling resistance is compounded of a number of plant responses and probably controlled by a number of different genes, any one screening technique may not select for all such genes. The extent to which different techniques select for different sets of genes is

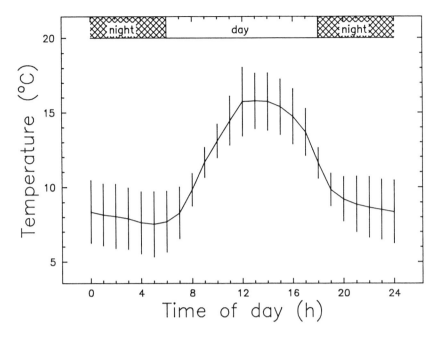

FIGURE 4. Temperatures recorded during selection for chilling resistance at 2000 m above sea level in Maui, Hawaii. The bars show the standard deviation about the mean during the 30 d of the experiment. (Modified from Patterson, B. D., Mutton, L., Paull, R. E., and Nguyen, V. Q., *Plant Cell Environ.*, 10, 363, 1987. With permission.)

not known. Despite this problem, selection techniques are all-important tools in breeding for chilling resistance. To be effective, they must be able to discriminate chilling-resistant genotypes and to be able to handle large numbers of plants at the relevant stage of development. For different kinds of chilling resistance there must be different selection techniques, for as yet it is not known how far particular genes for chilling resistance are common to different stages of development.

B. THE POTENTIAL OF GENE TECHNOLOGY

Gene technology is concerned with using laboratory techniques to transfer DNA between organisms and/or to modify genomic DNA. It offers the possibility of transferring useful genes between species that are not closely related. Using conventional plant breeding alone, there is little hope of improving the chilling resistance of many crops of tropical origin. This is because the crop either has no close relatives that are resistant to chilling, or because hybridization is unsuccessful. Even when hybridization is possible, the F_1 may be sterile. For instance, a useful source of resistance for the tropical fruit, papaya (*Carica papaya*), would be its close relative *Carica pubescens* (syn. *C. cundinamarcensis*), which originates from the andean highlands and is considerably more resistant to chilling. While an F_1 hybrid of these species has been reported,[71] it is not fertile, so that further progress has not been possible.

The techniques used for the isolation and transfer of genes are explained in a number of texts.[72,73] Before these techniques can be employed, the genes of interest must be identified in some way. While there are no simple methods for doing this, it may be possible to locate genes for chilling resistance to particular positions on the chromosome map of the species by reference to genetic markers. A second possibility is available if the enzymes or structural proteins controlled by the genes for chilling resistance could be identified. These possibilities are discussed below.

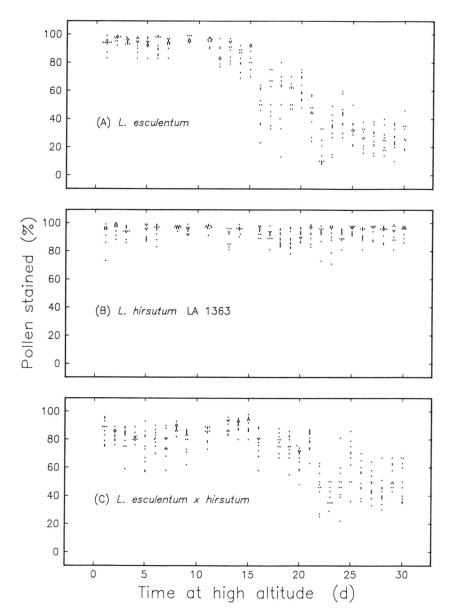

FIGURE 5.   Effect of low temperatures on pollen development in tomato. Plants of *Lycopersicon esculentum* cv. Flora-Dade (domestic tomato), *L. hirsutum* LA 1363 (wild tomato originally collected from 3000 m in Peru), and their F₁ hybrid were grown to the flowering stage on the island of Maui, Hawaii, then transported to 2000 m above sea level (see Figure 4 for temperatures). The percentage of pollen that was stained by acetyl carmine is plotted vs. time at the high altitude. (Modified from Patterson, B. D., Mutton, L., Paull, R. E., and Nguyen, V. Q., *Plant Cell Environ.*, 10, 363, 1987. With permission.)

1. Identification of Important Genes

When a chilling-sensitive species is crossed with a species that is a source of genes for chilling resistance, the progeny of their hybrid can be screened for chilling resistance. The resistant progeny will contain segments of the source chromosomes in which the genes for the chilling resistance are embedded. Although unwanted from the point of view of chilling resistance, these associated segments contain genes that can be used as markers as long as their map position is found.[74,75] The markers may be isozyme variants, polypeptides, or differences in DNA structure.

Molecular markers — In tomato, the genes for nearly 30 enzyme variants which can be separated by electrophoresis have been mapped on the 12 tomato chromosomes,[74] as well as more than 100 differences in DNA sequence[75] which have been detected using the specificity of restriction enzymes and DNA probes (RFLP analysis).[76]

Location of genes for chilling resistance — Judging from their association with chromosome markers, the genes for different kinds of chilling resistance appear to be scattered in different parts of the genome. For instance, genes governing adaptations to low temperatures in rice are distributed over several linkage groups.[77,78] In tomato, genes for the ability of pollen to fertilize at low temperatures (derived from *L. hirsutum*) are on chromosomes 6 and 12.[66] However, in any one tomato plant, successful fertilization at low temperatures depends either on the segments from chromosome 6 or chromosome 12 being present, rather than on them being present together. A different kind of chilling resistance in the tomato, the ability to grow at low temperatures as measured by the plastochron index[32] is associated with markers on chromosome 7.[(79,55)]

Biochemical identification — A further approach towards identifying genes for chilling resistance is possible if the metabolic or ultrastructural feature responsible for temperature sensitivity can be identified. A number of techniques are available for producing copy DNA (cDNA) from proteins for which the structure is known, or to which antibodies can be raised.[72] This can then be used as a probe for the gene. While little is known about the nature of the gene products (presumably enzymes, cytoskeletal proteins, or control sequences) whose structural differences result in the different degrees of chilling resistance, two possibilities are considered below.

An enzyme which controls the degree of unsaturation of phosphatidyl glycerol has been purified from squash,[80,81] following the hypothesis that the degree of unsaturation of phosphatidyl glycerol in chloroplast membranes is responsible for chilling resistance (see Chapter 11). The ability of tomato seeds to germinate at low temperatures is genetically associated with the degree of unsaturation of the storage lipid (triglyceride) in the seeds.[82] A possible explanation is that because the unsaturated lipid is more fluid at low temperatures, it is therefore more easily attacked by lipases during germination. There is considerable interest in manipulating the degree of unsaturation of seed lipids, and the methods of gene technology are already being applied to the enzymes governing their synthesis.[83] With greater control over the unsaturation of the storage lipid using such methods, it may be possible to decrease the minimum temperature of germination of seeds that have lipid as a major food reserve.

Microtubules may be important in adaptation to chilling temperatures.[14] They are destabilized to different degrees in the cold, depending on the species. The microtubule cytoskeleton is made up of two structurally related subunits, α- and β- tubulin, as well as a number of microtubule-associated proteins.[84] In a wide range of plants, the stability of microtubule structures at low temperatures is adapted to the environmental temperature.[(1,85-87)] In animals, there are apparently analogous adaptations, and indications that the genetic differences in cold stability are a function of the primary structure of the tubulin monomers rather than of the MAPS. One suggestion is that cold stability may be associated with a greater degree of amidation of glutamate side chains at the C terminal end of the α- and β-tubulins.[86] Given the strikingly close structural homology of plant and animal tubulins, it seems possible that analogous mechanisms of temperature adaptation will have evolved in the two groups.

DNA probes for animal tubulins hybridize with plant tubulin genes. They have been used to show that tubulin genes are arranged in the mung bean as tandem repeats of alternating α- and β- tubulin genes.[88] These results open the way to isolating tubulin genes from plants whose microtubules are more stable in the cold and transferring them to a chilling-sensitive plant using a suitable vector. This would create an experimental system for studying the role of microtubules in chilling resistance.

Quantitative nature of chilling resistance — A possible problem in using gene technology for transferring chilling resistance is that it appears to vary in a quantitative manner among related genotypes. For instance, survival at low temperatures[28] and photosynthetic capacity[54] increase gradually with altitude of origin in the species *Lycopersicon hirsutum* (wild tomato) and tuberous *Solanum* (wild potato) species. A family derived from a cross between *L. esculentum* (domestic tomato) and the more cold-resistant *Solanum lycopersicoides* also has a continuous variation in chilling resistance as determined by chlorophyll fluorescence.[61] One possibility is that a number of genes are interacting in a quantitative way to give the different levels of chilling resistance. The existing techniques of gene technology are concerned with the isolation of single genes that confer special properties. If a number of genes are required for resistance to chilling, they would all have to be identified in order to be transferred. So far, none has been identified.

Before gene technology can be applied to the improvement of chilling resistance, more needs to be known of the molecular genetics and biochemistry of the diverse kinds of chilling resistance. Once the relevant genes can be identified, it may be possible to isolate and transfer them between species. A very diverse gene pool could then be drawn on, free from unwanted parts of foreign genomes. While the difficulties of this approach should not be underestimated, it is likely to become increasingly important as techniques in gene technology improve.

IV. ENVIRONMENTAL EFFECTS

The nature of the environment profoundly affects the response of plants to chilling. Apart from some remarkable aroids, which generate heat so as to maintain themselves at nonchilling temperatures even when covered with snow,[89] plants tend to respond to environmental changes rather than modifying them. It is hardly surprising that vegetative plants respond to chilling quite differently at different times of the diurnal cycle. Plants are autotrophic only during the hours of daylight; at night their metabolism must switch to a wholly heterotrophic mode. Likewise, the capacity of plants to lose water varies diurnally, because most species only open their stomata during daylight. Fruits and storage organs are not so sensitive as leafy shoots to diurnal changes in response to chilling. Bearing these points in mind, some of the ways in which changes in the environment before, during and after a chilling stress can influence its effect is examined below.

A. EFFECTS BEFORE CHILLING
1. Hardening and Conditioning
Many, but not all, chilling-sensitive plants are able to be hardened against chilling injury. The hardening process is analogous to the freeze hardening that is typical of perennial plants from seasonal climates. In fact, with such plants it is often difficult to differentiate freeze hardening from chill hardening, or, indeed, from a hardening against stress in general. There are no absolute criteria for the amount a plant has been hardened. A "soft" plant is one that has been grown under conditions in which stress has been minimized, while a hardened plant is one that has reacted to stress in an adaptive way. At the one extreme, plants that have been grown aseptically on tissue culture media are stressed when they are transferred to a less protected environment. Such plants are adapted to an environment almost saturated with water vapor. They have little epicuticular wax, and are very sensitive to loss of water in undersaturated air.[90]

Hardening involving an increase in the level of abscisic acid (ABA) — The factors causing hardening do not necessarily have to include low temperature. Water stress alone is an effective treatment for many species.[11] Low temperatures (but not so low as to lead to injury) may also be effective. In some species, hardening is associated with an increase

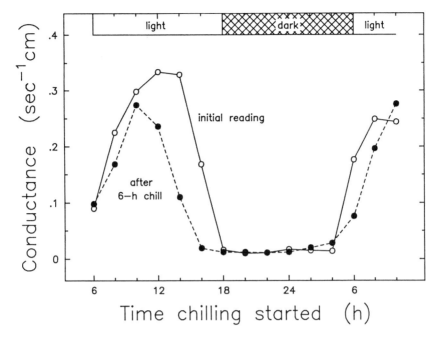

FIGURE 6.   The effect of chilling at 0°C on preventing stomatal opening and closure during the diurnal changes in conductance of mung bean leaves. Mung bean seedlings were grown in a cabinet at 22°/18°C day/night temperature until the first pair of leaves had expanded. At the times indicated, a batch of seedlings was divided into two; on one group initial conductance was measured with a porometer, and the other was kept at 0°C in the dark in 100% rH for 6 h before measuring the conductance. Each point is the mean from four leaves.

in ABA, as in chilling-sensitive tomato plants[91] or in chilling-resistant wheat plants.[92] Plants can be hardened against chilling by applying exogenous ABA to them.[93] It is therefore possible that much of these two hardening treatments are mediated through increases in ABA.

Hardening and stomatal action — Stomatal action in chilling-sensitive plants is sensitive to low temperature.[94] The sensitivity of stomata to chilling is illustrated by the results shown in Figure 6. If mung bean plants are chilled at different times of the day, when stomata are open or closed to different extents, the stomata become "locked" in their prechilling state. During the period of chilling, the stomata do not close at a temperature of 0°C if they are already open, nor do they open if they are already closed. As a consequence, leaves that are chilled from the middle of the day, when stomata are usually open, will be more easily damaged if water can continue to be lost through the open stomata. If the stomata are closed at the beginning of chilling, either because it is during the dark period or because the leaves have been previously given a water stress, the rate which water is lost will be reduced. Likewise, a hardening treatment that reduces stomatal aperture before chilling is started will reduce the loss of water and therefore damage during chilling. Part of the effect of ABA in reducing chilling is independent of its effect on stomata.

Low temperatures and lipid unsaturation — The effect of low temperatures in causing an increase in the unsaturation of membrane lipids is reviewed in Chapters 9 and 11. These changes have been associated with freezing resistance[92] as well as with chilling resistance.

Not only membrane lipids become less saturated at low temperatures. The degree of unsaturation of seed lipids (triglyceride) is increased by low temperatures in sunflower[95] and soybean.[96] This may be a hardening phenomenon, because in tomato, germination at low temperatures is associated with unsaturation of the triglyceride stored in the seeds.[82]

Hardening by photoinhibition — Photoinhibition is usually thought of as a chilling injury itself, but there may be circumstances where it is a hardening adaptation. The protective effect that ABA has toward chilling may derive at least partly from its inhibition of metabolism and growth. Photoinhibition can also inhibit growth by decreasing the availability of photosynthate. It therefore offers one way by which metabolism can be shut down. The photosynthesis of plants from warm climates is inhibited when they are exposed to high light at low temperatures (see Chapter 7). Recovery takes place slowly and is aided by moderately high temperatures. Even well into the seasonal tropics, night temperatures below 10°C can occur when the atmosphere is particularly dry, as is illustrated by Figure 2. Other tropical locations that are subject to very dry air are likely to experience similar kinds of chilling conditions. After a cold morning accompanied by bright sunlight, less photosynthate will be produced because of photoinhibition. With metabolism limited by the supply of photosynthate, the risk of further chilling damage during the subsequent cold nights could be decreased.

Temperature conditioning — Grapefruit have been found to suffer less chilling injury if they are held at relatively high temperatures (15 to 20°C) before storage at chilling temperatures.[97-99] One interpretation of this effect is that it allows the repair of wounds that would otherwise develop into larger lesions at the chilling temperature. It would then be analogous to the findings that "curing" (wound repair) of sweet potatoes at 32°C and high humidity reduces chilling injury during subsequent refrigerated storage.[100]

Cold shock as a conditioning treatment — The Japanese apricot *(Prunus mume)* suffers chilling injury on prolonged storage around 6°C, but this can be prevented if the fruit are given a cold shock by plunging them into ice water.[101] It is difficult to see any unifying explanation for the striking contrasts between the different hardening and conditioning treatments discussed above. Perhaps this diversity of effect should be seen as reflecting the physiological diversity of chilling injury itself and stand as a warning against too simple an explanation of its causes and prevention.

2. Diurnal Variations

The time at which a chilling-sensitive plant is put into a chilling temperature has a large influence on the extent to which it is damaged. Two aspects of this have been studied: the susceptibility to chilling injury where water loss is prevented, and the susceptibility where water loss is important.

At high humidities — Figure 7 illustrates the diurnal effect on the survival of tomato seedlings under humid conditions.[48,68,102] Seedlings chilled from just before the onset of the light period are about twice as sensitive as those chilled starting during the early part of the dark period. Changes analogous to those shown by tomatoes are common to a variety of mono- and dicotyledenous plants.[68,103] Diurnal changes in levels of metabolites through the day, and/or the changes in metabolic pathways between day and night, may be responsible for these changes in susceptibility. Levels of reduced glutathione have been found to vary on a diurnal basis in tomato seedlings. However, maximum levels are in the middle of the light period, and minimum levels are in the middle of the dark period.[104] As this is out of phase with the variation in chilling sensitivity, it seems unlikely that glutathione levels are a causative factor in the variation in chilling sensitivity.

At low humidities — Under conditions of high humidity, effects of chilling on the roots, which would alter the supply of water, or the degree of stomatal opening, which under drier conditions would influence the loss of water, do not change the form of diurnal variation in sensitivity.[102] A distinct kind of diurnal variation in susceptibility to chilling can be observed at lower humidities, however, and this is a consequence of stomatal "locking" at low temperatures, as illustrated by Figure 6. The conductance of the leaf to water, and hence their ability to lose water, rises to a maximum around the middle of the day,

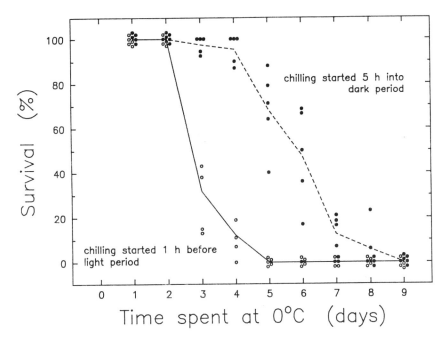

FIGURE 7.   Differences in survival of tomato seedlings after chilling from contrasting times in the diurnal cycle. (Modified from King, A. I., Reid, M. S., and Patterson, B. D., *Plant Physiol.*, 70, 211, 1982. With permission.)

while during the night it is low. In a natural environment, the "locking" of stomata in the prechilling state is likely to be an adaptive feature, because the onset of the most severe chilling temperatures is likely to occur during the dark period. At this time, the stomata will be shut, and the effect of chilling will ensure that the stomata do not open until temperatures become less stressful. A possible exception concerns CAM plants, for these plants tend to open their stomata at night rather than during the day.

Harvesting and storage practices should take diurnal variations in chilling sensitivity into account. If leafy tissues of tropical crops are cooled to chilling temperatures, they will be less likely to be injured if they are cooled during the first half of the dark period. At this time, stomata will normally be closed so that both kinds of diurnal susceptibility will be at a minimum. Alternatively, both kinds of diurnal increase in susceptibility can be supressed with ABA.[68]

B. EFFECTS DURING CHILLING

The conditions during chilling can have a large influence on the severity of symptoms. Some of the most important ones are discussed below.

1. Water Stress

With most plant tissues, damage during chilling is made worse by water loss, although some varieties of apple fruit that suffer from internal breakdown (low-temperature breakdown) are a notable exception to this generality.[105] Water loss before chilling tends to harden against subsequent injury, probably through the resulting increase in levels of ABA. However, at chilling temperatures, it is likely that the mechanisms of increasing ABA synthesis and transport in response to a water stress are not able to operate.

In leafy tissues — Leafy tissues are very susceptible to overall desiccation during chilling[106] because of their large surface area and concentration of stomata which can be "locked" open by chilling. If roots are chilled, the uptake of water may be impaired.

In bulky tissues — In some stored fruit and vegetables, pitting of the surface is a characteristic chilling injury, and this is made worse by water loss.[107] The cells around places with initially higher rates of water loss desiccate and collapse, leading to pit formation. The reason why internal breakdown (low-temperature breakdown) in apples is reduced by water loss seems to be that the existing water vapor entrains acetate esters. Acetate and compounds related to it increase the incidence of this chilling injury in apples.[105]

2. Chemical Treatments

A number of chemicals modify the response to chilling of sensitive plants. One of the major responses of chilled plants is that they tend to die through desiccation. Therefore, substances that close stomata (antitranspirants) will tend to reduce the effects of chilling.[106] However, some substances that reduce stomatal aperture or close stomata may be acting in other ways as well.

Effect of ABA — ABA reduces chilling injury[108] in addition to its effect in closing stomata. While the stomatal closure is probably a factor in the reduced injury, ABA also reduces chilling injury in tobacco callus, where there is no water stress and no stomata to be affected.[109] Accumulation of ABA seems to be associated with the hardening of plants against chilling, either with low temperature or with water stress.[91]

Choline — Although this substance reduces stomatal aperture, it reduces chilling susceptibility in addition to any antitranspirant effect.[110,111] Besides being a precursor of membrane lipids, amino alcohols such as choline can act as free radical acceptors. This class of compound has also been found to be effective at reducing some forms of chilling injury.[112,113]

Ethylene — When ethylene is applied to citrus fruits before harvest as the compound Ethrel®, it reduces the incidence of oleocellosis[16] and rind pitting,[17] both of which are increased by storage at chilling temperatures. The mechanism of action is not known, but both injuries are dependent on oil glands being ruptured,[17] and ethylene may render the oil glands less liable to mechanical damage.

Paclobutrazol — This is a dwarfing compound that inhibits kaurene oxidase, a key enzyme in the synthesis of gibberellins.[114] It also reduces stomatal opening, but it seems unlikely that this is the only reason for its effectiveness in reducing chilling injury. Some of its protective effect may be associated with its ability to reduce the loss of membrane lipid during chilling.[115,116] Any reduction in water loss will tend to add to its protective effect. It would be interesting to know whether any of the protective effect of paclobutrazol is a consequence of reducing carbon flux through metabolic pathways. Some chilling damage may result in the accumulation of toxic products of metabolism;[117] if carbon flux is reduced, the time taken to accumulate the toxic products would be expected to increase.[14] If this is a correct model, it would be expected that once growth had been stopped by treatment with ABA, paclobutrazol would have no additional effect.

Carbon source for growth — Tomato protoplasts die if they are exposed to a temperature of 7°C in the presence of glucose, which at higher temperatures supports cell growth. If the glucose is replaced by mannitol, which does not support cell growth, the protoplasts become resistant to the chilling stress.[118] Differences in the stage of growth or in the availability of carbon sources for growth may account for the striking differences in chilling resistance reported for cell cultures of tomato. While one group found that they were grossly injured in a few days at all temperatures below 12°C,[119] another found that cell cultures of tomato were apparently unharmed by being exposed to 0°C, so that even after 1 week, they could resume normal growth at 28°C.[120]

3. Cold Shock and Translocation

The effects of chilling to a given cold temperature are often influenced by the rate of cooling. The most extreme examples are in phloem transport, where the rate of cooling is often more important than the time spent at low temperature. Cold shock may be said to occur when the stress on a plant is increased at greater rates of cooling. Even this broad definition must sometimes be qualified, however. The survival of the protozoan *Tetrahymena pyriformis* to chilling can be critically dependent on the rate of cooling, with rates both greater and less than 2.5°C/min decreasing survival.[121] With this in mind, it should not be assumed that stress on a plant will necessarily always increase as the rate of cooling is increased. Work that has been done concerning the physiological effects of the rate of cooling suggests that it is an important variable that should not be neglected.

Translocation — Cold shock can have striking effects of translocation through the phloem.[122] As with so much of chilling, more than one mechanism may be involved. Even a sudden drop of 2.5°C from 30°C has been shown to stop phloem transport for about 3 min in stems of *Ipomoea alba* (morning glory), *Phaseolus vulgaris* (bean), and in the petiole of the aquatic *Nymphoides geminata*.[123] Similar effects are not produced by sudden rises in temperature, or by slow rates of cooling (<5°/min). In sugar beet *(Beta vulgaris)*, an initial slow cooling (0.25°C/min) does not change the rate of translocation.[124] It does have an effect on subsequent cycles of slow cooling, however, producing a stepwise reduction in the rate of translocation as the temperature is decreased. The lack of correlation of susceptibility of phloem transport to cold shock with other aspects of chilling sensitivity is underlined by a study[125] on 86 species from 50 families. A sudden drop in temperature from 25° to 3°C inhibited phloem transport in dicotyledons generally, but in only 30% of monocotyledons. In neither case was the effect related to the cultural requirements of the plant for warm conditions. For example, a cold shock stopped translocation in the chilling-resistant species creeping butttercup *(Ranunculus repens)*, but not in the chilling-sensitive corn *(Zea mays)*. In another chilling-sensitive monocotyledon, *Sorghum bicolor*, translocation stopped on the first chill, but was not affected by subsequent chills.

Protoplasmic streaming — Another possible result of cold shock concerns protoplasmic streaming, although here there seems to be a closer correlation with other responses to chilling. Lewis[126] found that streaming in the trichome cells of cucumber, tomato, watermelon, tobacco, and sweet potato (all chilling sensitive by varying criteria) stopped when the tissues (trichome hairs) were suddenly chilled by transferring them to cold rooms at 10°C or less. In contrast, the more chilling-resistant species radish, carrot, and filaree *(Erodium cicutarium*, Geraniaceae) could be transferred to a temperature at least as low as 2.5°C without stopping their streaming. Later studies have shown that with more gradual cooling (1°C/min or less), a number of chilling-sensitive species, such as tobacco,[127] tomato, and watermelon,[52,70] can continue streaming, albeit at low rates, to temperatures below 10°C. There are differences between species in the extent to which streaming rates are reduced at low temperatures[52] as well as in the rates at which they recover on rewarming.[70] However, sudden cessation of streaming on transfer to a lower temperature may be due more strictly to cold shock.

Ion uptake by roots — The uptake of ^{86}Ru, used as a tracer for potassium, is influenced by a cold shock.[128] The kinetics of uptake are similar in chilling-sensitive and chilling-resistant species if the cooling is gradual, with rates of uptake decreasing with decreases in temperature. However, if the roots are cooled suddenly from 25°C, the uptake by the chilling-sensitive species is increased at the lower temperatures. This effect of cold shock is abolished by divalent metal ions (Ca^{++}, Mg^{++}, Sr^{++}) in excess of $10^{-5} M$ in the medium.

Fruit storage — The finding that a cold shock protects fruit of the Japanese apricot *(Prunus mume)* against subsequent chilling injury during storage[101] suggests that this work should be extended to other species of *Prunus* (plums, peaches, cherries).

In summary, the effects of cold shock may be severe and quite different to the longer term effects of chilling. In the reporting of experiments on chilling, it is rare for rates of cooling to be specified. More attention needs to be paid to this aspect of technique, considering that varying the rate of cooling can result in such striking differences in effect.

V. CONCLUSIONS

While a genetic approach to the solution of chilling sensitivity no longer seems intractable, a number of problems remain. One is that there are many kinds of chilling sensitivity, each probably controlled by a number of genes. Evidence concerning tomato and rice suggests that these genes are scattered through the genome and so will be difficult to transfer in total. Many crops do not have a sexually compatible relative that can serve as a source of genes for chilling resistance. Even where a suitable source of genes is available, it may prove difficult to introduce more than a few of those genes with major effects into modern cultivars.

On the positive side, the finding that most forms of chilling resistance are expressed in F_1 hybrids offers a short-term solution in that a chilling-resistant line can be combined with one containing other genes of importance. With the knowledge gained through the genetics of chilling resistance, it will be possible to identify the genes responsible for chilling resistance, to isolate them, and transfer them even between sexually incompatible species. This approach shows great promise in the longer term. The priority at this stage is to identify the relevant genes.

The effects of environmental variables on the expression of chilling injury are important from two aspects. First, they influence the expression of genetic traits. As a consequence, genetic studies may be flawed if they do not take into account environmental effects such as diurnal variations in sensitivity and water relations. Second, environmental modification is a practical method which can be used to avoid or at least delay chilling injury.

ACKNOWLEDGMENTS

The authors are grateful for the advice of Mr. Peter Hind (Royal Botanic Gardens, Sydney) in preparing the data in Table 1 and Mr. Andrew Eddy for the preparation of the figures.

REFERENCES

1. **Burch, M. D. and Marchant, H. J.**, Motility and microtubule stability of antarctic algae at sub-zero temperatures, *Protoplasma,* 115, 240, 1983.
2. **Hoham, R. W.**, Optimum temperature and temperature ranges for growth of snow algae, *Arct. Alp. Res.,* 7, 13, 1975.
3. **Soeder, C. J. and Stengel, E.**, Physico-chemical factors affecting metabolism and growth rate, in *Algal Physiology and Biochemistry,* Stewart, W. P. D., Ed., Blackwell Scientific, Oxford, 1974.
4. **Bliss, L. C.**, A comparison of plant development in microenvironments of arctic and alpine tundras, *Ecol. Monogr.,* 26, 303, 1956.
5. **Chapin, F. S.**, Direct and indirect effects of temperature on arctic plants, *Polar Biol.,* 2, 47, 1983.
6. **Fujishita, N.**, Cytological, histological and physiological studies on the pollen degeneration. II. On the free amino acids with reference to pollen degeneration caused by low temperature in vegetable fruit crops, *J. Jpn. Soc. Hortic. Sci.,* 34, 113, 1965.
7. **Simmonds, N. W.**, *Bananas* Longman, London, 1982.
8. **Paull, R. E.**, Effect of storage duration and temperature on cut anthurium flowers, *HortScience,* 22, 459, 1987.
9. **Lyons, J. M. and Rappaport, L.**, Effect of temperature on respiration and quality of brussels sprouts during storage, *J. Am. Soc. Hortic. Sci.,* 73, 361, 1959.
10. **Smith, L. G.**, Cause and development of blackheart in pineapples, *Trop. Agric. (Trinidad),* 60, 31, 1983.
11. **Wilson, J. M.**, Drought resistance as related to low temperature stress, in *Low Temperature Stress in Crop Plants: The Role of the Membrane,* Lyons, J. M., Graham, D., and Raison, J. K., Eds., Academic Press, New York, 1979, 47.

12. **Smillie, R. M.,** A highly chilling-sensitive angiosperm, *Carlsberg Res. Commun.,* 49, 79, 1984.
13. **Chen, Y. Z. and Patterson, B. D.,** Ethylene and 1-aminocyclopropane-1-carboxylic acid as indicators of chilling sensitivity in various plant species, *Aust. J. Plant Physiol.,* 12, 377, 1985.
14. **Patterson, B. D. and Graham, D.,** Temperature and metabolism, in *The Biochemistry of Plants: A Comprehensive Treatise,* Vol. 12, Davies, D. D., Ed., Academic Press, New York, 1987, 153.
15. **Hetherington, S. E. and Öquist, G.,** Monitoring chilling injury: a comparison of chlorophyll fluorescence measurements, post-chilling growth and visible symptoms of injury in *Zea mays, Physiol. Plant.,* 72, 241, 1987.
16. **Erner, Y.,** Reduction of oleocellosis damage in 'Shamouti' orange peel with an ethephon preharvest spray, *J. Am. Soc. Hortic. Sci.,* 57, 129, 1982.
17. **Gilfillan, I. M. and Pelser, P. D. T.,** Reduction of rind pitting development in 'Marsh' grapefruit (*Citrus paradisis* MacF.) with ethephon, *Citrus J.,* 15, 1985.
18. **Eaks, I. L.,** The physiological breakdown of the rind of lime fruits after harvest, *Proc. Am. Soc. Hortic. Sci.,* 66, 141, 1955.
19. **Hall, E. G.,** Fruit fly: low temperature trials with oranges confirm greater susceptibility of navels to injury, *Citrus News,* 32, 27, 1956.
20. **Anon.,** Climatic Averages: South Australia and Northern Territory, Australian Government Publishing Service, Canberra, 1975.
21. **Schwertdfeger, W.,** *World Survey of Climatology,* Elsevier, New York, 1976.
22. **Rentoul, J. N.,** *Growing Orchids,* Book 4: The Australasian Families, Lothian Publishing, Melbourne, 1985.
23. **Nichols, W. H.,** *Orchids of Australia,* Nelson, Melbourne, 1969.
24. **Anon.,** Climatic Atlas of Australia, Map Ser. 1, Temperature, Bureau of Meteorology, Australian Government Publishing Service, Canberra, 1975.
25. **Rentoul, J. N.,** *Growing Orchids,* Book 2: Cattleyas and Other Epiphytes, Lothian Publishing, Melbourne, 1982.
26. **Patterson, B. D.,** Genes for cold resistance from wild tomatoes, *HortScience,* 23, 794 and 947, 1988.
27. **Rick, C. M.,** Potential genetic resources in tomato species: clues from observations in native habitats, in *Genes Enzymes and Populations,* Srb. A. M., Ed., Plenum Press, New York, 1973, 255.
28. **Patterson, B. D., Paull, R., and Smillie, R. M.,** Chilling resistance in *Lycopersicon hirsutum* Humb. & Bonpl., a wild tomato with a wide altitudinal distribution, *Aust. J. Plant Physiol.,* 5, 609, 1978.
29. **Scott, S. J. and Jones, R. A.,** Quantifying seed germination responses to low temperature: variation among *Lycopersicon* spp., *Environ. Exp. Bot.,* 25, 129, 1985.
30. **Patterson, B. D., Mutton, L., Paull, R. E., and Nguyen, V. Q.,** Tomato pollen development: stages sensitive to chilling and a natural environment for the selection of resistant genotypes, *Plant Cell Environ.,* 10, 363, 1987.
31. **Patterson, B. D. and Payne, L. A.,** Screening for chilling resistance in tomato seedlings, *HortScience,* 18, 340, 1983.
32. **Vallejos, C. E., Lyons, J. M., Breidenbach, J. W., and Miller, M. F.,** Characterization of a differential low-temperature growth response in two species of *Lycopersicon*: the plastochron as a tool, *Planta,* 159, 487, 1983.
33. **Vallejos, C. E. and Pearcy, R. W.,** Differential acclimation potential to low temperatures in two species of *Lycopersicon*: photosynthesis and growth, *Can. J. Bot.,* 65, 1303, 1987.
34. **Miltau, O., Zamir, D., and Rudich, J.,** Growth rates of *Lycopersicon* species at low temperature, *Z. Pflanzenzuecht.,* 96, 193, 1986.
35. **Polowick, P. L. and Sawhney, V. K.,** Temperature effects on male fertility and flower and fruit development in *Capsicum anuum* L., *Sci. Hortic.,* 25, 117, 1985.
36. **Pet, G.,** Verband tussen kwaliteit stuifmeel en vroegheit bij paprika, *Groenten Fruit,* 41, 42, 1985.
37. **Allan, P.,** Pollen studies in Carica papaya. I. Formation, development, morphology and production of pollen, *S. Afr. J. Agric. Sci.,* 6, 517, 1963.
38. **Goto, K. and Yamamoto, T.,** Studies on cold injury in pulses. III. The effect on pollen germination and fertilization of low temperatures preceding anthesis in soya bean, *Hokkaido Nogyo Shikenjo Iho,* 100, 14, 1972.
39. **Marshall, D. R., Thompson, N. J., Nicholls, G. H., and Patrick, C. M.,** Effects of temperature and day length on cytoplasmic male sterility in cotton (*Gossypium*), *Aust. J. Agric. Res.,* 25, 443, 1974.
40. **Maisonneuve, B.,** Action des basses températures nocturnes sur une collection varietale de tomate (*Lycopersicon esculentum* Mill.). II. Etude de la quantité et de qualité du pollen, *Agronomie,* 2, 453, 1982.
41. **Koike, S. and Satake, T.,** Sterility caused by cooling treatment at the flowering stage in rice plants. II. The abnormal digestion of starch in pollen grain and metabolic changes in anthers following cooling treatment, *Jpn. J. Crop Sci.,* 56, 666, 1987.
42. **Mashingaidze, K. and Muchena, S. C.,** The induction of floret sterility by low temperatures in pearl millet (*Pennisetum typhoides* (Burm.) Stapf. & Hubbard), *Zimbabwe J. Agric. Res.,* 20, 29, 1982.

43. **Berding, N.,** Improved flowering and pollen fertility under increased night temperatures, *Crop Sci.,* 21, 863, 1981.

44. **Brooking, I. R.,** Male sterility in *Sorghum bicolor* (L.) Moench induced by low night temperature. I. Timing of the stage of sensitivity, *Aust. J. Plant Physiol.,* 35, 589, 1976.

45. **Teplyakov, B. I., Maksimenko, V. P., and Chenkurov, V. M.,** The effect of reduced temperatures on aberrations during meiosis in spring bread wheat, *Tsitol. Genet.,* 8, 406, 1974.

46. **Maisonneuve, B.,** Effet d'un traitement à baisses températures, en conditions controlées, sur la qualité du pollen de tomate (*Lycopersicon esculentum* Mill.), *Agronomie,* 2, 755, 1982.

47. **Maisonneuve, B.,** Cold resistance of *Lycopersicon hirsutum* pollen, *Rep. Tomato Genet. Coop.,* 33, 8, 1983.

48. **Patterson, B. D., Paull, R., and Graham, D.,** Adaptation to chilling: survival, germination, respiration and protoplasmic dynamics, in *Low Temperature Stress in Crop Plants: The Role of the Membrane,* Lyons, J. M., Graham, D., and Raison, J. K., Eds., Academic Press, New York, 1979, 25.

49. **Zamir, D., Tanksley, S. D., and Jones, R. A.,** Low temperature effect on selective fertilization by pollen mixtures of wild and cultivated tomato species, *Theor. Appl. Genet.,* 59, 235, 1981.

50. **Maisonneuve, B. and Den Nijs, A. P. M.,** In vitro pollen germination and tube growth of tomato (*Lycopersicon esculentum* Mill.) and its relation with plant growth, *Euphytica,* 33, 833, 1984.

51. **Zamir, D. and Gadish, I.,** Pollen selection for low temperature adaptation in tomato, *Theor. Appl. Genet.,* 74, 545, 1987.

52. **Patterson, B. D. and Graham, D.,** Effect of chilling temperatures on the protoplasmic streaming of plants from different climates, *J. Exp. Bot.,* 28, 736, 1977.

53. **Paull, R. E.,** Leucine uptake by tomato leaf tissue under low temperature: preincubation dependence of uptake reduction, *Physiol. Plant.,* 56, 84, 1982.

54. **Hetherington, S. E. and Smillie, R. M.,** Practical applications of chlorophyll fluorescence in ecophysiology, physiology and plant breeding, *Adv. Photosyn. Res.,* 4, 447, 1984.

55. **Yakir, D., Rudich, J., and Bravdo, B.-A.,** Adaptation to chilling: photosynthetic characteristics of the cultivated tomato and a high altitude wild species, *Plant Cell Environ.,* 9, 477, 1986.

56. **Collins, J. L.,** *The Pineapple: Botany, Cultivation and Utilization,* InterScience, London, 1960.

57. **Vallejos, C. E.,** Genetic diversity of plants for response to low temperatures and its potential use in crop plants, in *Low Temperature Stress in Crop Plants: The Role of the Membrane,* Lyons, J. M., Graham, D., and Raison, J. K., Eds., Academic Press, New York, 1979, 473.

58. **Purseglove, J. W.,** *Tropical Crops - Dicotyledons,* Longman, London, 1968, 192.

59. **Smillie, R. M., Melchers, G., and von Wettstein, D.,** Chilling response of somatic hybrids of tomato and potato, *Carlsberg Res. Commun.,* 44, 127, 1979.

60. **McKersie, B. D., Pauls, K. P., and Walker, M. A.,** Fluorescence screening procedure for chilling tolerance in tomato, *Rep. Tomato Genet. Coop.,* 37, 55, 1987.

61. **Patterson, B. D., Murata, T., and Graham, D.,** Electrolyte leakage induced by chilling in *Passiflora* species tolerant to different climates, *Aust. J. Plant Physiol.,* 3, 435, 1976.

62. **Rentoul, J. N.,** *Growing Orchids,* Book 1: Vandas, Dendrobiums and Others, Lothian Publishing, Melbourne, 1982.

63. **Simmonds, N. W.,** *Evolution of Crop Plants,* Longman, London, 1976.

64. **Batten, D.,** Personal communication, 1987.

65. **Gavrilenko, T. A.,** Influence of temperature on crossing over in tomato (in Russian), *Tsitol. Genet.* 18, 347, 1984.

66. **Zamir, D., Tanksley, S. D., and Jones, R. A.,** Haploid selection for low temperature tolerance of tomato pollen, *Genetics,* 101, 129, 1982.

67. **Zamir, D. and Vallejos, E. C.,** Temperature effects on haploid selection of tomato microspores and pollen grains, in *Pollen: Biology and Implications for Plant Breeding,* Mulcahy, D. L. and Ottaviano, E., Eds., Elsevier, New York, 1983, 335.

68. **King, A. I., Reid, M. S., and Patterson, B. D.,** Diurnal changes in the chilling sensitivity of seedlings, *Plant Physiol.,* 70, 211, 1982.

69. **Britten, E. J.,** Hawaii as a natural laboratory for research on climate and plant response, *Pac. Sci.,* 16, 160, 1962.

70. **Woods, C. M., Reid, M. L., and Patterson, B. D.,** Responses to chilling stress in plant cells. I. Changes in cyclosis and cytoplasmic structure, *Protoplasma,* 121, 8, 1984.

71. **Horovitz, S. and Jimenez, H.,** Cruzamientos interespecificos e intergenericos en Caricaceas y sus implicaciones fitotecnicas, *Agron. Trop.,* 17, 323, 1967.

72. **Maniatis, T., Fritsch, E. F., and Sambrook, J.,** *Molecular Cloning — A Laboratory Manual,* Cold Spring Harbor Series, Cold Spring Harbor Laboratory, Cold Spring Harbor, NY, 1982.

73. **Dodds, J. H.,** *Plant Genetic Engineering,* Cambridge University Press, London, 1985.

74. **Tanksley, S. D. and Orton, T. J.,** *Isozymes in Plant Genetics and Breeding,* Elsevier, New York, 1983.

75. **Bernatsky, R. and Tanksley, S. D.,** Towards a saturated linkage map in tomato based on isozymes and random cDNA sequences, *Genetics,* 112, 887, 1986.

76. **Mutschler, M. A. and Rick, C. M.,** 1987 linkage maps of the tomato *(Lycopersicon esculentum), Rep. Tomato Genet. Coop.,* 37, 5, 1987.

77. **Chang, T. T. and Oka, H. I.,** Genetic variousness in the climatic adaptation of rice cultivars, in *Climate and Rice,* Chang, T. T., Vergara, B. S., and Yoshida, S., Eds., International Rice Research Institute, Los Banos, Philippines, 1976.

78. **Toriyama, K. and Futsuhara, Y.,** Genetic studies on cool tolerance in rice. I. Inheritance of cool tolerance, *Jpn. J. Breed.,* 11, 191, 1960.

79. **Vallejos, C. E. and Tanksley, S. D.,** Segregation of isozyme markers and cold tolerance in an interspecific backcross of tomato, *Theor. Appl. Genet.,* 66, 241, 1983.

80. **Frentzen, M., Nishida, I., and Murata, N.,** Properties of the plastidoyl acyl-(acyl carrier protein)-glycerol 3 phosphate acyl transferase from the chilling-sensitive plant squash *(Cucurbita moschata), Plant Cell Physiol.,* 28, 1195, 1987.

81. **Nishida, I., Frentzen, M., Ishizaki, O., and Murata, N.,** Purification of isometric forms of acyl-(acyl carrier protein)-glycero 3 phosphate acyl transferase from squash, *Plant Cell Physiol.,* 28, 1071, 1987.

82. **Maluf, W. R. and Tigchelaar, E. C.,** Relationship between fatty acid composition and low-temperature seed germination in tomato, *J. Am. Soc. Hortic. Sci.,* 107, 620, 1982.

83. **Anon.,** Calgene targets recombinant gene expression to plant seeds, *Genet. Technol. News,* 7, 2, 1987.

84. **Dawson, P. J. and Lloyd, C. W.,** Comparative biochemistry of plant and animal tubulins, in *The Biochemistry of Plants,* Vol. 12, Davies, D. D., Ed., Academic Press, New York, 1987, 3.

85. **Melkonian, M., Kroeger, K. H., and Marquardt, K. G.,** Cell shape and microtubules in zoospores of the green alga *Chlorosarcinopsis gelatinosa, Chlorosarcinales.* Effects of low temperature, *Protoplasma,* 104, 283, 1980.

86. **Detrich, H. W. and Overton, S. A.,** Heterogeneity and structure of brain tubulins from cold-adapted antarctic fishes. Comparison to brain tubulins from a temperate fish and a mammal, *J. Biol. Chem.,* 261, 10922, 1986.

87. **La Claire, J. W. I. I.,** Microtubule cytoskeleton in intact and wounded coenocytic green algae, *Planta,* 171, 30, 1987.

88. **Raha, D., Sen, K., and Biswas, B. B.,** cDNA cloning of beta-tubulin gene and organization of tubulin genes in *Vigna radiata* (mung bean) genome, *Plant Mol. Biol.,* 9, 565, 1987.

89. **Knutson, R. M.,** Heat production and temperature regulation in eastern skunk cabbage, *Science,* 186, 746, 1974.

90. **Dhawan, V. and Bhojwani, S. S.,** Hardening *in vitro* and morpho-physiological changes in the leaves during acclimatization of micropropagated plants of *Leucaena leucocephala* (Lam.) de Wit, *Plant Sci.,* 53, 65, 1987.

91. **Daie, J. and Campbell, W. F.,** Response of tomato plants to stressful temperatures: increase in abscisic acid concentrations, *Plant Physiol.,* 67, 26, 1981.

92. **Lalk, I. and Dörffling, K.,** Hardening, abscisic acid, proline and freezing resistance in two winter wheat varieties, *Physiol. Plant.,* 63, 287, 1985.

93. **Rikin, A., Atsmon, D., and Gitler, C.,** Chilling injury in cotton *(Gossypium hirsutum* L.): prevention by abscisic acid, *Plant Cell Physiol.,* 20, 1537, 1979.

94. **Guye, M. G. and Wilson, J. M.,** The effect of chilling and chill-hardening temperatures on stomatal behaviour in chill-sensitive species and cultivars, *Plant Physiol. Biochem.,* 25, 717, 1987.

95. **Silver, J. G., Rochester, C. P., Bishop, D. G., and Harris, H. C.,** Unsaturated fatty acid synthesis during the development of isolated sunflower *(Helianthus annuus* L.) seeds, *J. Exp. Bot.,* 5, 1507, 1984.

96. **Cherry, J. H., Bishop, L., Hasegawa, B. M., and Leffler, H. R.,** Differences in the fatty acid composition of soybean seed produced in northern and southern areas of the U.S.A., *Phytochemistry,* 24, 237, 1985.

97. **Hatton, T. T. and Cubbedge, R. H.,** Preferred temperature for prestorage conditioning of 'Marsh' grapefruit, *HortScience,* 18, 721, 1983.

98. **Chalutz, E., Waks, J., and Schiffmann-Nadel, M.,** Reduced susceptibility of grapefruit to chilling injury during cold treatment, *HortScience,* 20, 226, 1985.

99. **Ismail, M. A., Hatton, T. T., Dezman, D. J., and Miller, W. R.,** In transit cold treatment of Florida grapefruit shipped to Japan in refrigerated van containers: problems and recommendations., *Proc. Fla. State Hortic. Soc.,* 99, 117, 1986.

100. **Picha, D. H.,** Chilling injury, respiration and sugar changes in sweet potatoes stored at low temperature, *J. Am. Soc. Hortic. Sci.,* 112, 497, 1987.

101. **Goto, M., Minamide, T., Fujii, M., and Iwata, T.,** Preventative effect of cold-shock on chilling injury of Mume (Japanese apricot, *Prunus mume* Sieb. et Zucc.) fruits in relation to changes of permeability and fatty acid composition of membrane., *J. Jpn. Soc. Hortic. Sci.,* 53, 210, 1984.

102. **King, A. I. and Reid, M. S.,** Diurnal chilling sensitivity and desiccation in seedlings of tomato, *J. Am. Soc. Hortic. Sci.,* 112, 821, 1987.

103. **Pomeroy, M. K. and Mudd, J. B.,** Chilling sensitivity of cucumber cotyledon protoplasts and seedlings, *Plant Physiol.,* 84, 677, 1987.

104. **Koike, S. and Patterson, B. D.,** Diurnal variation of glutathione levels in tomato seedlings, *HortScience,* 23, 713, 1988.

105. **Wills, R. B. H. and Scott, K. J.,** Studies on the relationship between minerals and the development of storage breakdown in apples, *Aust. J. Agric. Res.,* 32, 331, 1981.

106. **Christianson, M. N. and Ashworth, E. N.,** Prevention of chilling injury to seedlings of cotton with antitranspirants, *Crop Sci.,* 18, 907, 1978.

107. **Purvis, A. C.,** Importance of water loss in the chilling injury of grapefruit stored at low temperature, *Sci. Hortic.,* 23, 261, 1984.

108. **Krizek, D. T., Semeniuk, P., Moline, H. E., Mirecki, R. M., and Abbot, J. A.,** Chilling injury in coleus as influenced by photosynthetically active radiation, temperature and abscisic acid treatments. I. Morphology and physiological responses, *Plant Cell Environ.,* 8, 135, 1985.

109. **Bornmann, C. H. and Jansson, E.,** *Nicotiana tabacum* callus studies. X. ABA increases resistance to cold damage, *Physiol. Plant.,* 48, 491, 1980.

110. **Horvath, I. and van Hasselt, P. R.,** Inhibition of chilling-induced photooxidative damage to leaves of *Cucumis sativus* L. by treatment with amino alcohols, *Planta,* 164, 83, 1985.

111. **Guye, M. G., Vigh, L., and Wilson, J. M.,** Choline-induced chill-tolerance in mung bean *Vigna radiata* L. Wilcz.), *Plant Sci.,* 53, 223, 1987.

112. **Wang, C. Y. and Baker, J. E.,** Effects of two free radical scavengers and intermittent warming on chilling injury and polar lipid composition of cucumber and sweet pepper fruits, *Plant Cell Physiol.,* 20, 243, 1979.

113. **Wills, R. B. H., Hopkirk, S. G., and Scott, K. J.,** Reduction of soft scald in apples with antioxidant, *J. Am. Soc. Hortic. Sci.,* 106, 569, 1981.

114. **Dalziel, J. and Lawrence, D. K.,** Biochemical and biological effects of kaurene oxidase inhibitors such as paclobutrazol, in *Biochemical Aspects of Synthetic and Naturally-Occurring Plant Growth Regulators,* Monogr. No. 11, Menhenett, R. and Lawrence, D. K., Eds., British Plant Growth Regulator Group, Wantage, 1984, 43.

115. **Wang, C. Y.,** Modification of chilling susceptibility in seedlings of cucumber and zucchini squash by a bioregulator, paclobutrazol (PP333), *Sci. Hortic.,* 26, 853, 1985.

116. **Whitaker, B. D. and Wang, C. Y.,** Effect of paclobutrazol and chilling on leaf membrane lipids in cucumber seedlings, *Physiol. Plant.,* 70, 404, 1987.

117. **Lyons, J. M.,** Chilling injury in plants, *Annu. Rev. Plant Physiol.,* 24, 445, 1973.

118. **Mühlbach, H. P. and Thiele, H.,** Response to chilling of tomato mesophyll protoplasts, *Planta,* 151, 399, 1981.

119. **Breidenbach, R. W. and Waring, A. J.,** Response to chilling of tomato seedlings and cells in suspension cultures, *Plant Physiol.,* 60, 190, 1977.

120. **Dupont, F. M., Staraci, B., Chou, B., Thomas, B., Williams, B. G., and Mudd, J. B.,** Effect of chilling temperatures on cell cultures of tomato, *Plant Physiol.,* 77, 64, 1984.

121. **Morris, G. J., Coulson, G. E., and Clarke, A.,** Cold shock injury in *Tetrahymena pyriformis, Cryobiology,* 21, 664, 1984.

122. **Geiger, D. R. and Sovonick, S. A.,** Effects of temperature, anoxia and other metabolic inhibitors on translocation, in *Encyclopedia of Plant Physiology* (New Series), Zimmermann, M. H. and Milburn, V. A., Eds., Springer-Verlag, Berlin, 1975, 256.

123. **Minchin, P. E. H. and Thorpe, M. R.,** A rate of cooling response in phloem translocation, *J. Exp. Bot.,* 34, 529, 1983.

124. **Grusak, M. A. and Lucas, W. J.,** Cold-inhibited phloem translocation in sugar beet. II. Characterization and localization of the slow-cooling response, *J. Exp. Bot.,* 36, 745, 1985.

125. **Lang, A. and Minchin, P. E. H.,** Phylogenetic distribution and mechanism of translocation inhibition by chilling, *J. Exp. Bot.,* 37, 389, 1986.

126. **Lewis, D. A.,** Protoplasmic streaming in plants sensitive and insensitive to chilling temperatures, *Science,* 124, 75, 1956.

127. **Das, T. M., Hilderbrandt, A. T., and Riker, A. J.,** Cinephotomicrography of low temperature effects on cytoplasmic streaming, nucleolar activity and mitosis in single tobacco cells in microculture, *Am. J. Bot.,* 53, 253, 1966.

128. **Zsoldos, F. and Karvaly, B.,** Cold-shock injury and its relation to ion transport by roots, in *Low Temperature Stress in Crop Plants: The Role of the Membrane,* Lyons, J. M., Graham, D., and Raison, J. K., Eds., Academic Press, New York, 1979, 123.

Chapter 7

LIGHT-INDUCED DAMAGE DURING CHILLING

P. R. van Hasselt

TABLE OF CONTENTS

I. INTRODUCTION

Sunlight is essential for the growth of all green plants and other photosynthetic organisms. Although light is necessary and beneficial for plants, it can also be damaging under some circumstances, e.g., when high-intensity light occurs combined with stress conditions which inhibit carbon metabolism.[1,2] Chilling temperature (0 to 12°C) is an important climatic factor affecting plant growth and distribution. Most chilling-sensitive plants are native to tropical or subtropical climates.[3] Many important horticultural and agricultural crops, such as bean, tomato, cucumber, maize, tea, and cotton, are chilling-sensitive plants. The increasing demand for food due to the growth of the world population, as well as the economic and agricultural developments during the last decades, have resulted in an increasing use of chilling-sensitive crops in areas of higher latitudes and altitudes.

Characteristic for these areas is the occurrence of large differences between day and night temperatures. For example, in western Europe, night temperatures around the freezing point can occur even in early summer. Cold, clear nights are likely to be followed by bright sunlight early in the morning when the temperature is still low. It was recently shown that under such conditions, light can cause substantial photoinhibition of photosynthesis in field-grown maize.[4] Light during chilling can be damaging to chilling-sensitive plants directly or indirectly, e.g., by enhancing chilling-induced water stress.[5,6] The most obvious symptom of direct damage during chilling is a light-induced degradation of leaf pigments, also called bleaching, chlorosis, or photooxidation, as can be observed in tropical grasses,[7] maize,[8] and cucumber.[9] Investigations during the last 10 years made it clear that prior to visible damage, several other injurious light-induced effects occur during chilling. This chapter will describe the main damaging effects of light during chilling and recent insights into the mechanisms underlying light-induced damage at chilling temperatures.

II. PRIMARY EVENTS

A. PRIMARY EFFECTS OF TEMPERATURE

It should be noted that light-induced damage during chilling is temperature dependent. When bean leaves were exposed to strong light at different temperatures, no damage to the photosynthetic apparatus was observed between 25 and 11°C. However, at 10°C an inhibition of 30% occurred, and at 5°C inhibition increased to 60%.[10] A similar dependence of temperature was observed for light-dependent leaf pigment degradation in cucumber leaf discs. At 10°C no damage was observed. Below that temperature, leaf pigment degradation occurred and the rate of damage increased at lower temperatures.[11] Apparently, damage by light only occurs below a certain threshold temperature and increases rapidly upon further lowering of the temperature. Therefore, it is evident that alterations induced by chilling temperature per se are a prerequisite for the damaging effect of light.

According to Raison and Lyons,[12] the overall process of chilling injury should be considered in two stages. First, a primary event which is followed by a series of secondary damaging events. In the case of light-dependent damage, the primary event might be a chilling-induced structural change of lipid domains[13,14] or a protein[15] in the chloroplast membrane. Also, a conformational change in some regulatory enzyme might underlie further damage.[16,17]

Although the exact nature of the primary temperature sensor is still an open question, there is circumstantial evidence that chilling temperatures induce structural changes in the chloroplast membranes. Comparison of temperature-dependent activity of coupled electron transport of chloroplasts from chilling-sensitive and chilling-resistant plants showed that abrupt changes in activation energy occurred below 12°C in the chilling-sensitive, but not in the chilling-resistant species.[18] Similar abrupt changes in activation energy for chloroplast functioning were observed in cucumber leaves by measuring temperature-dependent chlorophyll fluorescence.[19] The origin of the abrupt alterations in the activation energy of chloroplast membrane functioning is still open to debate. Lipid phase transitions have been implicated as initial events in chilling injury.[20] However, the degree of unsaturation of the lipids of the chloroplast membranes is so high that a phase transition of the bulk lipids from a liquid-crystalline to a gel phase does not occur until far below 0°C.[21] However, it cannot be excluded that small membrane areas undergo a phase change at chilling temperatures. Murata and Yamaya[22] reported that phosphatidyl glycerols from chilling-sensitive, but not from chilling-resistant plants showed a phase separation at room temperature. The observation that an increased phospholipid content of cucumber leaves induced by application of amino alcohols protected against photooxidation of chlorophyll during chilling suggests that alterations in membrane-lipid fluidity may be a primary cause for light-induced damage.[23] Finally, the strong correlation between the low-temperature-induced increase of the degree of unsaturation of monogalactosyl diglyceride, the main chloroplast lipid, and the overall electron transport capacity also indicate an effect of membrane fluidity on chloroplast functioning.[24] Changes in lipid fluidity might influence lipid-protein interactions in the membrane and might abruptly change the functioning of proteins involved in transfer of light energy. Also, temperature-induced weakening of hydrophobic interactions in a protein could cause conformational alterations which result in an abrupt change of its activation energy. Although it was shown that the critical temperature for changes in the activation temperature for electron transport of isolated chloroplasts did not correlate with the differences in chilling sensitivity of *Passiflora* species, these changes may reflect primary events underlying light-induced damage during chilling.[25] This hypothesis is supported by recent results showing that the phase transition in the polar lipids from thylakoids of chilling-sensitive plants occurred at a similar temperature as the start of the decrease in photosynthesis and the abrupt increase of chlorophyll fluorescence.[26,27]

It was found that chilling combined with light induced changes in the light-harvesting chlorophyll protein complex associated with photosystem II (LHCII).[28] These results suggest that a chilling-induced structural alteration of LHCII-protein might be the primary event for light-induced impair photosystem II (PSII) functioning.

There is ample evidence that electron transport of chloroplasts isolated from prechilled leaves of sensitive plants is inhibited. Many workers have identified the water splitting system as the chill-sensitive site.[29-31] The relevance of the inhibition of electron transport as the primary cause of the reduction of CO_2 fixation after chilling was questioned by Martin and Ort.[32] They found that the electron transfer capacity of chloroplast membranes isolated from dark-chilled tomato plants was in excess of that necessary to support light- and CO_2^--saturated photosynthesis of attached leaves. The results of their studies lead to the conclusion that the limiting factor for photosynthesis after chilling in the dark was not electron transport capacity, although diminished, but some as yet undefined step in the CO_2^- fixation mechanism outside the thylakoid membrane.[33]

The exact nature of the primary event or possibly a combination of primary events underlying chilling-induced damage by light is still unresolved. However, the present data suggest that the site of the primary event is in the chloroplast and that structural changes in proteins are involved.

B. PRIMARY EFFECTS OF LIGHT

Before light can cause an effect, it has to be absorbed. Chlorophyll is the main absorbing pigment in green plants. Absorption of light quanta by chlorophyll results in excitation to its singlet state. A small part of the singlet chlorophyll is converted to the long-lived triplet state. The energy level of the triplet state is lower than the energy level of the corresponding singlet state. Singlet-triplet transitions are only about 10^{-6} as probable as singlet-singlet transitions. Consequently, the lifetime of the triplet state is much longer than that of the excited singlet state. The excited electron will regain its ground state by releasing its excess energy in a deexcitation reaction. Three main ways of deexcitation may be discerned. First, the energy can be dissipated as heat. Second, the excitation may occur by emission of light as fluorescence (excited singlet to ground state) or as phosphorescence (excited triplet to ground state). The third way is a reaction between the excited triplet or singlet chlorophyll molecule and another molecule. By the last type of reaction, light energy is funneled by resonance transfer via several chlorophyll molecules into the reaction centers of the photosystem and used for photochemistry. Furthermore, chlorophyll triplet may react with molecules of atmospheric oxygen, giving rise to singlet excited oxygen which may initiate damaging photooxidative reactions.[34]

III. SECONDARY EVENTS

A. PHOTOINHIBITION OF PHOTOSYNTHESIS DURING CHILLING
1. Effects of Light

One of the first secondary events, which can be observed when sensitive plants are exposed to chilling in the light, is photoinhibition of photosynthesis. Chilling of tomato plants under strong illumination resulted in a fast and large inhibition of light- and CO_2-saturated photosynthesis, combined with a remarkable decline of the quantum efficiency of photosynthesis.[33] Similar results were obtained for several other chilling-sensitive species, e.g., cucumber,[35] bean,[10] olive,[36] and maize.[37] Several experiments showed that photoinhibition preceded leaf pigment degradation.[37-40] It can be concluded that photoinhibition of photosynthesis during chilling is not caused by a decrease of the leaf pigment concentration. Apparently, light-induced leaf pigment degradation only starts after prolonged exposure to light during chilling, much later than photoinhibition of photosynthesis (see Powles[1] for a review of photoinhibition).

The extent of the reduction of the quantum yield after chilling in the light depends on the light intensity during chilling. Little or no inhibition was observed at light intensities of 10% or less of full sunlight.[10] It was demonstrated that quantum yield of photosynthesis was hardly affected by chilling in the dark.[33] Therefore, a reduction of the quantum yield of photosynthesis appears to be a primary and specific manifestation of light-induced damage during chilling.

The state I-state II transition, a light-induced alteration of excitation spillover from PSII to photosystem I (PSI), occurs as a consequence of LHCII protein phosphorylation.[41] Recent results show that the state I-state II transition and the phosphorylation of LHCII are inhibited at 5°C in chilling-sensitive rice, but not in chilling-resistant barley, suggesting a chill-induced inhibition of the action of the protein kinases due to a lower enzyme activity or a structural change of the LHCII protein, which inhibits access of the kinase.[42] A light-induced inhibition of the structural changes underlying the state I-state II transition during chilling might be the origin of the observed decline of the quantum efficiency after chilling in the light.

The rate of photoinhibition of photosynthesis is not only dependent on the light intensity, which induces the damage, but also on the light intensity during the growth period preceding photoinhibitive stress. The effect of the light intensity during growth was demonstrated by Powles and Critchley,[43] who found that photoinhibition of low-light-adapted plants occurred after exposure to full sunlight at normal conditions. Light intensity during growth evidently also affects the sensitivity of plants to light-induced damage during chilling. It was observed that cucumber plants grown in a greenhouse under a screen of cheese cloth were more sensitive to photooxidative damage during chilling than were unscreened plants.[89] It seems likely that plants grown in the field during a period of low light intensity, for example, prevailing in the winter in mediterranean climates, are more sensitive to light-induced damage during chilling.[36,40]

2. Oxygen Dependency

A dependence on oxygen of photoinhibition at chilling temperatures was observed in intact plants[37,44] and in leaf segments.[39] Evidently, in the absence of oxygen, photoinhibition hardly occurred during chilling. It should be noted that, while a low oxygen concentration reduces photoinhibition at chilling temperatures, it enhances photoinhibition at nonchilling temperatures. At higher temperatures, a reduction of photoinhibition of photosynthesis by atmospheric oxygen was observed in C_3 plants in the absence of CO_2. This was ascribed to the effect of CO_2, generated by the photorespiratory pathway.[45] Furthermore, it was found that enrichment of the atmosphere with CO_2 protected plants against chilling-induced damage.[10,46] Apparently, some activity of carbon metabolism is essential to prevent photoinhibition caused by overenergetization of the photosystem reaction centers by maintaining an adequate supply of electronacceptors as sink for photochemically generated energy. Oxygen could also function as a direct sink by acting as electron acceptor in a Mehler-type reaction producing superoxide, which can be dismutated efficiently by enzymatic reactions at normal temperatures.[47] At chilling temperatures, however, both carbon metabolism and enzymatic superoxide dismutation will be reduced. As a consequence, overexcitation of the reaction centers will occur and, in the presence of oxygen, superoxide will be generated, resulting in photoinhibition. Recent results indicate that PSI photoinhibition, contrary to PSII photoinhibition, is oxygen dependent at normal temperatures.[48]

3. Role of Stomata

It is well established that chilling affects stomatal functioning.[49,50] It might be argued, therefore, whether the inhibition of photosynthesis after chilling in the light is caused by a reduction of stomatal conductance. Indeed, a decrease of stomatal conductance was observed after chilling in the light in maize and in tomato.[51,33] However, the calculated intercellular CO_2 partial pressure after return to 25°C of bean leaves chilled in strong light was higher than that of leaves chilled at low, nondamaging light intensity. It was therefore concluded that the reduction of CO_2 uptake following chilling in strong light primarily resulted from nonstomatal inhibition of photosynthetic reactions in the chloroplast.[10] Similar results were obtained with tomato plants after chilling in the light.[33] In olive, however, a decreased internal CO_2 concentration was observed after chilling in the light, suggesting that a reduction in stomatal conductance is an important factor in the chilling-induced photoinhibition of this crop.[36]

B. PHOTOOXIDATIVE DEGRADATION DURING CHILLING
1. Degradation of Leaf Pigments

Bleaching of leaf pigments has been frequently observed in chilling-sensitive plants upon prolonged chilling in strong light.[7-9,36] Several workers showed that this chilling-induced bleaching is oxygen and light dependent and, by consequence, is caused by a photooxidative process.[9,37,44] Figure 1 shows the effect of different oxygen concentrations on light-induced leaf pigment degradation in cucumber leaf discs during chilling. Maximal pigment degradation occurred at 100% oxygen, while at 0% oxygen, chlorophyll was hardly affected, even after 40 h chilling in the light. Another property of photooxidative degradation is also shown in Figure 2. The time course of pigment degradation is biphasic: a lag phase precedes a fast phase of degradation. The length of the lag phase appears to be proportional with the sensitivity of the pigments to photooxidation. In cucumber leaf discs, β-carotene showed the shortest lag phase and was the most sensitive to photooxidative degradation followed by xantophyll, chlorophyll a, and chlorophyll b.[9]

A light-dependent decrease of leaf pigment concentration during moderate chilling (10 to 15°C), often called chlorosis, is present in maize and other tropical grasses.[51] At these temperatures, there is some leaf extension, but the synthesis of chlorophyll seems to be

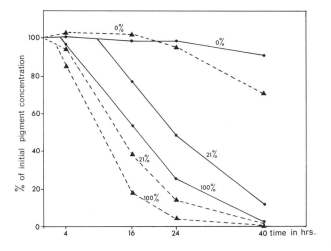

FIGURE 1. Time course of the light-induced degradation of chlorophyll (—●—) and carotene (.. △..) in cucumber leaf discs at different oxygen concentrations at 1°C and 22,000 lux.

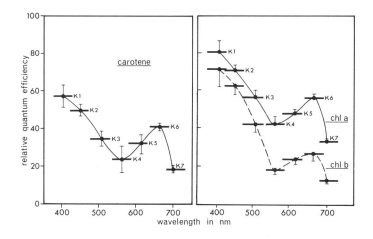

FIGURE 2. Action spectrum of the photooxidative degradation of carotene, chlorophyll a (chl a), and chlorophyll b (chl b) in cucumber leaf discs after 48 h exposure at 21,000 lux and 1°C. The mean of three experiments with their standard deviation is represented.

inhibited, or newly synthesized pigment is photooxidized.[8,53,54] Green leaves are resistant to chlorosis. Thus, when chlorophyll is incorporated in the chloroplast membrane it is evidently protected from destruction at moderate chilling. At temperatures below 10°C, however, photooxidative degradation of the leaf pigments in the chloroplast membranes may occur.

2. Degradation of Chloroplast Lipids

Chloroplast membranes contain, in addition to leaf pigments, a remarkably high content of unsaturated lipids. Photoperoxidation of chloroplast lipids, concurrent with leaf pigment degradation, was observed in cucumber leaf segments during chilling.[55,56] It was found that the most abundant unsaturated fatty acid, linolenic acid, was mainly affected by photooxidation and that the linolenic acid from the chloroplast glycolipids was mostly oxidized.[56,57] The action spectrum of the degradation resembled the absorption spectrum of chlorophyll, indicating a primary role of chlorophyll in the photooxidative degradation of chloroplast membrane lipid during chilling. The electron transport inhibitors dichlorophenyl dimethyl urea (DCMU) and atrazine inhibited the photoperoxidation of lipids in cucumber leaf discs during chilling, indicating an involvement of superoxide, which is generated by electron transfer to atmospheric oxygen, in the damaging process.[9,55] It should be noted that light-induced lipid peroxidation was also observed at elevated temperatures after prolonged chilling of tomato plants in the dark. This was attributed to an increase of the superoxide concentration in the chloroplast due to a chilling-induced, irreversible inhibition of superoxide dismutase.[58,59] Ethylene production by enzymatic oxidation has been measured as a result of damage by chilling in the dark.[60] Wang[61] observed, however, that the light-induced increase in chilling damage was not reflected in an increased production of ethylene and its precursors. Apparently, photooxidation of chloroplast lipids does not lead to enhanced ethylene production.

3. Structural Damage to the Chloroplast

Because leaf pigments and unsaturated lipids, which are both sensitive to photooxidation, are mainly localized in the chloroplast membranes, photooxidative damage to the chloroplast structure seems likely. Indeed, light-induced ultrastructural damage of the chloroplast membranes at low temperatures has been observed in several chilling-sensitive species, e.g., tomato, soy bean, cotton, bean, and cucumber.[55,62-66] In cucumber leaf discs, rupture and disintegration of the envelope occurred concurrent with progressive swelling of the inner chloroplast membranes and dilatation of the grana lamellae (Figure 3c).[67] Substantially less damage was observed in the absence of oxygen, indicating that photooxidation is underlying the structural damage (Figure 3d).

4. Damage to the Electron Transport

While chilling in the dark seems to affect only PSII, chilling in the light causes damage to both photosystems.[33,37] Much work has been done to elucidate the damaging effects of light on the electron transport chain by using measurements of chlorophyll fluorescence.[68,69] In cucumber leaf discs, a concomitant 60% decrease of both prompt and delayed fluorescence was observed after 2 h chilling in the light, while fluorescence in dark-chilled leaf discs was unaffected.[39] Because prompt fluorescence mainly depends, and delayed fluorescence totally depends, on PSII functioning, it was concluded that PSII was damaged by light during chilling. Baker et al.[38] demonstrated in maize leaves, after chilling in the light, a disturbance of the distribution of excitation energy occurred within the photochemical apparatus of the thylakoid membranes, which led to a modification of the energy distribution in favor of PSI. This would result in a reduction of the photochemical capacity of PSII and possibly in an enhancement of cyclic electron transport. In a related study, it was shown that light combined with chilling produced a reduction of energy transfer from the light-harvesting chlorophyll protein complex to PSII.[28] Such stress-induced alterations of PSII could underlie the observed reduction of the quantum rendement of photosynthesis. Recently, it was shown that in thylakoids of chloroplasts isolated from light-chilled tomato leaves, the activity of reactions involving PSI were inhibited to the same extent as those involving PSII. As the calculated rate of whole chain linear electron transport of the isolated chloroplasts was nearly a factor of two higher than required for supporting the measured rate of photosynthesis of intact leaves after chilling in the light, it was suggested that impaired electron transport capacity was not the rate-limiting factor for inhibition of photosynthesis after chilling in the light.[33] The relevance of the correlation between *in vitro* and *in vivo* electron transport capacity of the chloroplast needs further study before the exact site of the rate-limiting step of photosynthesis after chilling in the light can be established.

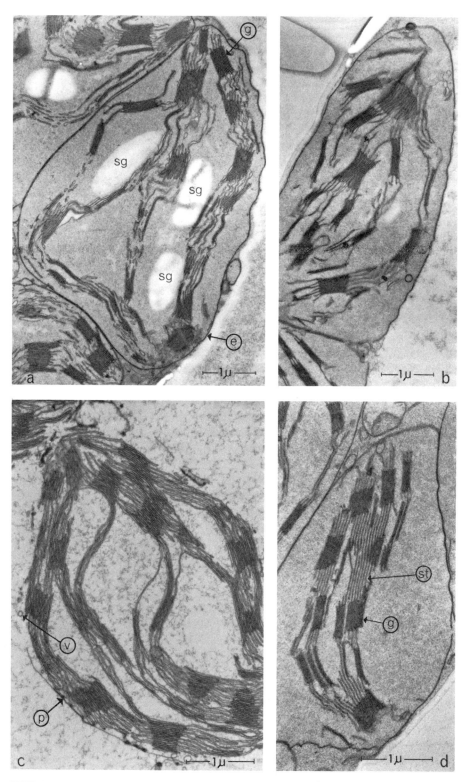

FIGURE 3. Effects of light and darkness on chloroplast ultrastructure at 1°C. (**a**) Chloroplast of an untreated leaf: (sg) starch grains; (e) envelope; (g) granum. (**b**) Chloroplast ultrastructure after 29 h darkness at 1°C. (**c**) Chloroplast ultrastructure after 29 h light (20,000 lux) in air at 1°C: (p) electron-dense particle; (V) vesicle. (**d**) Chloroplast ultrastructure after 29 h light in an atmosphere of nitrogen at 1°C: (st) stroma thylakoids; (g) grana thylakoids.

IV. CAUSES OF PHOTOOXIDATIVE DAMAGE

A. THE SENSITIZING PIGMENT

Since light has to be absorbed by a sensitizing pigment before it can have an effect, the question arose as to which pigment or pigments sensitized the photooxidative damage. Figure 2 shows action spectra of the light-induced pigment degradation in cucumber leaf discs. Chlorophyll a, chlorophyll b, and carotene showed a high degradation in the blue and the red wavelengths and a low degradation in the green part of the spectrum. The action spectra of the photooxidative degradation of the three pigments were similar and corresponded with the absorption spectrum of chlorophyll. It can therefore be concluded that the photooxidative degradation of carotene and chlorophyll is sensitized by chlorophyll. A similar action spectrum was observed for the photooxidative degradation of lipids and the photooxidative decrease of reductive capacity in cucumber leaf discs during chilling.[56,70] Apparently, chlorophyll plays a crucial role in the mechanism of photooxidative damage in leaves during chilling.

B. REACTIVE OXYGEN SPECIES

It is generally accepted that reactive oxygen species are involved in photooxidative reactions. Two potentially toxic oxygen species may be considered as a primary cause of photooxidative damage: singlet excited oxygen and superoxide.

1. Singlet Oxygen

As described under primary events, singlet oxygen can be generated by the transfer of energy from triplet excited chlorophyll, which has a sufficiently long lifetime to allow chemical reactions, to atmospheric oxygen. Evidence for the involvement of singlet oxygen in photooxidative reactions was obtained by the use of artificially generated singlet oxygen. It was shown that singlet oxygen can exist in solution and can initiate reactions which were previously thought to be light dependent.[71,72] A direct involvement of singlet oxygen in photooxidative damage in intact leaves is difficult to prove since it readily reacts with another singlet oxygen molecule to form superoxide. There is, however, ample indirect evidence that singlet oxygen may have damaging effects in plants. Substances which prevent the formation of triplet chlorophyll *in vitro*, like benzoquinone and benzidine, also inhibited photooxidative damage in cucumber leaf discs.[73] This result indicates that photooxidation is inhibited when triplet chlorophyll formation and, therefore, singlet oxygen generation is prevented. Furthermore, implication of singlet oxygen in photooxidative damage during chilling in intact leaves was suggested recently because photooxidation was inhibited by Dabco, a singlet oxygen quencher, while it was increased by D_2O, an enhancer of singlet oxygen.[74]

2. Superoxide

Superoxide can be generated in chloroplasts in two ways. First, in a so-called Mehler reaction where oxygen interacts directly with a component of the electron transport chain, probably with the primary electron acceptor of PSI. Second, superoxide may be generated via the autoxidation of ferridoxin, the primary soluble electron acceptor of PSI.[47] Superoxide itself is not very damaging, but by reaction with another superoxide molecule or by enzymatic dismutation by SOD it readily forms toxic H_2O_2. Toxic effects of H_2O_2 in the chloroplast are prevented by enzymatic peroxidations.[75] The inhibition of these reactions at low temperature in chilling-sensitive plants might result in photooxidative damage. There are several indications that superoxide is involved in the photooxidative mechanism during chilling. DCMU and atrazine, inhibitors of electron transport which decrease the formation of reduced electron acceptors of PSI and, as a consequence, the generation of superoxide, also inhibited photooxidation in cucumber and maize leaf tissues during chilling.[49,55,73]

Implication of superoxide in photooxidative damage during chilling is further indicated by the observation that application of two superoxide anion scavengers decreased chilling-induced photooxidation in detached cucumber leaves.[74] It appears that on account of our present information, both singlet oxygen and superoxide may play a primary role in the mechanism of photooxidation during chilling.

V. PROTECTIVE MECHANISMS AGAINST PHOTOOXIDATION

A. ENZYMATIC REACTIONS

Most plants can resist exposure to full sunlight even under stress conditions. Only after prolonged exposure to intense light may irreversible photooxidative damage occur. Therefore, plants must possess efficient protective mechanisms against photooxidation. As stated in Section III.A, primary protection is provided by enzymatic reactions, which prevent overexcitation of the reaction centers in the chloroplast by maintaining a certain level of carbon metabolism and by enzyme reactions which protect against reactive oxygen species, e.g., carotene de-epoxidase and superoxide dismutase.[76,47] At low temperatures, however, most enzymatic reactions will be reduced and enzymatic protection will be less effective. Under these conditions, structural changes in the chloroplast membranes might result in an increase of energy dissipation in the form of heat or light, and, in addition, intrinsic antioxidants of the chloroplast, such as β-carotene and α-tocopherol, may play a protective role.

B. STRUCTURAL CHANGES

Low-temperature-induced structural changes of chloroplast membrane proteins might be the origin of the observed changes in the light-harvesting complex of PSII after chilling in the light.[28,42] Such changes may not only cause an altered energy distribution over the two photosystems, as was observed in maize,[38] but may also cause an increase of the dissipation of possible harmful light energy as heat.

C. β-CAROTENE

It is remarkable that β-carotene is present in the chloroplasts of all green plants. A primary indication for a protecting role of carotene against photooxidation in plants came from the observation that mutants of maize, which were defective in carotene synthesis, were bleached in full sunlight, but remained green in low light.[77] Recent results show that β-carotene is exclusively bound to the core complexes of PSI and PSII.[78] These complexes contain the redox complexes functioning in the primary photochemical events. Protection against the damaging effects of oxygen at this site is essential for chloroplast functioning, and it seems that carotene by its unique properties is protecting the primary photochemical processes. Carotene appears to be a very effective antioxidant because it is protecting in two ways. First, it can effectively quench excited triplet chlorophyll and dissipate the surplus of energy as heat, preventing the formation of toxic singlet oxygen by triplet chlorophyll and atmospheric oxygen.[34] Second, carotene is an efficient quencher of singlet oxygen. As a consequence of this reaction, some of the carotene may be oxidized.[72] Under normal conditions, oxidized carotene is reduced enzymatically to carotene and can be used for protection again. When the enzymatic regeneration of carotene is inhibited, e.g., by low temperatures in chilling-sensitive plants, a decrease of carotene concentration may occur. It is shown in Figure 1 that photooxidative carotene degradation has a shorter lag-phase than photooxidative chlorophyll degradation. As a consequence, degradation of carotene precedes chlorophyll degradation in cucumber leaf discs during chilling in the light. This sequence of degradation suggests that chlorophyll may be protected by preferential photooxidation of carotene as a consequence of the reaction of carotene with singlet oxygen during the initial stage of photooxidative stress.

D. α-TOCOPHEROL

α-Tocopherol (vitamin E) is a lipophylic antioxidant which is localized within the leaf in the chloroplast membranes.[79] α-tocopherol is just as β-carotene (provitamin A), an effective scavenger of singlet oxygen *in vitro*, and it has been observed that α-tocopherol has an inhibitive effect on chlorophyll photooxidation in acetone and in liposomal suspensions.[80,81] *In vivo* it might protect the chloroplast membranes from photoperoxidation by using glutathione as a reductant.[75] An interesting observation was made by Wise and Naylor,[74] who measured a fast decrease in α-tocopherol concentration in the first 3 h of exposure to light during chilling of cucumber leaves, while no decrease was observed in chilling resistant pea leaves under the same conditions. They suggest that the thylakoid pool of α-tocopherol might be the first line of defense for protecting against photooxidative injuries in the chloroplast. As it was shown that application of α-tocopherol increased the phase transition temperature of rice root membranes and liposomes, α-tocopherol may also have an indirect protective effect by influencing the phase transition temperature of membrane lipids.[82,83]

VI. RECOVERY OF LIGHT-INDUCED DAMAGE DURING CHILLING

The overall success of plants in coping with stress depends not only on the amount of stress-induced damage, but also on their ability to recover and on the rate at which the stressed plants can attain prestress growth rate after normal conditions are restored. It appears that the combined effects of high light intensity and chilling temperatures result in a slower and often more incomplete recovery as compared with recovery after one of these stresses alone. Recovery of the decrease of variable chlorophyll fluorescence in rice induced by exposure to high light intensity was much slower after exposure to light at chilling temperatures than after exposure at 25°C.[42] Apparently, light combined with chilling temperatures results in a reduced capacity for recovery of PSII functioning. Recovery from chilling-induced photoinhibition of photosynthesis generally occurs rather slowly. Full reversal of the inhibition of apparent quantum yield of CO_2 uptake of bean leaves, which had been exposed to high light intensity at 6°C for 3 h, always was in excess of 3 h, and recovery took usually 4 to 8 h.[10] After a 45-h chilling exposure at 4°C in the dark, complete recovery from chill inhibition of CO_2-saturated photosynthesis of tomato leaves was observed after 12 h at 25°C in the dark. However, recovery of leaves after chilling under high irradiance was slower, and after 12- and 20-h chilling and light treatment, recovery was incomplete even after 48 h incubation at 25°C in the dark.[84] Another interesting observation of Martin and Ort[84] was that high-intensity illumination has an adverse effect on recovery of photosynthesis after chilling in the dark. Similar results were obtained by others.[10] It has been suggested that the adverse effect of light after chilling is caused by superoxide, which can not be dismutated in the normal way by superoxide dismutase because this enzyme is inhibited by chilling treatment.[58] Recovery from light-induced damage during chilling is also dependent on the duration of the stress. Quantum yield of CO_2 assimilation in olive leaves chilled for 8 h or less at 5°C in the light recovered at 26°C in dim light to the prechilled value within 20 h, but leaves, which had been chilled for 12 h in the light, showed no recovery in 20 h and had only recovered to 60% of the prechilled value after 90 h.[36] Furthermore, it has been shown that the capacity for recovery varies between plant species. Yakir et al.[85] observed that tomato plants of the high altitude species *Lycopersicum hirsutum*, when compared with the cultivated species *L. esculentum*, maintained not only a higher CO_2- and light-saturated photosnythesis at the chilling temperature of 10°C, but also showed a better capacity for rapid recovery after prolonged exposure to chilling stress at moderate light intensity. A correlation between the capacity for recovery at 25°C and the degree of inhibition of photosynthetic CO_2 fixation during chilling in the light was also observed by Taylor and Rowley

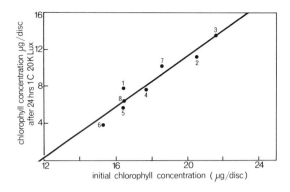

FIGURE 4.   Correlation between photooxidative chlorophyll degradation during chilling and the initial chlorophyll content of eight genotypes of tomato (*Lycopersicon esculentum*).

in tropical grasses.[7] The possibility that an impaired stomatal opening might be the limiting factor during recovery of photosynthesis after chilling was studied in tomato. Although an increase of stomatal resistance as a consequence of chilling occurred, the internal CO_2 concentration did not decrease or even increase during the recovery period.[84] Similar results were obtained earlier.[51] These results suggest that the limiting factor for recovery is not stomatal resistance, but probably the CO_2 fixation mechanism in the chloroplast.

VII. PREVENTION OF LIGHT-INDUCED DAMAGE DURING CHILLING

A. ARTIFICIAL SHADING
The level of light-induced damage during chilling appears to depend on the intensity of the irradiation.[9,10] Therefore, a reduction of damage by artificial shading of sensitive plants during the chilling period can be expected. Indeed, it was found that the photosynthetic rate of overwintering leaves was greater when field-grown tea plants were shaded using lawn cloth during winter, and that shading resulted in an increased yield of new shoots in the first flush.[86]

B. BREEDING FOR RESISTANCE
Genotypic variation in light-induced damage during chilling has been observed in a few studies. According to Miedema et al.[54] resistance to low temperatures and light-dependent chlorosis is a major selection criterium for the improvement of low-temperature tolerance of maize. This agrees with results of Eagles et al.,[87] who observed that the most chilling-resistant lines of a high-altitude tropical population of maize remained greener during growth at suboptimal temperature. Genotypic variation in susceptibility to chilling-dependent photoinhibition of photosynthesis was recently described in maize varieties.[88] Genotypic variation in light-induced damage during chilling was also evident in tomato lines. Figure 4 shows that there were differences in light-induced chlorophyll degradation between the eight lines. The level of the degradation correlated with the initial chlorophyll concentration of the leaf discs, suggesting that the chlorophyll concentration of the leaf could be a selection criterium for reduced sensitivity to light-induced damage during chilling in tomato.

VIII. CONCLUDING REMARKS

Most plant species are well protected against light-induced damage at low temperatures, e.g., wintergreen plants can endure full sunlight at temperatures far below the freezing point.

Chilling-sensitive plants appear to be exceptionally sensitive to light at low temperatures. The present data indicate that the sensitivity to light is due to a chilling-induced imbalance between the mechanisms, which dissipate possible harmful energy in the chloroplast membranes and in the chloroplast stroma.

The primary damaging event is as yet unidentified. It might be a chilling-induced inhibition of an enzyme in the stroma, which leads to a reduced concentration of the main sink for electron transport $NADP^+$. Light energy absorbed by chlorophyll will then accumulate in the photosystems and the electron transport chain and lead to photooxidative damage at different sites in the chloroplast.

Much information exists about the light-induced degradation of different chloroplast components, but a great deal remains to be learned about the underlying damage-inducing mechanisms. Particularly, more knowledge is needed about the specific damaging effects of reactive oxygen species in the chloroplast. Of special interest seems to be the unravelling of the role of chilling-induced structural alterations of proteins. Changes in the structure of proteins, which regulate the harmless dissipation of energy in the chloroplast, might underlie the exceptional susceptibility to light of chilling-sensitive plants.

Finally, the existence of genotypic variation in susceptibility to light-induced damage during chilling deserves further attention. Special breeding programs could be started to improve the resistance to photoinhibition during chilling and the yield of chilling-sensitive crops at higher altitudes and latitudes. However, more detailed information about the mechanisms which prevent light-induced damage during chilling will be required for choosing the most appropriate selection criteria.

ACKNOWLEDGMENT

The author thanks Prof. P. J. C. Kuiper and L. J. de Kok for valuable comments.

REFERENCES

1. **Powles, S. B.,** Photoinhibition of photosynthesis induced by visible light, *Annu. Rev. Plant Physiol.,* 35, 15, 1984.
2. **Powles, S. B. and Osmond, C. B.,** Photoinhibition of intact attached leaves of C3 plants illuminated in the absence of both carbon dioxide and of photorespiration, *Plant Physiol.,* 64, 982, 1979.
3. **Levitt, J.,** Chilling injury and resistance, in *Responses of Plants to Environmental Stresses,* Vol. 1, Academic Press, New York, 1980, 46.
4. **Farage, P. K., Long, S. P., Bradbury, M. R., and Baker, N. R.,** Damage to maize photosynthesis in the field during periods when chilling is combined with high photon fluxes, Proc. Meet. on Photosynthesis of the Agricultural and Food Research Counsil, Aberystwyth, Wales, 1986, 27.
5. **McWilliam, J. R., Kramer, P. J., and Musser, R. L.,** Temperature-induced water stress in chilling-sensitive plants, *Aust. J. Plant. Physiol.,* 9, 343, 1982.
6. **Bagnall, D., Wolfe, J., and King, R. W.,** Chill-induced wilting and hydraulic recovery in mung bean plants, *Plant Cell Environ.,* 5, 457, 1983.
7. **Taylor, A. O. and Rowley, J. A.,** Plants under climatic stress. I. Low temperature, high light effects on photosynthesis, *Plant Physiol.,* 47, 713, 1971.
8. **Miedema, P.,** The effect of low temperature on zea mays, *Adv. Agron.,* 35, 93, 1982.
9. **Van Hasselt, P. R.,** Photo-oxidation of leaf pigments in Cucumis leaf discs during chilling, *Acta Bot. Neerl.,* 21, 539, 1972.
10. **Powles, S. B., Berry, J. A., and Björkman, O.,** Interaction between light and chilling temperature on the inhibition of photosynthesis in chilling-sensitive plants, *Plant Cell Environ.,* 6, 117, 1983.
11. **Van Hasselt, P. R. and Strikwerda, J. T.,** Pigment degradation in discs of the thermophilic Cucumis sativus as affected by light, temperature, sugar application and inhibitors, *Physiol. Plant.,* 37, 253, 1976.

12. **Raison, J. K. and Lyons, J. M.,** Chilling injury: a plea for uniform terminology, *Plant Cell Environ.,* 9, 685, 1986.
13. **Raison, J. K. and Orr, G. R.,** Phase transitions in thylakoid polar lipids of chilling sensitive plants, *Plant Physiol.,* 80, 638, 1986.
14. **Lyons, J. M., Graham, D., and Raison, J. K., Eds.,** Epilogue, in *Low Temperature Stress in Crop Plants: The Role of the Membrane,* Academic Press, New York, 1979, 543.
15. **Wolfe, J.,** Chilling injury in plants-the role of membrane fluidity, *Plant Cell Environ.,* 1, 241, 1978.
16. **Taylor, A. O., Slack, C. R., and McPherson, H. G.,** Plants under climatic stress. VI. Chilling and light effects of photosynthetic enzymes of sorghum and maize, *Plant Physiol.,* 54, 696, 1974.
17. **Graham, D. and Patterson, B. D.,** Responses of plants to low, non freezing temperatures: proteins, metabolism, and acclimation, *Annu. Rev. Plant Physiol.,* 33, 347, 1982.
18. **Shneyour, A., Raison, J. K., and Smillie, R. M.,** The effect of temperature on the rate of photosynthetic electron transfer in chloroplasts of chilling-sensitive and chilling-resistant plants, *Biochim. Biophys. Acta,* 292, 152, 1973.
19. **Janssen, L. H. J. and Van Hasselt, P. R.,** Temperature-induced alterations of in vivo chlorophyll a fluorescence induction in cucumber as affected by DCMU, *Photosynth. Res.,* 1988, in press.
20. **Lyons, J. M.,** Chilling injury in plants, *Annu. Rev. Plant Physiol.,* 24, 445, 1973.
21. **Bishop, D. G., Kenrick, J. R., Bayston, J. H., MacPherson, A. S., Johns, S. R., and Willing, R. I.,** The influence of fatty acid unsaturation on fluidity and molecular packing of chloroplast membrane lipids, in *Low Temperature Stress in Crop Plants: The Role of the Membrane,* Lyons, J. M., Graham, D., and Raison, J. K., Eds., Academic Press, New York, 1979, 375.
22. **Murata, N. and Yamaya, J.,** Temperature dependent phase behaviour of phosphatidyglycerols from chilling sensitive and chilling resistant plants, *Plant Physiol.,* 74, 1016, 1984.
23. **Horvath, I. and Van Hasselt, P. R.,** Inhibition of chilling-induced photooxidative damage to leaves of Cucumis sativus by treatment with amino-alcohols, *Physiol. Plant.,* 164, 83, 1984.
24. **Öquist, G.,** Seasonally induced changes in acyl lipids and fatty acids of chloroplast thylakoids of Pinus sylvestris, *Plant Physiol.,* 69, 869, 1982.
25. **Critchley, C., Smillie, R. M., and Patterson, B. D.,** Effect of temperature on photoreproductive activity of chloroplasts from passion fruit species of different chilling sensitivity, *Aust. J. Plant Physiol.,* 5, 443, 1978.
26. **Hodgson, R. A. J., Orr, G. R., and Raison, J. K.,** Inhibition of photosynthesis by chilling in the light, *Plant Sci.,* 49, 75, 1987.
27. **Havaux, M. and Lannoye, R.,** Effects of chilling temperatures on prompt and delayed chlorophyll fluorescence in maize and barley leaves, *Photosynthetica,* 18, 117, 1984.
28. **Hayden, D. B., Baker, N. R., Percifal, M. P., and Beckwith, P. B.,** Modification of the Photosystem II light-harvesting chlorophyll a/b protein complex in maize during chill-induced photoinhibition, *Biochim. Biophys. Acta,* 851, 86, 1986.
29. **Margulies, M. M.,** Effect of cold-storage of bean leaves on photosynthetic reactions of isolated chloroplasts. Inability to donate electrons to photosystem II and relation to maganese content, *Biochim. Biophys. Acta,* 267, 96, 1972.
30. **Kaniuga, Z., Sochanowicz, B., Zabec, J., and Krzystyniak, K.,** Photosynthetic apparatus in chilling sensitive plants. I. Reactivation of Hill reaction activity inhibited on the cold and dark storage of detached leaves and intact plants, *Planta,* 140, 121, 1978.
31. **Smillie, R. M. and Nott, R.,** Assay of chilling injury in wild and domestic tomatoes based on photosystem activity of the chilled leaves, *Plant Physiol.,* 63, 796, 1979.
32. **Martin, B. and Ort, D. R.,** Insensitivity of water-oxidation and photosystem 2 activity in tomato to chilling temperatures, *Plant Physiol.,* 70, 689, 1982.
33. **Kee, S. C., Martin, B., and Ort, D. R.,** The effects of chilling in the dark and in the light on photosynthesis of tomato: electron transfer reactions, *Photosyn. Res.,* 8, 41, 1986.
34. **Fujimori, E. and Livingstone, R.,** Interactions of chlorophyll in its triplet state with oxygen, carotene, *Nature (London),* 180, 1036, 1957.
35. **Lasley, S. E., Garber, M. P., and Hodges, C. F.,** After effects of light and chilling temperatures on photosynthesis in excised cucumber cotyledons, *J. Am. Soc. Hortic. Sci.,* 104, 477, 1979.
36. **Bongi, G. and Long, S. P.,** Light-dependent damage to photosynthesis in olive leaves during chilling and high temperature stress, *Plant Cell Environ.,* 10, 241, 1987.
37. **Lindeman, W.,** Inhibition of photosynthesis in Lemna minor by illumination during chilling in the presence of oxygen, *Photosynthetica,* 13, 175, 1979.
38. **Baker, N. R., East, T. M., and Long, S. P.,** Chilling damage to photosynthesis in young Zea mays. II. photochemical function of thylakoids in vivo, *J. Exp. Bot.,* 34, 187, 1983.
39. **Van Hasselt, P. R. and Van Berlo, H. A. C.,** Photo-oxidative damage to the photosynthetic apparatus during chilling, *Physiol. Plant.,* 50, 52, 1980.

40. **Yakir, D., Rudich, J. Bravdo, B. A., and Malkin, S.,** Prolonged chilling under moderate light: effect on photosynthetic activity measured with the photoacoustic method, *Plant Cell Environ.,* 9, 581, 1986.

41. **Bennett, J.,** Regulation of photosynthesis by reversible phosphorylation of the light-harvesting chlorophyll a/b protein, *Biochem. J.,* 212, 1, 1983.

42. **Moll, B. A. and Steinback, K. E.,** Chilling sensitivity in Oryza sativa: the role of protein phosphorylation in protection against photoinhibition, *Plant Physiol.,* 80, 420, 1986.

43. **Powles, S. B. and Critchley, C.,** Effect of light intensity during growth on photoinhibition of intact attached bean leaves, *Plant Physiol.,* 65, 1181, 1980.

44. **Rowley, J. A. and Taylor, A. O.,** Plants under climatic stress IV. Effects of CO_2 and O_2 on photosynthesis under high-light low-temperature stress, *New Phytol.,* 71, 477, 1972.

45. **Powles, S. B. and Osmond, C. B.,** Inhibition of the capacity and the efficiency of photosynthesis in bean leaflets illuminated in a CO_2 free atmosphere at low O_2: a possible role for photorespiration, *Aust. J. Plant Physiol.,* 5, 619, 1978.

46. **Potvin, C.,** Amelioration of chilling effects by CO_2 enrichment, *Physiol. Veg.,* 23, 345, 1985.

47. **Badger, R. M.,** Photosynthetic oxygen exchange, *Annu. Rev. Plant Physiol.,* 36, 27, 1985.

48. **Chaturvedi, R. and Nilsen, S.,** Photoinhibition of photosynthesis: effect of O_2 and selective excitation of the photosystems in intact Lemna gibba plants, *Photosyn. Res.,* 12, 35, 1987.

49. **Mustardy, L. A., Sz-Rozsa, Z., and Faludi-Daniel, A.,** Chilling syndrome in light-exposed maize leaves and its easing by low doses of DCMU, *Physiol. Plant.,* 60, 572, 1984.

50. **Eamus, D. and Wilson, J. M.,** A model for the inter action of low temperature, ABA, IAA, and CO_2 in the control of stomatal behaviour, *J. Exp. Bot.,* 35, 91, 1984.

51. **Long, S. P., East, T., and Baker, N. R.,** Chilling damage to photosynthesis in young Zea mays. I. Effects of light and temperature variation on photosynthetic CO_2 assimilation, *J. Exp. Bot.,* 34, 177, 1983.

52. **Long, S. P.,** C4 photosynthesis at low temperatures, *Plant Cell Environ.,* 6, 345, 1983.

53. **Hodgins, R. and van Huystee, R. B.,** Chilling induced chlorosis in maize (Zea mays), *Can. J. Bot.,* 63, 711, 1984.

54. **Miedema, P., Post, J., and Groot, P. J.,** The effects of low temperature on seedling growth of maize genotypes, *Agric. Res. Rep.,* 926, 71, 1987.

55. **Wise, R. R. and Naylor, A. W.,** Chilling-enhanced photooxidation. The peroxidative destruction of lipids during chilling injury to photosynthesis and ultrastructure, *Plant Physiol.,* 83, 272, 1987.

56. **Van Hasselt, P. R.,** Photo-oxidation of unsaturated lipids in Cucumis leaf disks during chilling, *Acta Bot. Neerl.,* 23, 159, 1974.

57. **De Kok, L. J. and Kuiper, P. J. C.,** Glycolipid degradation in leaves of the thermophilic Cucumis sativus as affected by light and low temperature treatment, *Physiol. Plant.,* 39, 123, 1977.

58. **Kaniuga, Z., Zabek, J., and Michalski, W. P.,** Photosynthetic apparatus in chilling sensitive plants. VI. Cold and dark-induced changes in chloroplast superoxide dismutase activity in relation to loosely bound manganese content, *Planta,* 145, 145, 1979.

59. **Michalski, W. P. and Kaniuga, Z.,** Relationship between superoxide dismutase activity and photoperoxidation of chloroplast lipids, *Biochim. Biophys. Acta,* 637, 159, 1981.

60. **Wang, C. Y. and Adams, D. O.,** Chilling-induced ethylene production in cucumbers (Cucumis sativus L.), *Plant Physiol.,* 69, 424, 1982.

61. **Wang, C. Y.,** Effect of temperature and light on ACC and MACC in chilled cucumber seedlings, *Sci. Hortic.,* 30, 47, 1986.

62. **Ilker, R., Breidenbach, R. W., and Lyons, J. M.,** Sequence of ultrastructural changes in tomato cotyledons during short periods of chilling, in *Low Temperature Stress in Crop Plants: The Role of the Membrane,* Lyons, J. M., Graham, D., and Raison, J. K., Eds., Academic Press, New York, 1979, 97.

63. **Wise, R. R., McWilliam, J. R., and Naylor, A. N.,** A comparative study of low temperature-induced ultrastructural alterations of three species with different chilling sensitivity, *Plant Cell Environ.,* 6, 525, 1983.

64. **Musser, R. L., Thomas, S. A., Wise, R. R., Peeler, T. C., and Naylor, A. W.,** Chloroplast ultrastructure, chlorophyll fluorescence, and pigment composition in chilling-stressed soybeans, *Plant Physiol.,* 74, 749, 1984.

65. **Rikin, A., Gilter, C., and Atsmon, D.,** Chilling injury in cotton (Gossipium hirsutum L.): light requirement for the reduction of injury and the protective effect of abscissic acid, *Plant Cell Physiol.,* 22, 553, 1981.

66. **Taylor, A. O. and Craig, A. S.,** Plants under climatic stress. II. Low temperature, high light effects on chloroplast ultrastructure, *Plant Physiol.,* 47, 719, 1971.

67. **Van Hasselt, P. R.,** Photo-oxidative damage to the ultrastructure of Cucumis chloroplasts during chilling, *Proc. K. Ned. Akad. Wet.,* 77, 50, 1974.

68. **Critchley, C. and Smillie, R. M.,** Leaf chlorophyll fluorescence as an indicater of high light stress (photoinhibition) in Cucumis sativus L., *Aust. J. Plant Physiol.,* 8, 133, 1981.

69. **Yakir, D., Rudich, J., and Bravdo, B. A.,** Photoacoustic and fluorescence measurements of the chilling response and their relationship to carbon dioxide uptake in tomato plants, *Planta,* 164, 345, 1985.

70. **Van Hasselt, P. R.,** Photo-oxidative damage to triphenyltetrazoliumchloride (TTC) reducing capacity of Cucumis leaf disks during chilling, *Acta Bot. Neerl.,* 22, 546, 1973.

71. **Wilson, T. and Hastings, J. W.,** Chemical and biological aspects of singlet excited molecular oxygen, in *Photophysiology,* Vol. 5, Giese A. C., Ed., Academic Press, New York, 1970, 49.

72. **Foote, C. S. and Denny, R. W.,** Chemistry of singlet oxygen. VII. Quenching by carotene, *J. Am. Chem. Soc.,* 90, 6233, 1968.

73. **Van Hasselt, P. R.,** Protection of Cucumis leaf pigments against photo-oxidative degradation during chilling, *Acta Bot. Neerl.,* 25, 41, 1976.

74. **Wise, R. R. and Naylor, A. W.,** Chilling-enhanced photooxidation. Evidence for the role singlet oxygen and superoxide in the breakdown of pigments and endogenous antioxidants, *Plant Physiol.,* 83, 277, 1987.

75. **Jablonski, P. P. and Anderson, J. W.,** Light-dependent reduction of dehydroascorbate by ruptured pea chloroplasts, *Plant Physiol.,* 67, 1239, 1981.

76. **Krinsky, N. I.,** Protective function of carotenoid pigments, in *Photophysiology,* Vol. 3, Giese, A. C., Ed., Academic Press, New York, 1968, 123.

77. **Anderson, I. C. and Robertson, D. S.,** Role of carotenoids in protecting chlorophyll from photodestruction, *Plant Physiol.,* 35, 531, 1960.

78. **Thornber, J. P., Peter, G. F., and Nechushtai, R.,** Biohemical composition and structure of photosynthetic pigment-proteins from higher plants, *Physiol. Plant.,* 71, 236, 1987.

79. **Grumbach, K. H.,** Distribution of chlorophylls, carotenoids and qinones in chloroplasts of higher plants, *Z. Naturforsch.,* 38c, 996, 1983.

80. **Van Hasselt, P. R., De Kok, L. J., and Kuiper, P. J. C.,** Effect of α-tocopherol, β-carotene, monogalactosyldiglyceride and phosphatidylcholine on light-induced degradation of chlorophyl a in acetone, *Physiol. Plant.,* 45, 475, 1979.

81. **De Kok, L. J., Van Hasselt, P. R., and Kuiper, P. J. C.,** Photo-oxidative degradation of chlorophyll-a and unsaturated lipids in liposomal dispersions at low temperature, *Physiol. Plant.,* 43, 7, 1978.

82. **Tanczos, O. G., Erdei, L., Vigh, L., Kuipers, B., and Kuiper, P. J. C.,** Effect of α-tocopherol on growth, membrane-bound adenosine triphosphatase activity of the roots, membrane fluidity and potassium uptake in rice plants, *Physiol. Plant.,* 55, 289, 1982.

83. **Pohlmann, F. W. and Kuiper, P. J. C.,** Effect of α-tocopherol on the phase transition of phosphatidylcholine and on water transport through phosphatidylcholine liposome membranes. *Phytochemistry,* 20, 1525, 1981.

84. **Martin, B. and Ort, D. R.,** The recovery of photosynthesis in tomato subsequent to chilling exposure, *Photosyn. Res.,* 6, 121, 1985.

85. **Yakir, D., Rudich, J., and Bravdo, B. A.,** Adaptation to chilling photosynthetic characteristics of the cultivated tomato and a high altitude wild species, *Plant Cell Environ.,* 9, 477, 1986.

86. **Aoki, S.,** Interaction of light and low temperature in depression of photosynthesis in tea leaves, *Jpn. J. Crop Sci.,* 55, 496, 1986.

87. **Eagles, H. A., Hardacre, A. K., Brooking, I. R., Cameron, A. J., Smillie, R. M., and Hetherington, S. E.,** Evaluation of a high altitude tropical population of maize for agronomic performance and seedling growth at low temperature, *N. Z. J. Agric. Res.,* 26, 281, 1983.

88. **Long, S. P., Nugawela, A., Bongi, G., and Farage, P. K.,** Chilling dependent photoinhibition of photosynthetic CO_2 uptake, in *Progress in Photosynthesis Research,* Vol. IV.2, Biggins, J., Ed., Martinus Nijhoff, Dordrecht, Netherlands, 1987, 131.

89. **van Hasselt, P. R.,** Unpublished data.

Chapter 8

ASSESSMENT OF CHILLING SENSITIVITY BY CHLOROPHYLL FLUORESCENCE ANALYSIS

John M. Wilson and John A. Greaves

TABLE OF CONTENTS

I. INTRODUCTION

Even closely related species often differ not only morphologically, but physiologically and chemically as well. Given the great diversity of plant species, especially in the tropics, it is therefore not surprising that the symptoms of chilling injury and their rate of development in different tropical and subtropical species can vary enormously. Common symptoms of chilling injury to leaves include rapid wilting, the development of water-soaked patches, browning, photobleaching of chlorophyll, and the development of necrotic patches. Because the rate of development of such symptoms depends not only on the chilling temperature and the species selected, but also on other factors such as the previous growth temperature, age of plants, light intensity, and relative humidity, the quantification of these injuries to compare the chill sensitivity of even closely related species has been difficult. In addition, these comparisons of the rate of development of symptoms are complicated by the fact that the symptoms usually develop only after several days at the chilling temperature and are often accelerated by the return of the plants to the warmth. Physiological experiments to quantify chilling injury, such as changes in the rate of electrolyte leakage, lipid fluidity, respiration, or photosynthesis, have been shown to be either very time consuming or unreliable and destructive to the plant under investigation and, therefore, not suitable for use by plant breeders to screen large numbers of plants for chill tolerance.

Many of the most important crop plants of the world, such as rice, pearl millet, and sorghum, are susceptible to chilling temperatures; and researchers at international research centers, such as the International Rice Research Institute (IRRI) in the Philippines and at the International Centre for Research in the Semi-Arid Tropics at Hyderabad in India, are attempting to produce plants that are more tolerant of low-temperature conditions, as well as other environmental factors such as drought and salinity. For instance, in many areas of the world, high-yielding rice varieties cannot be grown due to their sensitivity to chilling; therefore research at IRRI is aimed at producing crosses between the high yielding *Indica* varieties and the more cold-tolerant *Japonica* varieties in an attempt to increase rice yields in countries where chilling conditions are encountered. In this type of breeding program, quantifying the effect of the chilling stress on the plant has proved to be difficult, time consuming, and expensive, as plants have to be chilled in water baths maintained at low temperatures for several days and damage estimated by visual assessment after several days of return to the warmth. However, the chlorophyll fluorescence technique offers considerable potential in accelerating the quantitative assessment of chilling injury as it is rapid, sensitive, nondestructive to the plant tissue, relatively low in cost, and able to detect injury before visible symptoms occur.

II. WHAT IS CHLOROPHYLL FLUORESCENCE?

Under optimal conditions, 85% of the light intercepted by a plant leaf is absorbed by the photosynthetic pigments and is used in photosynthesis. The remainder is lost as heat or is radiated as fluorescence. Most of the fluorescence is emitted from the chlorophyll of photosystem II (PSII). In particular, the variable chlorophyll fluorescence is very responsive to changes in PSII activity,[1] so that any stress applied to green plant tissue which directly or indirectly affects photosynthetic metabolism is likely to change the yield of this fluorescence. The yield is lowest when the electron acceptors are in the oxidized state, such as during the illumination of dark-adapted tissue. Thus, the yield of chlorophyll fluorescence will be reduced if there is an inhibition of the photooxidizing side of PSII and enhanced if the inhibition occurs on the photoreducing side of PSII.

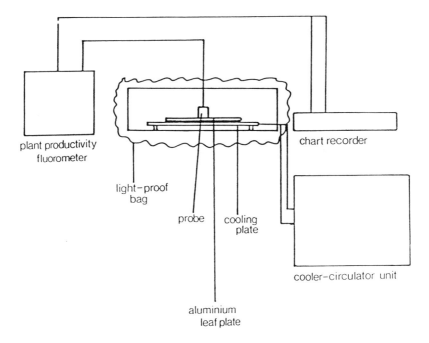

FIGURE 1.   Arrangement of the Brancker plant productivity fluorimeter in conjunction with
light-tight photographic bag, aluminium leaf plate, cooler-circulator unit, and chart recorder
for the measurement of continuous fluorescence.

III. MEASUREMENT OF CHLOROPHYLL FLUORESCENCE

A. CONTINUOUS FLUORESCENCE

The machine most commonly used to measure fluorescence is a plant productivity
fluorimeter (Richard Brancker Research Ltd., 27 Monk Street, Ottawa) in conjunction with
a chart pen recorder that has a fast paper speed of approximately 60 cm/min. The fluorimeter
is a miniaturized Kautsky apparatus developed for use on plants *in situ*. The instrument is
portable and consists of two main parts:

1.   Power and control unit containing the main electronic circuitry for light control, signal
      amplification, and connections for an oscilloscope or chart recorder (Figure 1).
2.   The fluorimeter probe which is placed directly on the surface of the leaf and incorporates
      a photodiode sensor and light-emitting diode (LED) which provides monochromatic
      illumination at a maximum wavelength of 670 nm at a variable light intensity. A cut-
      off filter (Corning® 769) is fitted behind the LED to absorb strong or reflected light
      and to exclude fluorescence at wavelengths below 710 nm. Therefore, this machine
      measures only PSII fluorescence (Figure 1).

Two models of the fluorimeter are currently available: the SF-10, which costs approximately US$1800, and the SF-20, which costs US$2800. The model SF-20 has additional
features such as digital readout of fluorescence intensity and a method for calibrating the
light intensity emitted by the probe.

Leaf material is usually dark-acclimated for 30 to 60 min before fluorescence is measured,
usually in a dark room with a green safety light. Whole plants can be dark acclimated and
the probe held firmly against the leaf surface while measurements are taken. However, the
most commonly used method is to place either detached whole leaves or parts of leaves
onto an aluminium plate covered in black card to reduce reflection and a layer of moist

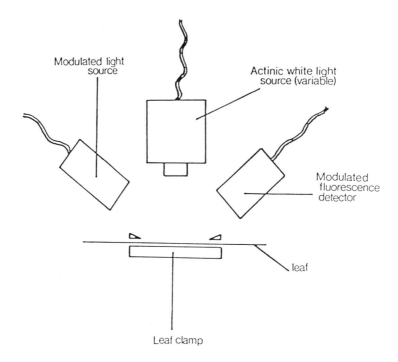

FIGURE 2.   Arrangement of the Hanstech fluorescence system with modulated light source, actinic white light source, and modulated fluorescence detector for the measurement of modulated fluorescence.

tissue paper to reduce leaf water loss. The leaf tissue is then covered in a layer of transparent film ("cling film" or similar) which is permeable to air, but not to water. Finally, on top of the film is placed a plastic grid with holes 3.3 cm in diameter through which the probe lead of the fluorimeter is placed. The leaves are arranged on the plate so that the LED of the probe does not sit on a main vein of the leaf. This method has several advantages over using whole plants, as it enables rapid measurements to be made in the dark, the plates can be easily chilled in crushed ice or an incubator, and it enables the probe lead to be located on the same part of the leaf if successive readings are required during a chilling treatment. Usually 16 replicate measurements are made per treatment.

If a dark room is not available for fluorescence measurements, an alternative method is to use a light-tight photographic bag positioned around a cooling plate supplied with refrigerant by a cooler-circulator unit (Figure 1). The prepared aluminium plate holding the leaves is then placed on top of the cooling plate in the light-tight bag, and the cooler-circulator can be adjusted to provide the desired chilling temperature which can be monitored by a digital thermocouple attached to one of the leaves.

B. MODULATED FLUORESCENCE

The most recent method developed for measuring chlorophyll fluorescence enables the fluorescence to be detected from leaves exposed to continuous, high-intensity white light. Ogren and Baker[2] have produced in conjunction with Hansatech Ltd. (Paxman Road, Hardwick Industrial Estate, Kings Lynn, Norfolk) an instrument for the generation and measurement of chlorophyll fluorescence signals from leaves exposed to continuous, high-intensity white light (Figure 2). The current cost of this machine, plus high-intensity white light source, is approximately US$3500. Modulated fluorescence is generated in the leaf by pulsed diodes emitting low-intensity yellow radiation, and this is detected with a photodiode where output is fed to an amplifier locked in to the frequency of the light-emitting diodes.

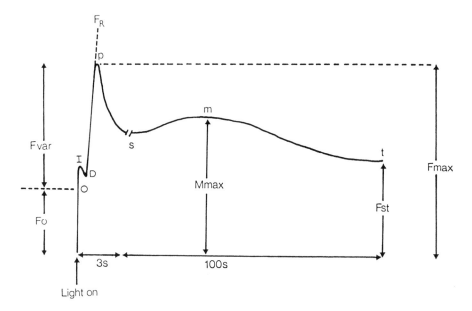

FIGURE 3. A typical fluorescence induction curve for a plant leaf. F_o represents the initial fluorescence level at 0, F_{max} the maximal level at p, M_{max} the maximal level of secondary fluorescence at m, and F_{st} the steady-state fluorescence obtained in time t.

A comparison of modulated fluorescence signals and the continuous signals emitted from dark-adapted leaf tissue showed only small differences between the two signals,[2] which are thought to be attributable to differences in the populations of chloroplasts being monitored in the two measurements as a result of differential penetration of the modulated and actinic light sources in the sample. The modulated fluorescence system will, therefore, expand the use of the chlorophyll fluorescence technique as a probe of photosynthesis both in the laboratory and in the field.

C. A TYPICAL CONTINUOUS FLUORESCENCE TRACE

When a leaf has been in the dark for a few minutes and the light is switched on, photosynthesis does not start immediately. Instead, several transitory "warm up" induction phases occur first (the Kautsky effect). Fluorescence induction follows these phases and can be related to fundamental processes within the chloroplast. The interpretation of these induction curves is somewhat controversial, and the reader is referred to three recent reviews[1,3,4] for further details.

A typical fluorescence curve is shown in Figure 3 and has five main components:

1. **Baseline to 0** — This is a very fast rise (<1 nsec) to the F_o level, called the nonvariable fluorescent component, and is not considered to be responsive to physiochemical events in the chloroplast thylakoids.
2. **F_o to p** — This is a slower rise and is termed the fluorescence of variable yield which reaches a maximum value at p. This fluorescence originates in the chlorophyll of PSII and is sensitive to changes in the rates of electron transfer and the ultrastructure of the thylakoid membranes. Plant physiologists have used mainly the rate of rise of chlorophyll fluorescence F_R (see Figure 4) to quantify the stress incurred by the plants. In some cases, the maximum fluorescent yield (i.e., distance from F_o to 0), called F_{var}, has been used, and there is usually a good correlation between F_R and F_{var}. Usually chilling and other environmental stresses decrease both the F_R and F_{var} values as the photooxidizing side of PSII is inhibited.

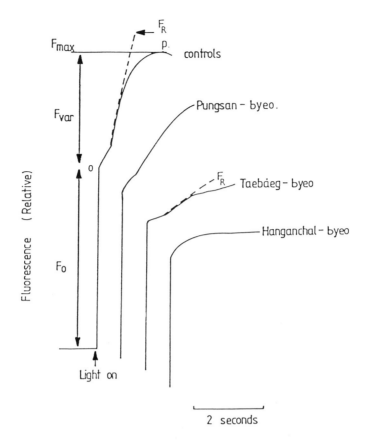

FIGURE 4.   Changes in the chlorophyll fluorescence kinetics of three Korean varieties
of rice leaves which differ in their sensitivity to chilling, *Pungsan-byeo, Taebaeg-byeo,*
and *Hanganchal-byeo,* showing the decrease in the rate of rice of F_R after 8 d of chilling
at 10°C and 85% rH. The results are the mean of 16 replicate measurements.

3.   **p to s** — Fluorescence is quenched as O_2 becomes available and NADP is reduced.
Q (primary electron acceptor of PSII) becomes oxidized and is a potent quencher of
fluorescence.

4.   **s-m-t transient** — This relates to the onset of carbon assimilation and is affected by
two quenching systems QUE (quenching system based on pH gradient across chloroplast membrane) and QUQ (quenching system based on Q). High ATP levels may
also cause reduced height of the peak. Both the decay p to s and the slow fluorescence
change s-m-t reflect processes in which the ultrastructure of the thylakoid membrane
plays a fundamental role, such as ion fluxes necessary for the establishment of the
prephosphorylation high-energy state.

5. F_{st} — Represents the final steady-state fluorescence and F_{st}-F_o, the quenching capacity
of the system.

It is important to note that at any phase of the fluorescence induction time course, the
flourescence yield is controlled by more than one photophysiological process. Therefore,
the assignments here refer only to those processes which are thought to exert the predominant
influences in each phase.

D. A TYPICAL MODULATED FLUORESCENCE TRACE
A demonstration of the measurement of the modulated fluorescence signal generated by
exciting a potato leaf with weak, modualated yellow light in the prescence of white actinic

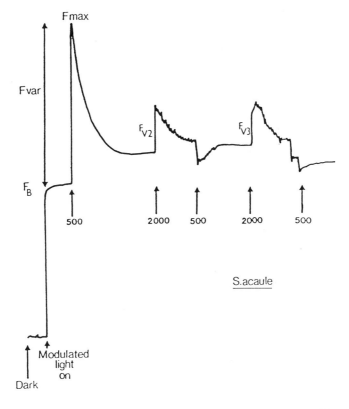

FIGURE 5. A typical modulated fluorescence signal generated by exciting a potato leaf with weaker, modulated yellow light in the presence of white actinic light of varying intensities. The photon flux density of the actinic light is given in μmol m$_{-2}$ s$_{-1}$ below each arrow.

light of varying intensities is shown in Figure 5. Exposure of the leaf to the modulated light produces a rise in fluorescence to a basal level F_B; however, no induction curve characteristic of variable fluorescence was observed. Exposure of the leaf, while being irradiated with the weak modulated light, to white actinic light produces a variable fluorescence emission which is observed as a component of the modulated fluorescence signal. The kinetics of the induction of the modulated variable fluorescence when the leaf is excited with 500 μmol m^{-2} s^{-1} of white light are shown in Figure 5. Under physiological conditions, the major factors determining the variable yield are (1) the redox state of "Q", and (2) the degree of thylakoid energization, which is dependent on the pH gradient across the thylakoid membranes.[5-7] The rapid rise in modulated fluorescence to a maximum is due to the rapid reduction of "Q" by the actinic light, while the fluorescence quenching from this maximum is attributable to the reoxidation of "Q" and increasing thylakoid energization. The rapid rise in modulated fluorescence, when the photon flux density of the actinic light reaching the leaf is increased from 500 to 2000 μmol m^{-2} s^{-1} (F_{V2}), is the result of increased "Q" reduction; the reduced maximal level attained by F_{V2} compared to that observed when the leaf is exposed to 500 μmol m^{-2} s^{-1} of actinic light (Figure 5) is probably a demonstration of the existence of a larger pH gradient across the thylakoids prior to exposure to 2000 μmol m^{-2} s^{-1} light that existed in the presence of only the weak, modulated light. The presence of quenching from the second maximum suggests that either reoxidation of "Q" and/or increased thylakoid energization occurs in this 2000-μmol m^{-2} s^{-1} light regime. Reduction of the actinic white light from 2000 back to 500 μmol m^{-2} s^{-1} rapidly reduces the modulated fluorescence level (Figure 5). On returning the leaf to a white light photon flux density of

2000 μmol m^{-2} s^{-1}, modulated fluorescence rose again (F_{V3}) to the previous level observed in F_{V2}, presumably indicating a similar degree of thylakoid energization. A further quenching was observed from F_{V3}.

This pulse modulation technique is based on the rationale (first used in the "light doubling" technique of Bradbury and Baker,[8] and further developed by Quick and Horton[9]) that at any time during the measurement of chlorophyll fluorescence, a short pulse of high light (L2) will transiently cause the complete reduction of the PSII electron acceptor Q_A and cause a further rise in variable fluorescence (F_{V2}), as shown in Figure 5. The extent of this second rise (F_{V2}) is dependent on the redox state of "Q" and the pH gradient across the thylakoid membrane. Any environmental stress such as chilling will have an effect on both of these and will consequently have an effect on F_{V2}, providing information on the state of health of the electron acceptor "Q", which is known to be affected by both high light and chilling temperatures. Although it is not the intention in this chapter to discuss the complex nature of these fluorescence transients, the data shown in Figure 5, together with the above interpretation of the reasons for the changes in the modulated fluorescence signal, demonstrate the potential of this modulated fluorescence technique and "light doubling" for studying *in vivo* photochemical events when leaves are exposed to chilling temperatures and high intensities of continuous white light.

As the modulated fluorescence technique is new, there are few reports in the literature of its use in assessing the chill sensitivity of plants. However, the present authors have investigated the use of modulated fluorescence to compare the chill sensitivity of the more cold-tolerant wild potatoes *Solanum acaule* and the cultivated species *S. tuberosum*. The wild species *S. acaule* showed little detectable change in F_{V2} after 18 d of chilling at 4/2°C (day/night) as compared to a marked reduction in the cultivated species *S. tuberosum*, indicting that substantial damage had occurred to the photosynthetic electron transport pathway (probably $Q_A \rightarrow Q_B$) in the cultivated species during chilling.[10]

IV. TWO EXAMPLES OF THE EFFECTS OF CHILLING TEMPERATURES IN CONTINUOUS CHLOROPHYLL FLUORESCENCE IN DIFFERENT SPECIES

A. RICE

Figure 4 shows the changes in the fluorescence after 8 d of chilling at 10°C and 85% rH of three rice varieties which differ in their sensitivity to chilling temperatures. It can be seen that both the rate of rise of chlorophyll fluorescence (F_R) and the maximum yield (peak height) was reduced after chilling, and these decreases correlated well with the known sensitivity of the plants to chilling.

Figure 6 compares the changes in F_R (as a percentage of control) of 11 different rice varieties with the visual assessment of chilling injury made on a 1-to-9 scale after 8 d of chilling at 10°C and 85% rH. The *Japonica* varieties *Samnam-byeo*[10] and *Sangpung-byeo*[11] are the most chill tolerant, showing an increase in F_R of 23 to 34% during chilling. In contrast, the *Indica/Japonica* varieties *Taebaeg-byeo*,[1] *Milyang 77*,[9] *Hanganchal-byeo*,[6] and *Chupung-byeo*[4] were the most susceptible to chilling and showed decreases in F_R of between 77 to 98%. The more tolerant *Indica/Japonica* varieties tested showed smaller decreases in chlorophyll fluorescence which agreed well with visual observations of injury (Figure 6).

B. POTATO

Although adapted for growth in cool climates, the common potato is still only moderately cold tolerant compared with other major crops. Below 5°C, the potato plant shows symptoms of mild chilling injury, such as little or no growth,[11,12] and in the tuber, increased respiration[11] and the conversion of starch to sugars.[13] In many cases, if cultivars were to possess 2 to

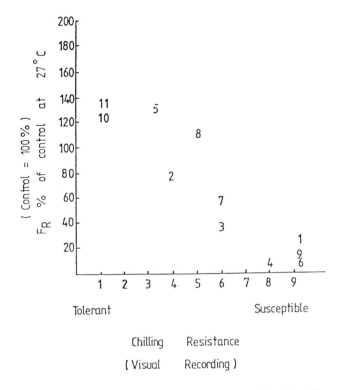

FIGURE 6. A comparison of the changes in the rate of rise of variable fluorescence (F_R) of the leaves of 11 rice varieties (expressed as a percentage of the control at 27°C) after 8 d of chilling at 10°C and 85% RH with a visual assessment of injury on a 1-to-9 scale. (1) *Taebaeg-byeo* (I/J), (2) *Chungchung-byeo* (I/J), (3) *Pungsan-byeo* (I/J), (4) *Chupung-byeo* (I/J), (5) *Milyang* 23 (I/J), (6) *Hangachal-byeo* (I/J), (7) *Manseog-byeo* (I/J), (8) *Baegyang-byeo* (I/J), (9) *Milyang* 77 (I/J), (10) *Samnan-byeo* (J), and (11) *Sangpung-byeo* (J). I or J indicates whether variety is *Indica* or *Japonica* or a cross between the two.

3°C more cold tolerance, potato production could be increased in the presently cultivated areas by simply extending the growing season. Burton[14] reported that the average commercial daily increment for potato growth is in the region of 400 kg ha^{-1}. Therefore, extending the growing period just a few days at each end of the season could greatly increase the average potato yield. However, suitable cold-tolerant varieties are not yet commercially available. The cold sensitivity of commercial varieties of potato in comparison to wild potatoes may be due, in part, to the very restricted gene pool from which our present-day commercial varieties were developed.[15]

Wild species of potato are endemic to extensive areas of South America and parts of North America and are found over an unusually wide range of altitudes, from sea level up to nearly 5000 m.[16,17] Because the wild species inhabit such a diverse range of habitats, they might be expected to possess different levels of genetically determined cold tolerance. In a survey of 18 genotypes of wild and cultivated potato (*Solanum* spp.) Greaves and Wilson[18] showed that their relative cold sensitivity could be measured rapidly by the decrease in the rate of rise in chlorophyll fluorescence of leaflets chilled at 0°C, measured as the time taken for a 50% decrease in F_R in comparison to the control (Table 1). The majority of the wild species showed a greater tolerance to the low temperature, and the results showed a large genotypic range within the *Solanum* species for chill tolerance at 0°C. *S. albicans* showed the greatest tolerance to the stress with a chill tolerance value of 78 h, in contrast

TABLE 1
The Assessment of the Chill Sensitivity of Wild and Cultivated Potato Species by the Time Taken for a 50% Decrease in F_R in Leaves Kept at 0°C in the Dark

Solanum	Ref. no.[a]	Geographical spread	Altitude range (m)	Time (h) for a 50% decrease in F_R in leaves kept at 0°C in the dark
albicans (Ochoa, Hawkes)	CPC 3712	Peru	3400—3500	78
chacoense (Bitt.)	CPC 3057	Bolivia, Argentina, Paraguay, Uruguay, Brazil	1600—2800	75
acaule (Bitt.)	CIP 13561	Peru, Bolivia, Argentina	2600—4650	73
sanctae-rosae (Hawkes)	CPC 2483	Argentina	2500—3800	72
albicans (Ochoa, Hawkes)	CIP 12062	Peru	3400—3500	72
acaule (Bitt.)	CPC 2440	Peru, Bolivia, Argentina	2600—4650	71.5
acaule (Bitt.)	CIP 11979	Peru, Bolivia, Argentina	2600—4650	71
tuberosum (L.)	CPC 5374	Chile	2000—3500	71
chomatophilum (Bitt.)	CIP 11398	Peru	2000—4000	70
acaule (Bitt.)	CIP 13145	Peru, Bolivia, Argentina	2600—4650	66
multidissectum (Hawkes)	CPC 2723	Peru	3300—4100	59
tuberosum (L.)	CPC 6052	Chile	2000—3500	58
tuberosum (L.)	CPC 5646	Chile	2000—3500	54
bukasovii (Juz.)	CIP 13581	Peru	3000—3900	51
tuberosum (L.)	SCRI cv. Moira	Europe	0—2000	41
tuberosum (L.)	SCRI cv. Cara	Europe	0—2000	29
tuberosum (L.)	SCRI cv. Desiree	Europe	0—2000	28
tuberosum (L.)	CPC 3294	Chile	2000—3500	24

[a]   CPC: Commonwealth Potato Collection; CIP: International Potato Centre, Lima; SCRI: Scottish Crops Research Institute.

with *S. tuberosum* (CPC 3294) with a value of only 24 h for a 50% decrease in F_R, the rate of the induced rise in chlorophyll fluorescence. A large genotypic range was shown to exist within *S. tuberosum* (Table 1), which could prove to be very important in breeding for more chill-tolerant clones. The Chilean varieties showed greater tolerance to the chilling temperature in contrast with the European varieties. However, the variety "Moira" was shown to be more chill tolerant than other European cultivars. The ability of chlorophyll fluorescence analysis to distinguish degrees of chill tolerance both inter- and intraspecifically among the wild and cultivated potato species indicates a potential use of the method to screen for chill tolerance in future potato breeding programs. In addition, the technique has been successfully used in determining the different levels of the frost hardiness of potato species[19] and the frost hardiness of different clover species.[20] The effectiveness of frost-hardening treatments in these plants can also be rapidly measured using the chlorophyll fluorescence technique.

V. CONCLUSION

For a long time, there has been a need for a technique which can give a rapid quantitative assessment of the response of plants to environmental stress such as chilling temperatures. The chlorophyll fluorescence technique can be used for measuring the sensitivity of plants to chilling and many other environmental stresses such as freezing, salinity, drought, heat, and nutrient deficiency (Table 2). In conclusion, the ability to detect chill damage to the thylakoid membranes by chlorophyll fluorescence analysis should not only be a useful tool for plant breeders, but also provide plant physiologists with valuable insight into the causes and mechanisms by which different environmental stresses affect plant tissues and photosynthesis.

TABLE 2
The Different Environmental Stresses and Species in which Chlorophyll Fluorescence Has Been Used to Assess the Stress Sensitivity of Plant Tissue

Environmental stress	Species	Ref.
Chilling	Potato sp.	18
	Rice	21
	Maize	22, 23
	Tomatoes	24
	Cucumber	25
	Soybeans	26
	Episcia reptans	27
Freezing	Clover sp.	20
	Potato sp.	19, 28
Nutrient deficiency	*Pinus radiata*	29
	Amaranthus tricolor	30
	Chloris gayana	30
	Echinychloa utilis	30
	Kochia childsii	30
Drought	*Pinus radiata*	29
	Salix sp.	31
	Cotton	32
High temperatures	Barley, pea, bean	33
	Tomato, maize, papaya	
	Spinach	34
	Passiflora sp.	35
	Potatoes	36
	Plantago gracilus	37
	Lettuce, papaya	
Salinity	Beans	38
	Cucumber	39
	Borya nitida	40

ADDENDUM

The authors would like to point out that since this chapter was written, there have been considerable advances in instrumentation designed for measuring chlorophyll fluorescence. The PAM chlorophyll fluorescence measuring system (Heinz Waltz, Effeltrich, FRG) has been developed for the measurement of pulse saturation kinetics, and a number of papers have subsequently been published using this method to study chilling injury (Havaux, M., *Plant Physiol. Biochem.*, 25, 735, 1987; Larcher, W. and Neuner, G., *Plant Physiol.*, 89, 740, 1989). There have also been advances in the measurement of continuous chlorophyll fluorescence with the development of the PSM microprocessor-operated instrument (BioMonitor, Sweden; PK Morgan Instruments, U.S.); this instrument was developed by Prof. G. Oquist (Oquist, G., and Wass, R., *Physiol. Plant.* 73, 211, 1988). The authors are continuing their work developing chlorophyll fluorescence techniques in stress physiology.

REFERENCES

1. **Papageorgiou, G.,** Chlorophyll fluorescence: an intrinsic probe of photosynthesis, in *Bioenergetics of Photosynthesis,* Govindjee, Ed., Academic Press, London, 1975, 319.
2. **Ogren, E. and Baker, N. R.,** Evaluation of a technique for the measurement of chlorophyll fluorescence from leaves exposed to continuous white light, *Plant Cell Environ.,* 8, 539, 1985.
3. **Sivak, M. N. and Walker, D. A.,** Some effects of CO_2 concentration and decreased O_2 concentration on induction fluorescence in leaves, *Proc. R. Soc. London Ser. B,* 217, 372, 1983.
4. **Horton, P.,** Relations between electron transport and carbon assimilation, simultaneous measurements of chlorophyll fluorescence, trans thylakoid pH gradient and O_2 evolution in isolated chloroplasts, *Proc. R. Soc. London Ser. B,* 217, 405, 1985.
5. **Krause, G. H., Vernotte, C., and Briantais, J. M.,** Photoinduced quenching in intact chloroplasts and algae. Resolution into two components, *Biochim. Biophys. Acta,* 679, 116, 1982.
6. **Horton, P.,** Relations between electron transport and carbon assimilation; simultaneous measurement of chlorophyll fluorescence, transthylakoid pH gradient and O_2 evolution in isolated chloroplasts, *Proc. R. Soc. London Ser. B,* 217, 405, 1983.
7. **Bradbury, M. and Baker, N. R.,** A quantitative determination of photochemical and non-photochemical quenching during the slow phase of the chlorophyll fluorescence induction curve of bean leaves, *Biochim. Biophys. Acta,* 765, 275, 1984.
8. **Bradbury, M. and Baker, N. R.,** Analysis of the slow phase of the *in vivo* chlorophyll fluorescence induction curve. Changes in the redox state of photosystem II electron acceptors and fluorescence emission from photosystems I and II, *Biochim. Biophys. Acta,* 635, 4542, 1981.
9. **Quick, W. P. and Horton, P.,** Studies on the induction of chlorophyll fluorescence in barley protoplasts. II. Resolution of fluorescence quenching by redox state and the transthylakoid pH gradient, *Proc. R. Soc. London Ser. B,* 220, 371, 1984.
10. **Greaves, J. A.,** Low Temperature Physiology of Potato Species, Ph.D. thesis, University of Wales, Cardiff, 1987.
11. **Burton, W. G.** *Th Potato,* Veenman & Zonen, Wageningen, 1966.
12. **Moorby, J. and Milnthorpe, F. L.,** Potato, in *Crop Physiology,* Evans, L. J., Ed., Cambridge University Press, London, 1975, 225.
13. **Burton, W. G.,** The physics and physiology of storage, in *The Potato Crop,* Harris, P. M., Ed., Chapman & Hall, London, 1978, 545.
14. **Burton, W. G.,** Challenges for stress physiology in potato, *Am. Potato J.,* 58, 3, 1981.
15. **Simmonds, N. W.,** Variability in crop plants, it's use and conservation, *Biol. Rev.,* 37, 442, 1962.
16. **Correll, D. S.,** *The Potato and its Wild Relatives,* Texas Research Foundation, Renner, TX, 1962.
17. **Hawkes, J. G.,** Biosystematics of the potato, in *The Potato Crop,* Harris, P. M., Ed., Chapman & Hall, London, 15, 1978.
18. **Greaves, J. A. and Wilson, J. M.,** Assessment of the non-freezing cold sensitivity of wild and cultivated potato species by chlorophyll fluorescence analysis, *Potato Res.,* 29, 509, 1986.
19. **Greaves, J. A. and Wilson, J. M.,** Assessment of the freezing sensitivity of wild and cultivated potato species by chlorophyll fluorescence analysis, *Potato Res.,* 30, 381, 1987.
20. **Barnes, J. D. and Wilson, J. M.,** Assessment of the frost sensitivity of *Trifolium* species by chlorophyll fluorescence analysis, *Ann. Appl. Biol.,* 105, 107, 1984.
21. **Rho, Y. D. and Wilson, J. M.,** Assessment of chilling injury in eleven Korean rice cultivars by chlorophyll fluorescence analysis, *Int. Rice Res. Inst. Newsl.,* 10(3), 14, 1985.
22. **Hetherington, S. E., Smillie, R. M., Hardacre, A. K., and Eagles, H. A.,** Using chlorophyl fluorescence *in vivo* to measure the chilling tolerances of different populations of maize, *Aust. J. Plant Physiol.,* 10, 247, 1983.
23. **Baker, N. R., East, T. M., and Long, S. P.,** Chilling damage to photosynthesis in young zea mays. II. Photochemical function of the thylaxoids *in vivo, J. Exp. Bot.,* 34, 189, 1983.
24. **Smillie, R. M. and Nott, R.,** Assay of chilling injury in wild and domestic tomatoes based on photosystem activity of chilled leaves, *Plant Physiol.,* 63, 796, 1979.
25. **Smillie, R. M. and Hetherington, S. E.,** Stress tolerance and stress induced injury in crop plants measured by chlorophyll fluorescence *in vivo,* (chilling, freezing, ice cover, heat and high light), *Plant Physiol.,* 72, 1043, 1983.
26. **Musser, R. L., Thomas, S. A., Wise, R. R., and Peeler, T. C.,** Chloroplast ultrastructure, chlorophyll fluorescence and pigment composition in chilling stressed soybeans, *Plant Physiol.,* 74, 749, 1984.
27. **Smillie, R. M.,** A highly chilling sensitive angiosperm, *Carlsberg Res. Commun.,* 49, 75, 1984.
28. **Sundbom, E., Strand, M., and Hallgren, J. E.,** Temperature induced fluorescence changes. A screening method for frost tolerance of potato (*Solarum* spp.), *Plant Physiol.,* 70, 1299, 1982.

29. **Conroy, J. P., Smillie, R. M., Kuppers, M., Bevege, D. I., and Barlow, E. W.,** Chlorophyll a fluorescence and photosynthetic and growth responses of *Pinus radiata* to phosphorus deficiency, drought stress and high CO_2, *Plant Physiol.*, 81, 423, 1986.

30. **Groff, C. P. L., Johnston, M., and Brownell, P. F.,** *In vivo* chlorophyll a fluorescence in sodium-deficient C_4 plants, *Aust. J. Plant Physiol.*, 13, 589, 1986.

31. **Ogren, H. and Oquist, G.,** Effects of drought on photosynthesis, chlorophyll fluorescence and photo-inhibition susceptibility in intact willow leaves, *Planta*, 166, 380, 1985.

32. **Genty, B., Briantias, J. M., and Da Silva, J. B. V.,** Effects of drought on primary photosynthetic processes of cotton leaves, *Plant Physiol.*, 83, 360, 1987.

33. **Smillie, R. M. and Gibbons, G. C.,** Heat tolerance and heat hardening in crop plants measured by chlorophyll fluorescence, *Carlsberg Res. Commun.*, 46, 395, 1981.

34. **Santarius, K. A. and Muller, M.,** Investigations in heat resistance of spinach leaves, *Planta*, 146, 529, 1979.

35. **Smillie, R. M.,** Coloured components of chloroplast membranes as intrinsic membrane probes for monitoring the development of heat injury in intact tissues, *Aust. J. Plant Physiol.*, 6, 121, 1979.

36. **Hetherington, S. E., Smillie, R. M., Malagamba, P., and Huaman, Z.,** Heat tolerance and cold tolerance of cultivated potatoes measured by the chlorophyll fluorescence method, *Planta*, 159, 119, 1983.

37. **Smillie, R. M. and Nott, R.,** Heat injury in leaves of alpine, temperte and tropical plants, *Aust. J. Plant Physiol.*, 6, 135, 1979.

38. **Smillie, R. M. and Nott, R.,** Salt tolerance in crop plants monitored by chlorophyll fluorescence *in vivo*, *Plant Physiol.*, 70, 1049, 1982.

39. **Critchley, C. and Smillie, R. M.,** Leaf chlorophyll fluorescence as an inductor of high light stress (photoinhibition) in *Cucumis sativus*, *Aust. J. Plant Physiol.*, 8, 133, 1981.

40. **Hetherington, S. E. and Smillie, R. M.,** Tolerance of *Borya nitida*, a poikilohydrous angiosperm to heat, cold and high light stress in the hydrated state, *Planta*, 155, 76, 1982.

Part III. Biochemical Changes, Molecular Basis, and Concepts of Chilling Injury

Chapter 9

PROPOSALS FOR A BETTER UNDERSTANDING OF THE MOLECULAR BASIS OF CHILLING INJURY*

John K. Raison and Glenda R. Orr

TABLE OF CONTENTS

---

*   The thoughts and concepts outlined in this chapter are based on data and reviews published up to December 1987.

I. INTRODUCTION

The term "chilling injury" describes the visual manifestation of the cellular dysfunction in tropical plants when exposed to chilling temperatures, usually in the range of 0 to 15°C. The symptoms of chilling injury vary depending on the type of tissue, its state of maturity and metabolic status (active or dormant) immediately before the chilling, and on a variety of environmental factors as well as on the subjectivity of the assessor. Given these variables and the variety of horticultural crops and products which are sensitive to chilling, it is little wonder there is a plethora of descriptions detailing the visible symptoms. This is, however, in marked contrast with the dearth of information about the basic, temperature-induced lesion which initiates the disorder and details of the metabolic events which lead to the visible symptoms.

Chilling injury is of significant economic importance to world agriculture and has been recognized, described, and studied for over 100 years.[1,2] It is thus difficult to understand the reason for the present lack of definitive information about the basic lesion. There is little doubt that the visible symptoms of the injury result from a disruption of normal metabolic processes,[3] as well as a number of degradative processes initiated or accelerated by chilling.[3] It is thus not surprising that the symptoms of chilling injury have been described as "...elusive as well as conspicuous"[4] and prompted one author, after having read the relevant literature, to comment that the only message discernible was "that the dysfunction is so deep-seated and so fundamental that any search for ways to alleviate chilling injury is doomed to failure."[5] This view is understandable considering the current confusion and ambiguity in the terminology used to describe the various stages of chilling injury. Furthermore, much of the research on chilling injury has been concerned with the horticultural and postharvest aspects, with particular emphasis on descriptions of the symptoms, rather than attempting to detail the sequence of events leading to the symptoms. The main problem appears to be that most of the investigations have failed to separate the primary lesion, which initiates the injurious processes, from the events which lead to the development of the visible symptoms. It is thus easy to understand why reviews on chilling injury generally conclude that the phenomenon is a "complex process" and highlight a general inability to describe the cause(s) in molecular terms or even come to some consensus about the nature of the temperature-induced lesion.

Our understanding of the events contributing to the overall process of chilling injury and our ability to communicate our observations and views to others could be improved considerably if a clear differentiation could be established between the stages in the development of chilling injury. As a first step, a simple recognition that the immediate events can be separated from the later events, at least conceptually, would reduce the apparent complexity of the overall process.

This chapter discusses the rationale for dividing the overall process of chilling injury into two identifiable stages and provides terminology to define and describe the two stages. It also discusses the significance, interpretation, and relevance of data relating to hypotheses concerning the primary lesion with special reference to the hypothesis which considers injury results from a phase change in the lipids of cell membranes.

II. THE TWO STAGES OF CHILLING INJURY

A. CONCEPT AND JUSTIFICATION

For the purposes of discussion, the process or change which initiates chilling injury will be referred to as the "primary event"; that is, the temperature-induced lesion which initiates the metabolic dysfunctions leading to tissue damage. The second stage will be referred to as the "secondary events". These include the multitude of metabolic processes which are adversely affected as a consequence of the primary event and lead to the visible symptoms and cell death. The reason for the subdivision is not arbitrary. It is proposed because it allows the time-dependent secondary events to be conceptually separated from the more instantaneous primary event.

The subdivision is also logical, and the rationale for its proposal is based on the many observations of the development of chilling injury in tropical plants and fruits. A characteristic feature of chilling injury is that, for a particular plant or tissue, the injury becomes apparent only when the tissue is held below some critical temperature.[6-9] A good example of this abrupt demarcation in sensitivity can be seen in the results of experiments assessing the growth rate of mung bean seedlings over a period of 14 d. It was found that 50% of the plants held at 9.9°C grew, but plants held at 8.8°C, not only failed to grow, but died.[9] These data indicate that the temperature-induced lesion is an "all-or-nothing" event occurring at a particular critical temperature and suggests that it is a physical event. Another feature of chilling injury is that, in the short term, the overall process is reversible.[10] For example, the storage life of many tropical fruits can be extended by periodic warming above the critical temperature.[5] The time threshold, beyond which reversibility is not possible, can be explained with reference to the schematic shown in Figure 1. This shows the relation between the temperature-dependent primary event and the time-dependent secondary events. The primary event is initiated when the temperature of the tissue falls below the particular critical temperature. This, in turn, causes a disproportionate inhibition or acceleration in reactions associated with a number of metabolic processes which, if the chilling is prolonged, leads to an imbalance in metabolism and loss of cellular integrity. The reactions which have been shown to be adversely affected or accelerated by chilling include chlorophyll synthesis,[11] photosynthesis,[12] oxidative activity of mitochondria,[13] amino acid incorporation into protein,[14] ion leakage,[3] amino acid uptake,[15] and ethylene production.[16]

If the chilling treatment is terminated and the temperature raised above the critical temperature, the tissue either recovers from the metabolic imbalance, or the higher temperature accelerates the development of visible symptoms of injury.[10] Thus, in the short term, both the primary event and the secondary events are reversible. During the time period of reversibility, raising the temperature results in a burst of activity, assumed to be associated with the metabolism of intermediates accumulated during chilling.[10,16] After a short period, balanced metabolism is restored.[10] During the course of the secondary events, cell structures, such as membranes and organelles, are degraded.[17-19] If the temperature of the tissue is raised, after these degradative processes have progressed to the stage where synthesis cannot keep pace with catabolism, the cells rapidly lose their integrity, and this becomes manifest as the visible symptoms of injury. Exposing fruit to ripening temperatures after chilling beyond the time for reversibility, accelerates the development of the visible symptoms of injury, and this provides a useful technique for rapidly assessing injury.[20]

The other practical observation which must be considered in formulating a hypothesis on the mechanism for chilling injury is the variability in the time course of the development of injury. This variation depends on the type of tissue,[21] its metabolic state prior to chilling,[22] the time in the day-night cycle in which chilling commences,[23] the chilling temperature,[10] and a number of environmental factors such as humidity.[24] Given the variability in the development of the visible symptoms, it is easy to understand the difficulties in formulating hypotheses to describe the fundamental molecular mechanism underlying the injury. However, by separating chilling injury into two stages, it is possible to differentiate "cause" (the primary event) from "effect" (the secondary events). This provides an opportunity to design experiments aimed at providing definitive information about the more fundamental aspects of each of the events.

The ability to manipulate the genome of chilling-sensitive plants, such that the new plants are better able to tolerate or avoid the stress of chilling, is a prospect of significant

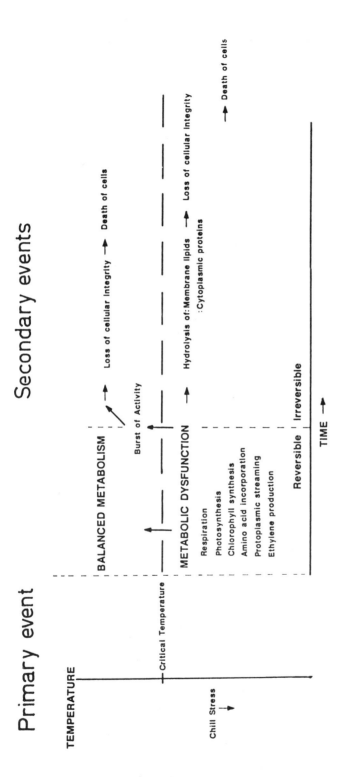

FIGURE 1.   A schematic representation of the relationship between the "primary" and "secondary events" of chilling injury. The "primary event" is initiated when the tissue is exposed to temperatures below the "critical temperature". The "secondary events", which include the processes indicated, are a consequence of the primary event. The severity of the secondary events (injury) increases with time at the chilling temperature. If rewarmed above the "critical temperature", after a brief chilling exposure, balanced metabolism can be restored, i.e., the effect of chilling is "reversible". If exposure to the chilling temperature is prolonged, a stage is reached where rewarming accelerates the development of injury and death of the tissue; this is the stage of "irreversibility".

horticultural and economic importance. Application of genetic techniques requires identification of the cause or causes of the injury. If there is a single cause, there is a reasonable likelihood that a gene could be manipulated to change the susceptibility of plants and their products to chilling. If there are a number of causes, or a multitude of lesions which become evident as the temperature is reduced, then the prospect of altering the genome becomes daunting, if not impossible, given the present techniques. Genetic manipulation to modify or prevent the development of injury, once initiated by the primary event, would be an equally daunting, if not impossible, task considering the number of metabolic processes known to be involved. Clearly, the approach most likely to succeed is that in which the molecular components of the primary event are identified and modified such that the critical temperature is abolished or at least lowered.

Before proceeding to the physiological, biochemical, and molecular aspects of chilling injury it is necessary to give the background and reasons for the terminology which will be adopted to provide more precise descriptions of the various aspects of chilling injury.

B. TERMINOLOGY

Using the concept that chilling injury can be subdivided into two separate stages, it is possible to more clearly define terms which describe the various aspects of chilling injury. The definitions used here follow those described by Raison and Lyons.[25] Terms such as *chilling sensitive* and *sensitive to chilling* are used to describe plants and tissue which suffer damage, and ultimately die, when exposed to chilling temperatures. These terms can be used in both a comparative and quantitative sense to compare and describe the susceptibility of tissue to chilling with reference to the temperature below which injury develops, i.e., the temperature of the primary event. Plants described as "very sensitive to chilling" would imply that the primary event is initiated at a relatively high temperature. The term *sensitive to chilling* should not be used to describe the time course of secondary events. Thus, if the visible injury takes longer to develop in one chilling-sensitive plant, compared with another, the latter should not be described as *more sensitive to chilling*. Comparisons of plants, showing variations in the time-course of the secondary events, should be described in terms of differences in *response* or *tolerance* to chilling.

The term *insensitive to chilling* will be used to describe plants which are not adversely affected by chilling and can complete most of their life cycle at or near 0°C. It is recognized that for some chilling-insensitive plants, the vegetative tissue can survive chilling, but the plant cannot complete its life cycle if maintained near 0°C. This could occur because a particular stage in the growth cycle, which in perennial plants is usually seasonally related, is particularly sensitive to chilling damage. Thus, the term *chilling-insensitive* will be used here to describe the effect of chilling on the vegetative part of a plant unless the study is specifically directed toward a particular stage in the life cycle which might be susceptible to chilling, e.g., pollen growth and fertilization.

The terms *resistant to chilling* and *tolerant to chilling* will be used to describe the response of chilling-sensitive plants and tissues to the stress of chilling. These are comparative terms, used in reference to the time course of the secondary events, and should be accompanied by relevant details about the temperature, time, humidity, the metabolic state of the tissue before the chilling stress, and the method used to assess the injury. An important point in this regard is that tissue which shows resistance or tolerance to chilling must, by definition, be sensitive to chilling. Conversely, if a plant is insensitive to chilling, it does not suffer stress at chilling temperatures and, therefore, is neither "resisting" nor "tolerating" the chilling.

The literature often describes comparative experiments of the thermal response of plants, and in many of these reports the terms *sensitive* and *resistant* to chilling, are used to differentiate between the behavior of the two groups of plants. However, in most of these comparisons both plants usually succumb to the treatment and were, therefore, sensitive to chilling. However, because one of the plants showed a longer time for the development of the symptom, or the symptom did not develop within the time course of the experiment, it was considered, erroneously, to be insensitive to chilling. Failure to appreciate that in this type of experiment both plants were sensitive to chilling has led to a large number of invalid comparisons and considerable confusion about events related to the cause of chilling sensitivity. For example, a number of studies have used altitudinal ecotypes of the wild tomato *Lycopersicon hirsutum* and variants of the commercial tomato (usually *L. esculentum*) in comparative studies on the effect of chilling.[26-38] The rationale for these comparisons was that the high-altitudinal ecotypes were adapted to grow at lower temperatures and were, therefore, insensitive to chilling. There is, however, no definitive evidence to indicate that the high-altitudinal ecotypes of *L. hirsutum*, are insensitive to chilling. Young seedlings of the high-altidudinal ecotype die when exposed to 5°C for 14 d,[26] showing they are sensitive to chilling. Measurements of the time for 50% germination of the *L. hirsutum* seeds held at various temperatures and the minimum temperature for chlorophyll synthesis show that there is only 2C degrees variation in the critical temperature of the high-and low-altitudinal ecotypes.[28] Thus, in terms of sensitivity to chilling, there appears to be little difference between the various ecotypes. As discussed later (see Section III.D.1. and Figure 3A) the high-altitudinal ecotype LA1361 has a transition and critical temperature of 14°C, 2°C degrees higher than that of *L. esculentum* cv. Grosse Lisse.[29] There are, however, marked differences in how the different ecotypes respond to the stress of chilling, e.g., survival after 7 d at 0°C[28] and variable fluorescence of leaves after exposure to 0°C.[30] These data show that all of the tomato strains and ecotypes tested are chilling sensitive. Some are obviously better able to cope with the strain imposed by the stress of chilling,[28] indicating there are differences in the response of the various ecotypes and commercial strains to chilling. However, it should be emphasized that the data obtained from these experiments show differences in response of the plants to chilling, not differences in their sensitivity. Thus, it would be inappropriate to use the different tomato ecotypes in experiments designed to compare the thermal behavior of tissue or cellular components of chilling-sensitive and chilling-insensitive plants. Studies using these ecotypes would provide comparisons of a response to chilling.

By clearly defining and explaining the meaning of these terms, it is anticipated that much of the confusion which has arisen regarding the descriptive aspects of chilling can be avoided. A particularly good example is the ambiguity concerning the use of the term "very sensitive to chilling". This has been used to describe a plant which suffers severe damage in a short exposure to a chilling temperature, usually between 0 and 5°C.[31] However, this term should be reserved to describe a plant which shows symptoms of chilling injury, within a defined time period, at a higher temperature, e.g., when held at 15°C, compared with one held at 12°C. For example, *Episcia* sp. has been described as "extremely chill sensitive"[31] and, indeed, exhibits symptoms such as wilting and necrosis after only 2 h at 5°C.[31] Unfortunately, it is not known whether the critical temperature for *Episcia* sp. is particularly high, thus providing a greater chilling differential when tissue is held at 5°C. An increase in the temperature differential below the critical temperature increases the severity of the injury for the same period.[10] Differentiating between these two conditions is important, particularly when describing and/or explaining alterations in the severity of the symptoms of injury for tissue exposed to the same chilling stress for the same time. If it is found that there is a decrease in the critical temperature, the change should be described as an alteration in the sensitivity of the tissue to chilling. A result of this type implies that in seeking an understanding of the molecular events involved, it would be logical to look for an alteration in the temperature of the primary event. On the other hand, if it is found that the time course for development of the injury is altered, an answer would best be sought by looking at the rate of metabolic processes associated with the secondary events which lead to the visible symptoms. Most of the studies which describe "amelioration" or "hardening" to chilling are usually referring to a decrease in the magnitude of some particular symptom for a given time, or to a longer time to reach the same degree of injury.[32] They are thus also concerned with the time-course of the secondary events.

Many recent reveiws and articles on the subject of chilling injury have failed to recognize the distinction between these terms and have perpetuated the confusion and ambiguity.[33-36]

III. THE CONCEPT OF AND CANDIDATES FOR A PRIMARY EVENT

Having evoked the concept that chilling injury is initiated by a single primary event, it is necessary to identify and describe the features of the event envisaged. It is also necessary to justify the concept in terms of providing a more logical and useful basis for formulating experiments aimed at providing definitive data which might increase our understanding of the molecular events underlying chilling injury.

From horticultural and postharvest studies, it is known that chilling injury occurs when tissue is held below a particular critical temperature and in the short term is reversible. To be consistent with these observations, the primary event envisaged must be virtually instantaneous and occur at a temperature which correlates with the critical temperature of the tissue. Furthermore, the event would have to be reversible, at least in the short term. The types of events envisaged or proposed as possible candidates for the primary cause of chilling injury are (1) an increase in the concentration of cytosolic calcium $[Ca^{2+}]_{cyt}$;[36] (2) a conformational change in some key regulatory protein (enzyme);[37] (3) a marked or abrupt decrease in the rate of cyclosis (protoplasmic streaming) and an alteration in cytoskeletal structure;[38] and (4) a temperature-induced change in the molecular ordering of membrane lipids.[13]

A. CONCENTRATION OF CYTOPLASMIC CALCIUM

Minorsky[36] has suggested that an increase in $[Ca^{2+}]_{cyt}$ might serve as the "primary physiological transducer of chilling injury in plants." The $[Ca^{2+}]_{cyt}$ plays an important role in many cellular processes, and it is envisaged that $[Ca^{2+}]_{cyt}$ may increase as much as two orders of magnitude in response to chilling.[36] It is difficult to envisage a mechanism whereby the $[Ca^{2+}]_{cyt}$ changes as a direct, instantaneous and reversible effect of temperature, and which causes a change in concentration sufficient to drastically alter metabolism at the critical temperature. Furthermore, a change in $[Ca^{2+}]_{cyt}$ is the nett result of changes in the active transport of Ca^{2+} and passive leakage through membranes. Thus, although a change in $[Ca^{2+}]_{cyt}$ is an attractive hypothesis and would be consistent with the metabolic dysfunction initiated by chilling, the phenomenon would be more appropriately classified as an early stage of the secondary events rather than a primary event. This classification does not imply that changes in $[Ca^{2+}]_{cyt}$ are not an important or significant aspect of chilling injury. As proposed by Minorsky,[36] it could be the immediate physiological transducer between the primary lesion and the secondary events.

B. COLD LABILE PROTEIN

The tertiary structure of many proteins is reversibly altered at low temperatures.[39] Thus, it has been proposed that an alteration in the kinetics and/or substrate specificity (K_m) of a regulatory enzyme could be a key lesion in the induction of chilling injury.[37] Such a change would qualify as a primary event. The enzymes proposed include pyruvate orthophosphate dikinase,[40] and phosphoenolpyruvate (PEP)-carboxylase.[37]

The dikinase from maize shows an abrupt increase in Arrhenius activation energy (E_a) at 12°C[40] but it has not been shown that this temperature-induced change is absent in the same enzyme from a chilling-insensitive plant or that the change has any effect on metabolism. Thus, there is no evidence that the abrupt change in E_a of this enzyme correlates with the sensitivity of the tissue to chilling or has any physiological implications with regard to chilling injury. Cold lability of the dikinase has also been measured in a variety of C_4 species and maize cultivars, but in only two species has lability of the enzyme correlated with sensitivity of photosynthesis to chilling.[41] Similarly, there was no clear distinction in the cold lability of this enzyme and the various cultivars of maize from northern and southern regions of Japan.[41] Although no data was presented, it was assumed that the cultivars from northern Japan were less sensitive to chilling than those from southern Japan.[42] The difference might, however, have been one of "response" rather than "sensitivity", as with the tomato ecotypes.[26]

The effect of chilling on the rate (V_{max}) and substrate affinity (K_m) of PEP-carboxylase has been studied in several species of temperate and tropical plants. Differences were noted in comparisons of the properties of the enzyme from a chilling-insensitive plant (wheat) with that from a chilling-sensitive plant (tomato).[37] However, the PEP-carboxylase from high-altitudinal ecotypes of *L. hirsutum*, plants adapted to warm days and cool nights, and from various species of *Passiflora*, which originated in cool climates, all show properties characteristic of the chilling-sensitive plants.[34] Differences have also been noted in the effect of chilling on the K_m (PEP) of this enzyme from various species. For pea, a chilling-insensitive plant, K_m (PEP) is relatively constant over the temperature range of 0 to 40°C, but for tomato, a chilling-sensitive plant, the K_m (PEP) increased from 0.1 mM, at temperatures below 10°C, reaching 0.5 mM at 5°C.[34] While it is assumed that the increase in K_m would result in a decrease in rate of carboxylation at chilling temperatures,[34] it has not been demonstrated. Other factors could compensate for the increase in K_m. In the absence of a correlation between V_{max} for PEP-carboxylase and the sensitivity of the plant to chilling, it is unlikely the thermal response of V_{max} for this enzyme has any implication in the initiation of chilling injury.

The proposition that some key regulatory enzyme could be the primary lesion in chilling injury is an interesting hypothesis. However, given the absence of a correlation between the sensitivity of plants to chilling and the cold lability of PEP-carboxylase, it must be concluded that it is unlikely that this enzyme has any implication in the initiation of chilling injury. The minimum requirement for this hypothesis to remain tenable would be a demonstration of a correlation between the temperature for the change in the kinetics or K_m of an enzyme and the critical temperature for the tissue. Since this has not been demonstrated, the hypotheses proposing a key role for cold labile proteins in the initiation of chilling injury must be treated as speculation.

C. STRUCTURAL CHANGES IN THE CYTOSKELETON

Temperature-induced changes occur in the cytoskeletal structure of cells and the dynamics of cyclosis which are either unique to or dominant in chilling-sensitive plants, compared with chilling-insensitive plants.[43] Such changes have been proposed as an early stage of chilling injury, even as the primary event.[38] One such change, observed in cells of watermelon, is the disappearance of the cytoplasmic strands which span the vacuole when tissue is chilled. This occurs within minutes of chilling, and they reappear if the cells are warmed, showing that in the short term the changes are reversible.[43] Cyclosis is also reported to cease[44] or be disproportionately reduced[43] at a temperature close to, or coincident with, the critical temperature for the tissue. Cyclosis has also been reported to be reversible when the tissue was warmed after chilling.[38] Thus, in all respects, these structural and dynamic changes would qualify as primary events.

Structural alterations in the cytoplasmic strands and a reduction in active streaming is thought to be mediated by the organization of actin filaments, and this is sensitive to the

$[Ca^{2+}]_{cyt}$.[45] It has been proposed that the effect of chilling on cytoskeletal structure is a response to a change in the $[Ca^{2+}]_{cyt}$, mediated by changes in the permeability of membranes.[46] This follows from the observation that the changes in the dynamics and structure of the cytoplasmic strands, induced by chilling, are mimicked by increasing $[Ca^{2+}]_{cyt}$ from 10^{-8} M external $[Ca^{2+}]$ to 5×10^{-6} M.[46] If the change in the dynamics and structure of the cytoplasm is dependent on changes in $[Ca^{2+}]_{cyt}$, then it must be considered an early secondary event; the preceeding alteration in the $[Ca^{2+}]_{cyt}$ would be more appropriately considered the primary event. Other structural changes have been noted in cells in response to chilling, but they take some time to develop and are undoubtably manifestations of secondary events. These include the vacuolization of cytoplasm and loss of definition of cellular membranes, noted in tomato mesophyll cells chilled at 5°C for 4 h,[19] and the separation of protein particles of the plasma membrane of avocado mesophyll cells after 7 d at 6°C in the presence of ethylene.[47]

D. TRANSITION IN MEMBRANE LIPIDS

The hypothesis that the primary event causing chilling injury is a transition in the molecular ordering of membrane lipids was first proposed in 1970.[13] It was predicated on two findings. One showed a correlation between the flexibility of mitochondrial membranes, as determined by their capacity to swell, and the chilling sensitivity of the tissue from which they were derived.[48] The other study showed that abrupt changes occurred in the function of mitochondria from chilling-sensitive plants at temperatures coincident with the critical temperature of the tissue.[13] The temperature for the functional changes were detected as "breaks" or "discontinuities" in Arrhenius-type plots of succinate oxidation.[13] Spin-label studies 1 year later indicated a change in the molecular ordering of lipids in mitochondrial membranes at temperatures coincident with the changes in function.[49] Thus, the change in function was considered the consequence of a "phase change in the membrane lipids."[49] No "break" in the Arrhenius plots of enzyme activity or spin-label motion was evident with mitochondria from chilling-insensitive plants above 0°C.[13,49] A phase change in the membrane lipids, postulated in this hypothesis, is consistent with the characteristics suggested for a primary event; it is an instantaneous physical change which is reversible and occurs at temperatures which can be directly related to the critical temperature of the tissue.

Within the limits of the definitions outlined here for a primary event, the only tenable proposal for the molecular event underlying chilling injury is that suggesting a transition in the membrane lipids. If it is shown that chilling has a direct effect on cytoskeletal structure, this proposal would also qualify. The proposal that the structural changes are mediated by a change in $[Ca^{2+}]_{cyt}$ precludes its classification as a primary event at present.

Advocation of a phase change in membrane lipids as the primary event in chilling does not imply that it is the basis of all low-temperature-induced disorders. Although the membrane lipid hypothesis has been proposed as the basic cause of chilling injury for 18 years, it would be an understatement to say it had not been universally accepted. It has, however, attracted much attention and discussion, as outlined in the following section.

1. Criticisms of the Membrane Lipid Hypothesis

The hypothesis relating chilling sensitivity to a phase transition in membrane lipids has been the subject of much criticism. The main criticisms are (1) the use of "breaks" in Arrhenius-type plots of enzyme activity to infer phase transitions in membrane lipids; (2) that it is unlikely a heterogeneous mixture of membrane lipids containing predominantly polyunsaturated fatty acids would undergo an abrupt transition above 0°C; (3) phase transitions have not been demonstrated in native membranes; and (4) even if a small proportion of the membrane lipids phase separate at chilling temperatures, it is unlikely this would disrupt metabolism and account for the injury.

In the early 1970s, the first of these criticisms was valid. However, subsequent studies of the physical properties of membranes and membrane lipids, starting in 1971, have demonstrated abrupt changes in the molecular ordering,[49] phase separation,[50,51] and an exothermic transition[29,52,53] in isolated membrane polar lipids at the same temperature as the change in enzyme function. These later results obviated the necessity to rely on Arrhenius-type plots of function to infer structural changes in the membrane lipids, as proposed in 1970.[13] Even so, reviews[33-36,54-58] published over the past few years continue to point out weaknesses in the hypothesis based on the argument that "breaks" in Arrhenius-type plots are not invariably synonomous with a phase change in the membrane lipids. This point is obvious; the proponents of the hypothesis never contended that "breaks" in such plots of enzyme function were proof of a phase change in the membrane lipids. They noted, for the tissue examined, that the "breaks" occurred only with mitochondria from tissue of chilling-sensitive plants and were indicative of a phase change.[13] Reviewers discussing this aspect of chilling injury have also pointed out that linear Arrhenius plots have been reported for the effect of temperature on the rate of oxidation of pollen grains from tomato[22] and seeds of mung bean and cucumber,[59] all chilling-sensitive plants. Because of the lack of a "break" in the respiration rate of these systems, the authors conclude that "breaks" of respiration in Arrhenius plots are not invariably associated with chilling sensitivity. They are undoubtedly correct. However, the hypothesis relating chilling sensitivity to a phase change in the membrane lipids was not predicated on such results. It was based on results obtained using freshly isolated, intact, coupled mitochondria oxidizing succinate.[13] It was never suggested that the thermal response of the respiration rate of tissues would be similar. The original paper,[13] in fact, pointed out differences between the effect of chilling on the respiration of tissue and that of the mitochondria isolated from nonchilled tissue. The rate of respiration of tissue could be limited by available substrates or accelerated by loss of cellular integrity induced by chilling. In addition, there are some reports of Arrhenius plots obtained with isolated mitochondria which appear to be inconsistent with the differences previously reported for chilling-sensitive and chilling-insensitive tissues.[60,61] In this regard, it should be pointed out that the effect of temperature on the rate of succinate oxidation, even by isolated mitochondria, is influenced by the concentration of mitochondria in the reaction mixture[62] and whether the mitochondria have been "conditioned" to overcome an initial inhibition of the state 3 rate of succinate oxidation.[62,63] Thus, in evaluating the validity of criticisms of the membrane lipid hypothesis, based on "breaks" in Arrhenius-type plots, care should be taken to ensure that the plots being compared were obtained with systems which are similar, if not identical, to that used in the development of the hypothesis.

While the criticisms based on the use of Arrhenius-type plots were pertinent to the early studies, they are of less significance now that exothermic transitions have been shown to occur in the membrane lipids of chilling-sensitive plants at a temperature coincident with that for the critical temperature of the tissue and involving about 6% of the polar lipids.[52]

There are other major criticisms of the membrane lipid hypothesis. One is that there is no direct evidence that a phase transition occurs in native membranes of chilling-sensitive plants at the critical temperature for injury.[27,35,57] Another criticism proposes that even if a transition which involved less than 10% of the lipid occurred, it is unlikely this would adversely affect metabolism and lead to the loss of cellular integrity associated with chilling injury.[57]

The first of these criticisms is justified. There is no unequivocal evidence for a phase transition in the lipid bilayer of thylakoids or mitochondrial membranes. The view that a transition occurs at the critical temperature for chilling injury derives from the correlation between the temperature of the transition detected in isolated membrane polar lipids and the critical temperature for the tissue[64] (see also Section III.D.1. and Figure 3). Although it has been argued that membrane proteins and neutral lipids could alter the temperature of the transition in native membranes,[37,65,66] no data have been presented to show this occurs in chloroplasts or mitochondrial membranes. This argument is based on experiments where a hydrophobic protein is reconstituted with vesicles prepared from pure lipids or binary mixtures of lipids. However, results obtained using polar lipids from thylakoids show that the addition of neutral lipids does not alter the temperature of the transition detected by DSC.[67] Furthermore, transitions detected in the lipids extracted from rat intestine enterocytes were also detected in the native membranes, by DSC, at the same temperature.[68] Thus, in the absence of specific data to the contrary, it must be considered equally probable that the transition observed in liposomes formed from the polar lipids of thylakoids and mitochondrial membranes also occur in the membranes at the same temperature. Indirect evidence supports this view. For example, phase separation, as detected by the fluorescence of chlorophyll a, has been noted for isolated active chloroplasts from corn at 16 to 19°C for decreasing and increasing temperature, respectively, and from tomato at 9 to 13°C.[69] For the polar lipids extracted from the chloroplasts, the transition, detected by fluorescence of *trans*-parinaric acid, was at 16 and 9°C for corn and tomato, respectively.[69] In addition, studies with intact chloroplasts show that the reflection coefficient (a measure of membrane permeability), for chilling-sensitive bean and tomato, shows an abrupt change at about 12°C.[70] In contrast with chloroplasts from pea and spinach, chilling-insensitive plants, the reflection coefficient was a monotonic function of temperature from 0 to 27°C.[70] The change in thermal response of permeability for the bean and tomato chloroplasts at about 12°C is similar to the temperature for the phase change in the membrane lipids of these plants[29,71] and is consistent with the view that the increase in permeability is a result of a phase change in the membrane lipids.[70]

In opposition to this view, Low et al.[27] reported a similarity in the thermograms of thylakoids from a variety of plants of varying sensitivity to chilling. All showed a broad, reversible endotherm, centered at about 20°C and finishing at about 40°C, which the authors state "almost certainly originates from a membrane lipid state transition."[27] This conclusion was based on the reversibility of the endotherm after heating the membranes, and the similarity of the endotherm for the membranes and that of the extracted polar lipids.[27] In this study the thermograms start at about 9°C. Thus, it would be difficult to detect an endotherm for the thylakoids of the high-altitudinal ecotype of *L. hirsutum*, LA 1361, one of the species examined.[27] For this ecotype, the endotherm would be complete by 14°C, the temperature for the transition for chloroplast lipids of this species and the critical temperature for the susceptibility of the leaves to chill-induced photoinhibition (see Figure 3). Furthermore, from the value of the enthalpy of the transition reported by Low et al.[27] (4 to 8 mcal/mg of polar lipid), it can be calculated that 40 to 70% of the membrane lipid is involved in the endotherm they detected between 0 and 40°C. This is a remarkably high value, considering that about 80% of the thylakoid polar lipids contain polyunsaturated fatty acids, and these would be in the liquid-crystalline phase at temperatures above 0°C and could not contribute to the observed transition.

That the broad endothermic transition observed by Low et al.[27] is that of melting lipid was also predicated on the similarity of the thermograms of the membranes and that of the isolated lipids. These thermograms are not similar. The data presented show that the thermogram for the isolated lipids extends from 5 to 18°C, while that for the thylakoids ranges from 10 to 35°C (see Figure 2 in Reference 27). An estimate of the specific enthalpy for the lipid transition, by the authors, was 0.5 mcal/mg lipid.[27] Based on this value and an enthalpy for mixed glycerolipids of 25 kJ/mol,[52] it can be calculated that only 6% of the polar lipid is involved in the transition of the isolated polar lipids, between 5 and 18°C, a value strikingly similar to the 7% reported for thylakoid lipids by Raison and Wright.[52] If only 6% of the polar lipids are involved in the transition observed with the membranes, the question must be asked, what is the derivation of the heat, equivalent to 40 to 70% of the lipid, shown in the thermograms of the thylakoids? The answer must be protein. The

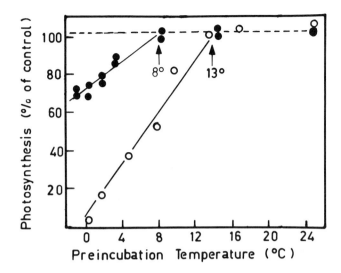

FIGURE 2. The effect of chilling on photosynthetic activity. Leaf slices from cucumber (○—○—○) and oleander grown at 45°C (●—●—●) were preincubated at the temperatures indicated at a light intensity of 450 μmol m^{-2} s^{-1} for 2.5 h. Photosynthetic activity was measured as CO_2-dependent O_2 evolution at 25°C as described by Hodgson et al.,[12] (Redrawn from data of Hodgson, R.A.J., Orr, G. R., and Raison, J. K., *Plant Sci.*, 49, 75, 1987.)

reversibility of the thermograms, after heating the thylakoids at 36°C for 3 min, is the main reason given for assuming the transition detected in the membranes between 10 and 50°C is due to lipid.[27] It is obvious from the thermograms presented[27] that the heat of denaturation of protein does not start until about 45°C and peaks at 70°C.[27] Heating at 36°C for 3 min would not, therefore, irreversibly denature much protein. The transition in the 10 to 40°C range would thus more likely represent the reversible denaturation of some proteins. The similarity of the thermograms for the various plants would therefore reflect the similarity in the thermal properties of thylakoid proteins rather than lipids. Given this possibility, it cannot be concluded that the thermograms demonstrate a lack of correlation between the sensitivity of the plants to chilling and a phase transition in lipids of the chloroplast lamella membrane, as claimed by the authors.[27]

Another objection to the membrane lipid hypothesis is that a phase transition involving only about 6% of the polar lipids would be unlikely to cause a major disruption in cellular metabolism.[57] The evidence, however, is to the contrary. One of the most sensitive responses to chilling is photosynthesis.[11] To determine one of the immediate effects of chilling, measurements were made of CO_2-dependent O_2 evolution at 25°C, using leaf slices which had been preincubated at chilling temperatures in moderate light for 2.5 h.[12] The result of such treatment is shown in Figure 2. Inhibition of photosynthesis was apparent for cucumber, a chilling-sensitive plant with a critical temperature of 12°C,[72] and *Nerium oleander* grown at 45°C (45°-oleander). For both plants the degree of inhibition increased as the increment of chilling increased, i.e., with decreasing temperature (Figure 2). No inhibition was observed for leaf slices from spinach, a chilling-insensitive plant, or from oleander, grown at 20°C.[12] When the data points for the progressive inhibition of cucumber and 45°-oleander were fitted by a straight line, it intersected the line representing no inhibition at 13 and 7°C, for cucumber and 45°-oleander, respectively. The phase transition in the polar lipids from the thylakoids of cucumber and 45°-oleander, determined by both calorimetery and spin labeling, is at 12 and 7°C, respectively.[12] This result shows that under the conditions of moderate light, photosynthesis is inhibited only at chilling temperatures, and the threshold temperature is

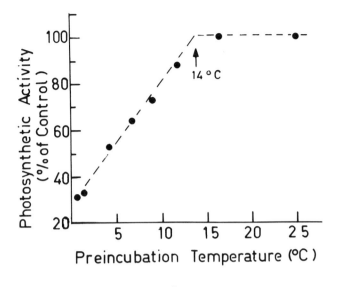

A

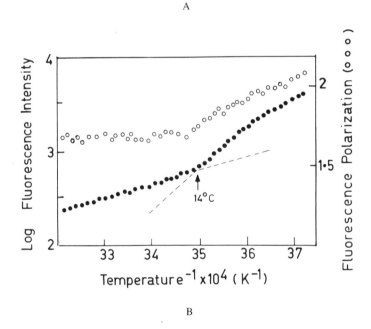

B

FIGURE 3. (A) The effect of chilling on the photosynthetic activity of leaf slices from *Lycopersicon hersutum*. For experimental details, see caption to Figure 2. (B) The fluorescence intensity and polarization of *trans*-parinaric acid intercalated with an aqueous dispersion of the polar lipids from *L. hirsutum*, LA1361. The temperature for the change in the temperature coefficient of intensity and that for a change in the polarization ratio is indicative of the temperature for the phase change in the polar lipids.[91]

coincident with that of the phase transition in the polar lipids from the thylakoids. The inhibition is dependent on both light and chilling.[12] Similar results have been obtained using leaf slices from the high-altitudinal ecotype of the wild tomato *L. hirsutum*, LA 1361, as shown in Figure 3A. For this ecotype, the critical temperature for the inhibition of photosynthesis is at 14°C, and this is coincident with the temperature for the transition detected by fluorescence intensity and fluorescence polarization (Figure 3B).

These data demonstrate two important features. The first is that a major disruption to a physiological process occurs at temperatures coincident with that of the phase transition in the polar lipids from thylakoids, and supports the view that even though the transition might involve only a small proportion of the polar lipid, it has a marked effect on cellular function. The second is that the high-altitudinal ecotype of wild tomato, LA1361, is a chilling-sensitive plant with a critical temperature for the inhibition of photosynthesis at about 14°C (Figure 3A) and with a transition temperature similar to that of the commercial tomato *L. esculentum*[29] (Figure 3B). The ability of the LA1361 ecotype to grow at altitudes of 1500 m is most probably related to the response of the tissue to chilling during the night rather than its sensitivity to chilling, where sensitivity is defined in terms of the temperature of the phase transition. Therefore, the use of this ecotype as an example of a chilling-insensitive species in comparative studies with commercial or low-altitudinal ecotypes would be fallacious. Given the similarity in the transition temperatures for the isolated lipids from the high-altitudinal ecotype and the commercial tomato, the similarity in the exotherms of the thylakoids reported by Low et al.[27] would be expected.

2. Misconceptions Regarding the Membrane Lipid Hypothesis

Early studies of Lyons and Asmundson,[73] using mixtures of fatty acids, showed that a small change in the proportion of saturated fatty acids caused a large shift in the temperature at which the melting transition of the mixture was initiated. It had also been noted that the mitochondria from chilling-sensitive plants contain a higher proportion of saturated fatty acids than those from chilling-insensitive plants.[73] Thus, in 1970, chilling sensitivity was equated with more "rigid" membranes containing a greater proportion of saturated fatty acids.[13] A year later it was obvious that this was an oversimplification, since reports at this time showed examples of mitochondria of chilling-sensitive plants which contained less saturated lipids than those of chilling-insensitive plants.[74] Furthermore, there are no definitive physical studies which show that the membrane lipids of chilling-sensitive plants are more "rigid" than those of chilling-insensitive plants. Thus, it is incorrect to equate an increase in the unsaturation of fatty acids with increased "fluidity" or decreased "rigidity" of a membrane, or a decrease in the transition temperature. It is also mere speculation to suggest that sensitivity to chilling is related to the fluidity of cell membranes.[75] These properties need to be measured before they are evoked to explain or relate the sensitivity of the tissue to chilling with physical properties of membranes.

When exposed to low temperatures, the membrane lipids of most plants show an increase in the proportion of unsaturated lipids,[76] and it is often inferred that this increases "fluidity" and/or lowers the transition temperature as well as increasing cold hardiness.[77] This view appears to be an extrapolation from studies on the adaptation of *Escherichia coli* to growth at different temperatures.[78] With bacteria, the composition and unsaturation of the membrane lipids are altered to precisely compensate for the change in growth temperature such that the fluidity of the membrane lipids, at the respective growth temperatures, remains constant, a process known as homeoviscous adaptation.[78] Similar effects have been noted for some plants. The transition temperature of the desert perennials decreases with decreasing growth temperature,[50,79] while fluidity of the polar lipids increases.[80] For these plants, the compensation in fluidity and transition temperature of the membrane lipids is only partial; e.g., with *N. oleander*, the equivalent of a shift of 10C in both fluidity and transition temperature for a 25C shift in growth temperature.[80] Similar shifts in either the transition temperature or fluidity have not been conclusively demonstrated in chilling-sensitive tropical plants. A chilling-"sensitive" genotype of cucumber has been reported to shift its transition temperature from 12 to 5°C when grown at 23/20°C and 20/10°C (day/night), respectively.[81] However, the same study also showed that three genotypes of cucumber, which were described as more "resistant" to chilling, had transition temperatures 4 to 5C degrees higher than the so-called "sensitive" genotypes.[81]

There are many reports of plants or fruits being "hardened" to chilling by exposure to a low but nonchilling temperature or by lowering water availability (drought hardening) before the chilling treatment.[24] With the low-temperature treatment, the membrane lipids usually show an increase in unsaturation.[76,77] This does not occur with drought hardening.[24] In these studies, the measure of the effectiveness of the "hardening" treatment is usually assessed as the time for the development of a particular symptom, i.e., the rate of some secondary events which lead to the measured symptom. For example, *Phaseolus vulgaris* seedlings, chill hardened at 12°C and 85% relative humidity (RH) for 4 d or drought hardened at 25°C and 40% RH, showed less injury after 9 d chilling or 5°C than plants grown at 25°C and 85% RH.[24] Injury was measured as electrolyte leakage at 5°C.[24] In terms of the definitions used here (see Section II.B.), the experiments described above measure the response of the tissue to chilling, i.e., electrolyte leakage as a consequence of chilling injury. It is, therefore, predicable and understandable that the conditions used for "hardening" could alter the amount of electrolyte leaked in a given time at 5°C by altering the amount of ions available to leak or the water status of the cells. This, however, is an alteration in the rate of the secondary events. There is no evidence that the "hardening" altered the critical temperature for chilling injury, i.e., it did not alter the sensitivity of the plant to chilling. In these experiments, the level of unsaturation of phosphatidylcholine increased during the chill hardening,[77] and it was assumed there was an alteration in the temperature of the phase transition.[24] Drought hardening did not increase the unsaturation to the same extent as the chill hardening, and it was assumed that the plants which were drought hardened would be more sensitive to chilling since the phase transition would not have been lowered to the same extent. The drought-hardened plants showed no visible injury in the time course of the experiment, and it was thus concluded that the phase change is not the primary cause of chilling injury.[24] It should be stressed that in this work the transition was not measured; it was assumed to change because unsaturation of the lipids changed. It is not, therefore, possible to conclude that these results refute the proposal that the primary event in chilling injury is a phase change in the membrane lipids, as claimed by the author.[24]

Similarly, most of the literature dealing with the "amelioration" of chilling injury refers to an increase in the time course of the secondary events. For example, the relative leakage of electrolytes from cucumber seedlings decreased and the water content increased when the seedlings were treated with abscisic acid (ABA) before chilling at 1.5°C for 15 h.[32] Since the critical temperature for cucumber is about 12°C,[16] (see also Figure 2), the experiment measured the effect of the secondary events only. The experiment could not determine if the ABA had a direct effect on the thermal response of the membrane lipids or on the water balance of the tissue which, in turn, affected the rate of ion leakage.

3. Characteristics of the Lipid Phase Transition

There is insufficient data available to provide a precise physical description of the membrane lipid transition. Since it is detected in liposomes formed from membrane polar lipids by both spin labeling and fluorescence polarization, it most likely involves a phase separation of some lipids. Although a change in heat capacity occurs, as evidenced by the exothermic transition detected by calorimetry,[29,52,53] there are no definitive data to indicate that it is a transition from a liquid-crystalline to a gel phase. The increase in thermal energy could be the result of a loss in entropy due to phase separation and immiscibility of some lipids.[29] From the information available, the transition is viewed as a solidification and phase separation involving about 6% of the lipids.[52] For the lipids from thylakoids, it is most probably initiated by solidification of some of the relatively high-melting-point molecular species of phosphatidylglycerol (PG) and/or sulfoquinovosyldiacylglycerol (SQDG).[52,53,82] Phosphatidylglycerol contains both dipalmitoyl and 1-palmitoyl-2-(*trans*)'Δ^3-hexadecenoyl species.[83,84] For three chilling-sensitive plants, the melting point of the PG fraction was

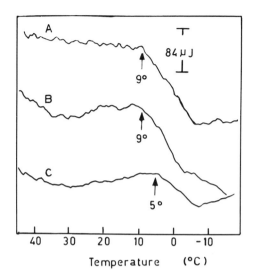

FIGURE 4.   Thermograms showing the temperature of the exothermic transition for cucumber polar lipids (A), cucumber polar lipids reconstituted with the phosphatidylglycerol (PG) from the polar lipids of cucumber (B), and cucumber polar lipids reconstituted with PG from cucumber supplemented (25%) with an unsaturated molecular species of PG from *Passiflora* sp.[90]

found to be between 32 and 29°C, and for three chilling-insensitive plants, between 14 and less than 5°C.[85] For *N. oleander* grown at 45°C, the PG exotherm occurs at 26°C, while for oleander grown at 20°C, it occurs at 16°C.[83] The transition for the polar lipids from the thylakoids of oleander occurs at 7 and -2°C for growth at 45 and 20°C, respectively.[29] It is thus obvious that the transition in the membrane polar lipids is not simply a solidification and phase separation of the PG. The PG must combine with some other low-melting-point lipids, and it is this mixture which presumably phase separates from the bulk membrane lipids. The temperature of the transition for the bulk polar lipids will depend on the relative proportion of the two components in the mixture. Under these conditions, it is not unexpected that the thylakoids of chilling-sensitive plants will, in general, contain a relatively higher proportion of saturated molecular species of PG (i.e., the sum of 16:0, t-16:1, and 18:0) than chilling-insensitive plants.[84]

The composition of the mixture which phase separates at the critical temperature of chilling is not known. Thus, there is no basis for any expectation that there would be a precise relationship between the total of the saturated molecular species of PG and the transition temperature of either the PG or the polar lipids and, hence, the sensitivity of the plant to chilling.[84] However, for each plant, it would be anticipated that the temperature of the transition for the polar lipids would be lowered if the proportion of unsaturated molecular species of PG increased. This has been investigated by removing the PG from the polar lipids obtained from the thylakoids of cucumber, adding 25% of an unsaturated molecular species of PG from *Passiflora* sp., and reconstituting the PG with the polar lipids and determining the transition temperature. The transition temperature of the PG from cucumber is 33°C, while that of the mixed PG is 25°C.[90] As shown in Figure 4, the transition for the bulk polar lipids from cucumber, with the PG from cucumber, was at 9°C, while that containing the PG mixture was at 5°C. Thus, the transition temperature of the polar lipids can be lowered by increasing the unsaturation of the PG.

This is a significant observation as it has a direct bearing on the feasibility of altering the thermal response of plant membrane lipids. The proportion of unsaturated molecular species of PG is controlled by the relative proportion of oleic acid (18:1) esterified to position 1 of glycerol-3-phosphate, compared with that of palmitic acid (16:0)[85] by the enzyme acyl-[acyl-carrier-protein]:glycerol-3-phosphate acyltransferase (AT).[85] Three isomeric forms of this enzyme have been identified and purified from greening squash cotyledons,[84] and one, AT1, shows a kinetic preference for esterifying oleoyl groups over palmitoyl at position 1.[85] Since it has been shown that the transition temperature of the bulk polar lipids can be reduced by increasing the proportion of unsaturated molecular species of PG, genetic manipulation to reduce the sensitivity of the thylakoid membranes to chilling injury would appear possible.

Most of the discussion about phase transitions has been in relation to the membrane lipids of thylakoids and, to a lesser extent, mitochondria. Transitions have also been observed in lipids from other cell membranes, such as the plasma membrane of mung bean seedlings[87] and the membrane lipids of glyoxysomes and proplastids of castor bean.[88] There are insufficient data to comment on the role of these transitions in chilling, except to say that they occur at temperatures below which injury occurs in these plants.

IV. EPILOGUE

It is almost 20 years since the membrane lipid hypothesis was proposed. Much has been written about the hypothesis and it is clear that, like the *Origin Of The Species* by Charles Darwin, it is "widely misunderstood, misquoted and misapplied."[89] It took 80 years for the theory of natural selection to be understood and accepted as the basis of evolution. It is to be hoped that definitive experiments over the next few years will either refute or substantiate the membrane lipid hypthesis so it does not suffer the same gestation period as did the theory of natural selection.

REFERENCES

1. **von Sachs, J.,** *Handbuch der Experimental Physiologie der Pflanzen,* Engelman, Leipzig, 1865.
2. **Molisch, H.,** *Untersuchungen uber das Erfrieren der Pflanzen,* Fischer, Jena, 1897.
3. **Lyons, J. M.,** Chilling injury in plants, *Annu. Rev. Plant Physiol.,* 4, 445, 1973.
4. **Ryall, A. L. and Lipton, N. J.,** *Handling, Transportation and Storage of Fruits and Vegetables,* Vol. 1, 2nd ed., AVI Publishing, Westport, CT, 1979.
5. **Couey, H. M.,** Chilling injury in crops of tropical and subtropical origin, *HortScience,* 17, 162, 1982.
6. **Wright, M. and Simon, E. W.,** Chilling injury in cucumber leaves, *J. Exp. Bot.,* 24, 400, 1973.
7. **Lutz, J. and Hardenburg, R. E.,** The Commercial Storage of Fruits and Vegetables and Florist and Nursery Stocks, Handb. No. 66, Department of Agriculture, Washington, D.C., 1968.
8. **McGlasson, W. B., Scott, K. J., and Mendoza, D. B.,** The refrigerated storage of tropical and subtropical products, *Int. J. Refrig.,* 2, 199, 1979.
9. **Bagnall, D. J. and Wolfe, J. A.,** Chilling sensitivity in plants. Do activation energies of growth processes show an abrupt change at a critical temperature?, *J. Exp. Bot.,* 29, 1231, 1978.
10. **Watada, A. E. and Morris, L. L.,** Effect of chilling and non-chilling temperatures on snap bean fruits, *Proc. Am. Soc. Hortic. Sci.,* 89, 368, 1966.
11. **McWilliam, J. R. and Naylor, A. W.,** Temperature and plant adaptation. I. Interaction of temperature and light in the synthesis of chlorophyll in corn, *Plant Physiol.,* 42, 1711, 1967.
12. **Hodgson, R. A. J., Orr, G. R., and Raison, J. K.,** Inhibition of photosynthesis by chilling in the light, *Plant Sci.,* 49, 75, 1987.
13. **Lyons, J. M. and Raison, J. K.,** Oxidative activity of mitochondria isolated from plant tissues sensitive and resistant to chilling injury, *Plant Physiol.,* 45, 386, 1970.
14. **Towers, N. R., Kellerman, G. M., Raison, J. K., and Linane, A. W.,** The biogenesis of mitochondria 29. Effects of temperature-induced phase changes in membranes on protein synthesis by mitochondria, *Biochim. Biophys. Acta,* 299, 153, 1973.

15. **Paull, R. E., Patterson, B. D., and Graham, D.,** Amino acid uptake by tomato leaf tissue under chilling stress: uptake conditions and a comparison of *Lycopersicon* species of different chilling resistance, *Aust. J. Plant Physiol.*, 6, 475, 1979.
16. **Wang, C. Y. and Adams, D. O.,** Chilling-induced ethylene production in cucumbers (*Cucumis sativus* L.), *Plant Physiol.*, 69, 424, 1982.
17. **Taylor, A. D. and Craig, A. S.,** Plants under climatic stress. II. Low temperature high light effects on chloroplast ultrastructure, *Plant Physiol.*, 47, 719, 1971.
18. **Wise, R. R., McWilliam, J. R., and Naylor, A. W.,** A comparative study of low-temperature-induced ultrastructural alterations of three species differing in chilling sensitivities, *Plant Cell Environ.*, 6, 525, 1983.
19. **Ilker, R., Breidenbach, R. W., and Lyons, J. M.,** Sequence of ultrastructural changes in tomato cotyledons during short periods of chilling, in *Low Temperature Stress in Crop Plants: The Role of the Membrane,* Lyons, J. M., Graham, D., and Raison, J. K., Eds., Academic Press, New York, 1979, 97.
20. **Chan, H. T., Sanxter, S., and Couey, H. M.,** Electrolyte leakage and ethylene production induced by chilling injury of papayas, *HortScience,* 20, 1070, 1985.
21. **Wheaton, T. A.,** Physiological Comparisons of Plants Sensitive and Insensitive to Chilling Temperatures, Ph.D. thesis, University of California, Davis, 1963.
22. **Kosiyachinda, S. and Young, R. E.,** Chilling sensitivity of avocado fruit at different stages of respiratory climacteric, *J. Am. Soc. Hortic. Sci.,* 101, 665, 1976.
23. **Patterson, B. D., Graham, D., and Paull, R.,** Adaptation to chilling: survival, germination, respiration and protoplasmic dynamics, in *Low Temperature Stress in Crop Plants: The Role of the Membrane,* Lyons, J. M., Graham, D., and Raison, J. K., Eds., Academic Press, New York, 1979, 25.
24. **Wilson, J. M.,** The mechanism of chill- and drought-hardening of *Phaseolus vulgaris* leaves, *New Phytol.,* 9, 685, 1986.
25. **Raison, J. K. and Lyons, J. M.,** Chilling injury: a plea for uniform terminology, *Plant Cell Environ.,* 9, 685, 1986.
26. **Dalziel, A. W. and Breidenbach, R. W.,** Physical properties of mitochondrial lipids from *Lycopersicon hirsutum, Plant Physiol.,* 70, 376, 1983.
27. **Low, P. S., Ort, D. R., Cramer, W. A., Whitmarsh, J., and Martin, B.,** Search for an endotherm in chloroplast lamella membranes associated with chilling-inhibition of photosynthesis, *Arch. Biochim. Biophys.,* 231, 336, 1984.
28. **Patterson, B. D., Paull, R., and Smillie, R. M.,** Chilling resistance in *Lycopersicon hirsutum* Humb. & Bonpl., a wild tomato with a wide altitudinal distribution, *Aust. J. Plant Physiol.,* 5, 609, 1978.
29. **Raison, J. K. and Orr, G. R.,** Phase transitions in thylakoid polar lipids of chilling-sensitive plants. A comparison of detection methods, *Plant Physiol.,* 80, 638, 1986.
30. **Smillie, R. M.,** The useful chloroplast: a new approach for investigating chilling stress in plants, in *Low Temperature Stress in Crop Plants: The Role of the Membrane,* Lyons, J. M., Graham, D., and Raison, J. K., Eds., Academic Press, New York, 1979, 187.
31. **Wilson, J. M.,** Drought resistance as related to low temperature stress, in *Low Temperature Stress in Crop Plants: The Role of the Membrane,* Lyons, J. M., Graham, D., and Raison, J. K., Eds., Academic Press, New York, 1979, 47.
32. **Rikin, A. and Richmond, A. E.,** Amelioration of chilling injuries in cucumber seedlings by abscisic acid, *Physiol. Plant.,* 38, 95, 1976.
33. **Markhart, A. H.,** Chilling injury: a review of possible causes, *Hortic. Sci.,* 21, 1329, 1986.
34. **Patterson, B. D. and Graham, D.,** Temperature and metabolism, in *The Biochmistry of Plants. A Comprehensive Treatise,* Vol. 12, Davies, D. D., Ed., Academic Press, New York, 1987, 153.
35. **Couey, H. M.,** Chilling injury of crops of tropical and subtropical origin, *HortScience,* 17, 162, 1982.
36. **Minorsky, P. V.,** An heuristic hypothesis of chilling injury in plants: a role for calcium as the primary physiological transducer of injury, *Plant Cell Environ.,* 8, 75, 1985.
37. **Graham, D., Hockley, D. G., and Patterson, B. D.,** Temperature effects on phosphoenolpyruvate carboxylase from chilling-sensitive and chilling-resistant plants, in *Low Temperature Stress in Crop Plants: The Role of the Membrane,* Lyons, J. M., Graham, D., and Raison, J. K., Eds., Academic Press, New York, 1979, 453.
38. **Woods, C. M., Reid, M. S., and Patterson, B. D.,** Response to chilling stress in plants cells. I. Changes in cyclosis and cytoplasmic structure, *Protoplasma,* 121, 8, 1984.
39. **Linderstrom-Lang, K. U. and Schellman, J. A.,** Protein structure and enzyme activity, in *The Enzymes,* Vol. 1, Boyer, P. D., Lardy, H., and Myrback, K., Eds., Academic Press, New York, 1959, 443.
40. **Shirahashi, K., Hayakawa, S., and Sugiyama, T.,** Cold lability of pyruvate, orthophosphate dikinase in the maize leaf, *Plant Physiol.,* 62, 826, 1978.
41. **Sugiyama, T., Schmitt, M. R., Ku, S. B., and Edwards, G. E.,** Differences in cold lability of pyruvate, Pi dikinase among C_4 species, *Plant Cell Physiol.,* 20, 965, 1979.

42. **Sugiyama, T. and Boku, K.,** Differing sensitivity of pyruvate orthophosphate dikinase to low temperature in maize cultivars, *Plant Cell Physiol.,* 17, 851, 1976.

43. **Patterson, B. D. and Graham, D.,** Effect of chilling temperatures on the protoplasmic streaming of plants from different climates, *J. Exp. Bot.,* 28, 736, 1977.

44. **Lewis, D. A.,** Protoplasmic streaming in plants sensitive and insensitive to chilling temperatures, *Science,* 124, 75, 1956.

45. **Taylor, D. L., Condeelis, J. S., Moore, P. L., and Allen, R. D.,** The contractile basis of amoeboid movement. I. The chemical control of mobility in isolated cytoplasm, *J. Cell Biol.,* 59, 378, 1973.

46. **Woods, C. M., Polito, V. S. and Reid, M. S.,** Response to chilling stress in plant cells. II. Redistribution of intracellular calcium, *Protoplasma,* 121, 17, 1984.

47. **Platt-Aloia, K. A. and Thomson, W. W.,** Freeze fracture evidence for lateral separations in the plasmalemma of chilling-injured avocado fruit, *Protoplasma,* 136, 71, 1987.

48. **Lyons, J. M., Wheaton, T. A., and Pratt, H. K.,** Relationship between the physical nature of mitochondrial membranes and chilling sensitivity in plants, *Plant Physiol.,* 39, 262, 1964.

49. **Raison, J. K., Lyons, J. M., Mehlhorn, R. J., and Keith, A. D.,** Temperature-induced phase changes in mitochondrial membranes detected by spin labeling, *J. Biol. Chem.,* 246, 4036, 1971.

50. **Pike, C. S. and Berry, J. A.,** Membrane phospholipid phase separations in plants adapted to or acclimated to different thermal regions, *Plant Physiol.,* 66, 238, 1980.

51. **Raison, J. K., Pike, C. S., and Berry, J. A.,** Growth temperature-induced alterations in the thermotropic properties of *Nerium oleander* membrane lipids, *Plant Physiol.,* 70, 215, 1982.

52. **Raison, J. K. and Wright, L. C.,** Thermal phase transitions in the polar lipids of plant membranes. Their induction by disaturated phospholipids and their possible relation to chilling injury, *Biochim. Biophys. Acta,* 731, 69, 1983.

53. **Raison, J. K. and Orr, G. R.,** Phase transitions in liposomes formed from the polar lipids of mitochondria from chilling-sensitive plants, *Plant Physiol.,* 81, 807, 1986.

54. **Graham, D. and Patterson, B. D.,** Response of plants to low, non-freezing temperatures: proteins, metabolism and acclimation, *Annu. Rev. Plant Physiol.,* 33, 347, 1982.

55. **Steponkus, P. L.,** Responses to extreme temperatures. Cellular and sub-cellular basis, in *Encyclopedia of Plant Physiology,* Vol. 12A, Lange, O. L., Nobel, P. S., Osmond, C. B., and Ziegler, H., Eds., Springer-Verlag, Berlin, 1981, 371.

56. **Quinn, P. J.,** The fluidity of cell membranes and its regulation, *Prog. Biophys. Mol. Biol.,* 33, 1, 1981.

57. **Martin, B.,** Arrhenius plots and the involvement of thermotropic phase transitions of the thylakoid membranes in chilling impairment of photosynthesis in thermophylic higher plants, *Plant Cell Environ.,* 9, 323, 1986.

58. **Oquist, G. and Martin, B.,** Cold climates, in *Photosynthesis in Contrasting Environments,* Baker, N. R. and Long, S. P., Eds., Elsevier, Amsterdam, 1986, 237.

59. **Simon, E. W., Minchin, A., McMenamin, M. M., and Smith, J. M.,** Low temperature limit of seed germination, *New Phytol.,* 77, 301, 1976.

60. **Pomeroy, M. K. and Andrews, C. J.,** Effect of temperature on respiration of mitochondrial and shoot segments from cold hardened and non hardened wheat and rye seedlings, *Plant Physiol.,* 56, 703, 1974.

61. **Bligny, R., Rebeille, F., and Douce, R.,** O_2-triggered changes of membrane fatty acid composition have no effect on Arrhenius discontinuities of respiration in Sycamore (*Acer pseudopatanus* L.) cells, *J. Biol. Chem.,* 260, 9166, 1985.

62. **Raison, J. K., Lyons, J. M., and Campbell, L. C.** Inhiition of state 3 respiration of isolated mitochondria and its implication in comparative studies, *Bioenergetics,* 4, 397, 1973.

63. **Raison, J. K., Laties, G. G., and Crompton, M.,** The role of state 4 electron transport in the activation of state 3 respiration in potato mitochondria, *Bioenergetics,* 4, 409, 1973.

64. **Raison, J. K., Chapman, E. A., Wright, L. C., and Jacobs, W. L.,** Membrane lipid transitions: their correlation with the climatic distribution of plants, in *Low Temperature Stress in Crop Plants: The Role of the Membrane,* Lyons, J. M., Graham, D., and Raison, J. K., Eds., Academic Press, New York, 1979, 177.

65. **McElhaney, R. N.,** Differential scanning calorimetric studies of lipid-protein interactions in model membrane systems, *Biochim. Biophys. Acta,* 864, 361, 1986.

66. **Yamamoto, H. Y. and Bangham, A. D.,** Carotenoid organization in membranes. Thermal transitions and spectral properties of carotenoid-containing liposomes, *Biochim. Biophys. Acta,* 507, 119, 1978.

67. **Brown, M. A. and Raison, J. K.,** The influence of carotenoids on the oxidative stability and phase behavour of plant polar lipids, *Phytochemistry,* 26, 961, 1987.

68. **Brasitus, T. A., Tall, A. R., and Schachter, D.,** Thermotropic transitions in rat intestinal plasma membranes studied by differential scanning calorimetry and fluorescence polarization, *Biochemistry,* 19, 1256, 1980.

69. **Fork, D. C. van Ginkel, G., and Harvey, G.,** Phase transition temperatures determined for chloroplast thylakoid membranes and for liposomes prepared from charged lipids extracted from thylakoid membranes of higher plants, *Plant Cell Physiol.,* 22, 1035, 1981.

70. **Nobel, P.,** Temperature dependence of the permeability of chloroplasts from chilling-sensitive and chilling-resistant plants, *Planta,* 115, 369, 1974.

71. **Shneyour, A., Raison, J. K., and Smillie, R. M.,** Effect of temperature on the rate of photosynthetic electron transfer in chloroplasts of chilling-sensitive and chilling-resistant plants, *Biochim. Biophys. Acta,* 292, 152, 1973.

72. **Whitaker, B. D. and Wang, C. Y.,** Effect of paclobutrazol and chilling on leaf membrane lipids in cucumber seedlings, *Physiol. Plant.,* 70, 404, 1987.

73. **Lyons, J. M. and Asmundson, C. M.,** Solidification of saturated/unsaturated fatty acid mixtures and its relationship to chilling sensitivity in plants, *J. Am. Oil Chem. Soc.,* 42, 1056, 1965.

74. **Yamaki, S. and Uritani, I.,** Mechanism of chilling injury in sweet potatoes. V. Biochemical mechanism of chilling injury with special reference to mitochondrial lipid components, *Agric. Biol. Chem.,* 36, 47, 1972.

75. **Wolfe, J. A.,** Chilling injury in plants — the role of membrane lipid fluidity, *Plant Cell Environ.,* 1, 241, 1978.

76. **Harwood, J. L.,** The synthesis of acyl lipids in plants and microbes, *Prog. Lipid Res.,* 18, 55, 1979.

77. **Wilson, J. M. and Crawford, R. M. M.,** Leaf fatty-acid content in relation to hardening and chilling injury, *J. Exp. Bot.,* 25, 121, 1974.

78. **Sinensky, M.,** Homeoviscous adaptation — a homeostatic process that regulates the viscosity of membrane lipids in *Escherichia coli, Proc. Natl. Acad. Sci. U.S.A.,* 71, 522, 1974.

79. **Raison, J. K., Pike, C. A., and Berry, J. A.,** Growth temperature-induced alterations in the thermotropic properties of *Nerium oleander* membrane lipids, *Plant Physiol.,* 70, 215, 1982.

80. **Raison, J. K., Roberts, J. K. M., and Berry, J. A.,** Correlations between the thermal stability of chloroplast (thylakoid) membranes and the composition and fluidity of their polar lipids upon acclimation of the higher plant *Nerium oleander* to growth temperature, *Biochim. Biophys. Acta,* 688, 218, 1982.

81. **Horvath, I., Vigh, I., Woltjes, J., Van Hasselt, P. R., and Kuiper, P. J. C.,** The physical state of lipids of the leaves of cucumber genotypes as affected by temperature, in *Biochemistry and Metabolism of Plant Lipids,* Wintermans, J. F. G. M. and Kuiper, P. J. C., Eds., Elsevier, Amsterdam, 1982, 427.

82. **Murata, N. and Yamaya, J.,** Temperature-dependent phase behavior of phosphatidylglycerols from chilling-sensitive and chilling-resistant plants, *Plant Physiol.,* 74, 1016, 1984.

83. **Orr, G. R. and Raison, J. K.,** Compositional and thermal properties of thylakoid polar lipids of *Nerium oleander* L. in relation to chilling sensitivity, *Plant Physiol.,* 84, 88, 1987.

84. **Murata, N., Sato, N., Takahashi, N., and Hamazaki, Y.,** Compositions and positional distributions of fatty acids in phospholipids from leaves of chilling-sensitive and chilling-resistant plants, *Plant Cell Physiol.,* 23, 1071, 1982.

85. **Frentzen, M., Nishida, I., and Murata, N.,** Properties of the plastidial acyl-[acyl-carrier-protein]:glycerol-3-phosphate acyltransferase from the chilling-sensitive plant squash *(Cucurbita moschata), Plant Cell Physiol.,* 28, 1195, 1987.

86. **Nishida, I., Frentzen, M., Ishzaki, O. and Murata, N.,** Purification of isomeric forms of acyl-[acyl-carrier-protein]: glycerol-3-phosphate acyltransferase from greening squash cotyledons, *Plant Cell Physiol.,* 28, 1071, 1987.

87. **Yoshida, S., Kawata, T., Uemura, M., and Niki, T.,** Properties of plasma membrane isolated from chilling-sensitive etiolated seedlings of *Vigna radiata* L., *Plant Physiol.,* 80, 152, 1986.

88. **Wade, N. L., Breidenbach, R. W., and Lyons, J. M.,** Temperature-induced phase changes in the membranes of glyoxysomes, mitochondria and proplastids from germinating castor bean endosperm, *Plant Physiol.,* 54, 320, 1974.

89. **Gould, S. J.,** *Ever Since Darwin, Reflections in Natural Historye,* W. W. Norton, New York, 1977.

90. **Orr, G. and Raison, J.,** Unpublished data.

91. **Raison, J. K.,** Unpublished data.

Chapter 10

PLANT ENZYMES IN RELATION TO CHILLING SENSITIVITY

Charles R. Caldwell

TABLE OF CONTENTS

I. INTRODUCTION

The ability of plants to acclimate or adapt to either low or high temperatures has been attributed to a number of processes, including the synthesis of temperature-shock proteins,[1] modification of membrane lipid composition,[2,3] and alterations in the structure or function of enzymes involved in key metabolic reactions.[4,5] In contrast to the study of thermal acclimation in poikilothermic animals,[6,7] the role of temperature-induced changes in membrane composition and structure has dominated the study of the low-temperature responses of plants.[8] Changes in membrane structure may be required for plant acclimation to temperature extremes. However, compensation for temperature-induced variations in the activities of soluble enzymes involved in key metabolic processes should also be necessary for acclimation.

Plants and poikilothermic animals have evolved metabolic processes which function optimally under the thermal conditions found in their habitats. This genetic adaptation to temperature results in enzymatic reactions which can compensate for the normal range in temperatures found in a particular environment. Plants adapted for growth under relatively stable temperature regimes, such as tropical or alpine regions, may have evolved processes which are unable to compensate for unseasonable temperature variations. Conversely, plants from more temperate regions must often respond to wide diurnal and seasonal variations in temperature. Therefore, one must differentiate between plants which have been genetically adapted to a particular temperature range and plants which also have the ability to acclimate in response to environmental changes. Thermal acclimation represents those temperature-induced changes in key biophysical or biochemical processes which permit plants to function under new temperature conditions. In this case, thermal acclimation is equivalent to the expression "photosynthetic acclimation", presented by Berry and Björkman,[4] "to denote environmentally induced changes in photosynthetic characteristics that result in an improved performance under the new growth regime." Furthermore, in the case of low-temperature stress, one must distinguish between processes related to cold acclimation or freeze/frost hardening and those involved in chilling acclimation. This is often difficult, since both chilling and cold acclimation can be induced by the same temperature conditions. For example, winter wheat can cold acclimate in response to growth at chilling temperatures and, as expected, is considered to be chilling resistant. However, chilling-resistant spring wheat is unable to cold acclimate.[9]

Plants and poikilothermic animals which evolved in temperate regions generally survive chilling temperatures (0 to 12°C).[7] As defined by Raison and co-workers,[10,11] these organisms would be considered to be chilling insensitive (CI), if they can complete their life cycle below 12°C, or chilling resistant (CR), if chilling results in some modification which reduces or prevents the adverse effects of chilling. Since CR plants must respond to chilling, they may also be defined as being chilling sensitive (CS).[10] However, it is often not possible to determine from the literature whether a particular process either has a predisposition to chilling insensitivity (i.e., genotypic adaptation) or is modified in response to chilling temperatures (i.e., phenotypic adaptation or acclimation). Therefore, in this review, CR has been used to describe plants which are not injured by chilling temperatures; CS plants are irreversibly damaged by low temperatures. These definitions depend in part upon the magnitude and rapidity of the chilling stress, as well as the biochemical or physiological process used to assess chilling damage. For example, organisms which can acclimate to seasonal variations in temperature may be damaged if the speed of the temperature change is too rapid to permit thermal compensation. Therefore, one must consider diurnal and seasonal changes in temperature in any evaluation of potential mechanisms for biochemical acclimation to temperature.

It is the purpose of this review to examine possible molecular mechanisms for plant acclimation to chilling temperatures (0 to 12°C), with particular emphasis on those processes which could permit rapid responses to temperature variations. Since considerations of recent research related to chilling injury and resistance in plants can be found elsewhere in this volume, this review is primarily concerned with the conservation of enzyme structure and function during plant acclimation to chilling. The effects of low temperatures upon photosynthesis, respiration, metabolism, and development have been extensively reviewed.[4,8,12-16] Since detailed evaluations of the potential mechanisms for temperature adaptation of enzymes from poikilothermic animals have been presented by Somero and co-workers,[6,7] many of the concepts presented below follow their ideas.

II. EFFECT OF CHILLING ON ENZYME STRUCTURE AND FUNCTION

A. TEMPERATURE SENSITIVITY OF ENZYME PROPERTIES
1. Chemical Bonding
Most of the structural and functional properties of proteins are temperature sensitive. This thermal sensitivity can often be attributed to the type of "weak" chemical bonds involved in such processes as the maintenance of the optimal protein secondary and tertiary structure, the interactions between an enzyme and its substrate or cofactors, protein-lipid and lipid-lipid interactions, the formation of protein oligomeric structures, and the structure of water around the protein.[6,7] Since the enthalpies of weak bond formation are low, minor changes in cellular temperature may be sufficient to disrupt protein structure and function. Furthermore, the sign of the enthalpy change for bond formation differs among the classes of weak bonds.[7] Since the formation of electrostatic and hydrogen bonds is exothermic, these chemical bonds would be stabilized by low temperatures. Conversely, the endothermic hydrophobic interactions of proteins are weakened by low temperatures. Both the stabilization and destabilization of weak bonds by low temperatures could have adverse effects on protein structure and function. Alexandrov[17] has proposed that the conservation of a biologically optimal flexibility may be characteristic of both lipid and protein adaptation to temperature. Therefore, both increased protein structural flexibility produced by disruption of hydrophobic interactions and increased protein structural rigidity resulting from strengthened electrostatic and hydrogen bonding could account for alterations in protein conformation and function at low temperatures.

Levitt[15] has emphasized the importance of protein disulfide bonds in plant responses to both high temperatures and freezing stress. The disruption of protein disulfide bonds and subsequent changes in protein conformation could result in the exposure of sulfhydryl groups. These sulfhydryl groups may facilitate protein aggregation or permit a protein to refold in an inactive configuration. Although there is little information available concerning the role of disulfide bonds in plant responses to chilling, the reversible cold inactivation of tobacco (*Nicotiana tabacum*) RuBP carboxylase apparently alters protein conformation, exposing sulfhydryl groups.[18]

2. Low-Temperature Lability of Plant Enzymes
Many enzymes are oligomers composed of interacting subunits. These enzymes are often allosteric, yielding sigmoidal velocity curves. The regulation of enzyme activity by allosteric modifiers probably results from conformational changes in protein structure mediated through subunit interactions. The subunit association of oligomeric enzymes is controlled by the weak chemical bonds described above. Allosteric or multisubunit enzymes are highly regulated, since they often catalyze key reactions involved in metabolic control. A number of oligomeric enzymes from both animals and plants are reversibly inhibited by cold.[19] The cold inhibition of these enzymes has often been attributed to the disruption of hydrophobic

bonds required for subunit interaction.[20] For example, maize (*Zea mays*) leaf pyruvate, P_i dikinase was reversibly inactivated by incubation at temperatures below 10°C.[21] The cold inactivation was accompanied by the apparent dissociation of the tetrameric enzyme into subunits. Since the dikinase may be the rate-limiting step in photosynthesis by C_4 plants at low temperatures,[22] the cold response of this enzyme may be correlated to the chilling sensitivity of maize cultivars. As mentioned above, tobacco RuBP carboxylase was also reversibly cold inhibited.[18] Although this octameric enzyme apparently does not completely dissociate at low temperatures, there is some evidence for the partial disruption of hydrophobic interactions between the subunits.

3. Enzyme Kinetics
a. Reaction Rates

In general, chemical reactions proceed at different rates as the temperature is varied. An increase in temperature imparts more kinetic energy to the reactant molecules, resulting in more productive collisions per unit time. Obviously, lowering the temperature would have the inverse effect. Enzyme-catalyzed reactions often behave similarily. A doubling of the enzyme reaction velocity with each increase in temperature of 10°C ($Q_{10} = 2$) is common at saturating substrate concentrations. However, enzyme proteins require a highly ordered tertiary structure which positions specific amino acid groups to form the stereospecific substrate and/or cofactor binding sites and the catalytic center. Furthermore, specific changes in protein tertiary structure may be required during the enzymatic reation. Therefore, any chilling-induced disruption of the weak chemical bonds involved in the maintenance of enzyme tertiary (conformation) and quaternary structure (subunit association), or in the actual binding of a substrate at the enzyme catalytic site, could markedly influence enzyme-substrate interactions, resulting in an increase in K_m. As a result of such increases in K_m, coupled with the reduced reaction rates induced by the lower kinetic energy, chilling can produce large changes in enzyme Q_{10} at physiological substrate concentrations.[6] Relative to CR wheat (*Triticum aestivum*), low temperatures induced a significant increase in Q_{10} for CS tomato (*Lycopersicon esculentum*) phosphoenolpyruvate (PEP)-carboxylase at nonsaturating substrate levels.[23] Somero[6] has suggested that such Q_{10} effects may establish the lower thermal limits of enzyme function.

b. Arrhenius Plot Nonlinearities

Many investigations of plant responses to chilling have involved the measurement of temperature-dependent changes in enzyme reaction velocities at a single saturating substrate concentration. Arrhenius plots of the activities of membrane-bound enzymes isolated from CS plants are often nonlinear with "breaks" or abrupt changes in slope at temperatures around 12°C.[8] This critical temperature usually coincides with temperature-induced alterations in the structure of the membrane lipids measured by various biophysical methods.[2,24] Since growth temperatures below 12°C are necessary to produce chilling injury in CS temperate and subtropical plants, it was proposed that nonlinear Arrhenius plots of essential enzyme activities with slope discontinuities at about 12°C are characteristic of CS plant species and result from temperature-induced changes in membrane bulk lipid structure.[3] With the exception of those soluble enzymes which have associated lipids, nonmembranous enzymes, as well as enzymes from CR plants, should have linear Arrhenius plots at temperatures below about 30°C.

There is little doubt that the large increases in activation energy for many enzymes and biochemical processes of CS plants at temperatures below about 12°C would have significant effects on metabolism.[25] However, the concept that membrane lipid composition and structure is the primary determinant of plant chilling sensitivity has been the subject of considerable debate.[10,26,27] Nonlinear Arrhenius plots for photosynthetic activity, respiration, and membrane-bound enzyme activities have been obtained with CR plants.[12,25, 28-30] These results have been interpreted as indicating either little correlation between the structure of membrane lipids[16] and chilling sensitivity, or temperature effects upon intrinsic, lipid-independent properties of the enzyme systems. However, in several cases, the experimental material was obtained from plants which had not been chilled.[28-30] If temperate-zone CR plants have the ability to acclimate to different temperatures, then there is no reason to expect differences between CS and CR plants in the absence of inductive low temperatures.

The biophysical and biochemical significance of nonlinear Arrhenius plots of membrane-bound enzymes is by no means clear. Silvius et al.[31] have noted that temperature-dependent changes in K_m can produce artifactual breaks in Arrhenius plots prepared from enzyme velocities obtained using single fixed substrate concentrations. Silvius and McElhaney[32] have presented a detailed theoretical analysis of processes which could result in nonlinear Arrhenius plots of membrane-associated reactions. They reached three major conclusions. First, Arrhenius plots with sharp "breaks" can occur in systems which do not undergo lipid phase transitions. Therefore, in the absence of supporting physical evidence, Arrhenius plot discontinuities should not be interpreted as indicating cooperative changes in lipid structure. Second, thermotropic processes which take place over a wide temperature range can produce sharp "breaks" in Arrhenius plots. This may account for the apparent discrepancies between the broad lipid phase changes measured by calorimetry and the abrupt changes in the temperature-dependent motion of membrane-associated probes presented as Arrhenius plots. Third, the temperature of an Arrhenius plot "break" represents the midpoint of the temperature-dependent process. Therefore, the width of the thermotropic process cannot be determined from an Arrhenius plot.

Caldwell and co-workers[28,33,34] have analyzed the temperature dependences of spin probe motion, intrinsic fluorescence, and adenosine triphosphatase (ATPase) activity of plasma membranes isolated from the roots of nonacclimated barley (*Hordeum vulgare*). Although barley is considered to be CR, an Arrhenius plot of the high-affinity ATPase maximal velocity was nonlinear with a "break" at about 22°C. However, EPR and fluorescence measurements of both lipid and membrane protein molecular motion in the intact membrane indicated no major change in membrane structure at 22°C. The ATPase K_m was relatively constant over the temperature range of 12 to 32°C, with significant increases in K_m at temperatures below 12°C and above 32°C. The change in the temperature dependence of the ATPase K_m at about 12°C coincided with altered spin probe motion and rate of spin probe lateral diffusion. Considering the concepts of Silvius and McElhaney detailed above and the oligomeric nature of ATPases, it is possible that the barley ATPase Arrhenius plot "break" at 22°C results from lipid-mediated, protein-protein interactions which occur at temperatures above 12°C and are disrupted at temperatures above about 32°C. At temperatures below 12°C, the restricted motion of the membrane lipids could reduce tha probability of any protein-protein association.

Using plasma membrane isolated from CS mung bean (*Vigna radiata*), Yoshida et al.[35] obtained nonlinear Arrhenius plots for the membrane-bound ATPase activity with "breaks" at 19 and 7°C. The "break" at 7°C may result from changes in the structure of the membrane lipids. The results presented above for CR barley ATPase[28] and the concepts of Silvius and co-workers[31,32] suggest that the change in mung bean ATPase activation energy at 7°C results from a modification in enzyme K_m. Therefore, the temperature dependences of CS mung bean and nonacclimated CR barley plasma membrane-bound ATPase activity may be similar and indicate lipid-mediated enzyme dysfunction at chilling temperatures.

c. *Enzyme-Substrate Affinities*

Considering the potential importance of enzyme kinetic properties in plant responses to chilling, it is necessary to consider the temperature dependences of both enzyme K_m and

V_{max}. Based upon studies of marine organisms, Somero[6] has proposed three major attributes of temperature-dependent alterations in enzyme K_m values. First, the K_m values of all metabolically important enzymes should not vary widely over the range of temperatures experienced by the cell. Second, when the K_m does vary with temperature, the trend is for enhanced enzyme-substrate binding as the temperature is lowered. Third, when rapid changes in K_m occur over a narrow temperature range, these are almost invariably at temperatures near the lethal limits of the species. "U-shaped" plots of enzyme K_m as a function of temperature have been obtained for enzymes from poikilothermic marine animals adapted for growth at specific temperatures.[6,7]

In contrast to studies of poikilothermic animals, there have been few detailed examinations of the thermal responses of plant enzyme-substrate interactions.[12] The K_m values of sweet potato (*Ipomoea batatas*) root phenylalanine ammonia lyase[36] and *Bryophyllum fedtschenkoi* PEP-carboxylase[37] increased at temperatures above or below the physiological temperature range. Graham et al.[23] have demonstrated that the K_m for PEP of PEP-carboxylase from tropical plants is more sensitive to low temperatures than that of plants from temperate or alpine environments. Using isolated plasma membrane preparations from the roots of barley seedlings grown at 20°C, Caldwell and Haug[28] demonstrated large increases in the K_m of the divalent cation-dependent ATPase at temperatures below about 12°C and above 32°C. Over the temperature range of 12 to 32°C, the K_m for the high-affinity ATPase was relatively constant. Considering the general characteristics of the thermal responses of enzyme K_m in relation to growth temperature detailed above, the increase in ATPase K_m at temperatures above 33°C and below 12°C suggests that these temperatures may represent the upper and lower limits for optimal root function in nonacclimated barley.

B.   CHILLING-INDUCED METABOLIC DYSFUNCTION

Considering the temperature sensitivities of enzyme properties described above, it is not surprising that most plant metabolic processes are influenced by low temperatures.[12,38-40] Although generally thought to be secondary effects of low temperature,[8,10] there is considerable evidence that chilling disrupts the biochemical processes of CS plants.[12,16,38] Unlike CR plants, chilling in the light significantly inhibited photosynthesis of CS plants.[16,41,42] Low-temperature photoinhibition is oxygen dependent and ultimately results in photooxidation.[41] Powles[41] indicated that these processes may result from the formation of singlet oxygen, superoxide, or hydroxyl radical. Considering the low-temperature sensitivities of catalase and superoxide dismutase of CS plants,[43,44] the accumulation of toxic oxygen compounds might result from the inability of the plant to detoxify these compounds at low temperatures.

Considering the different rates of glycolysis and oxidative respiration at low temperatures, Raison[45] has suggested that the production of toxic compounds in chill-stressed tissues could account for some chilling injuries. However, as noted by Wade,[46] it is often impossible to determine whether increases in metabolites at low temperatures are the cause rather than the effect of chilling injury. Furthermore, since injury is often observed only after rewarming of the chilled tissue, differences in chilling sensitivity can result from both the levels of toxic metabolites formed at low temperatures and the ability of plants to detoxify or assimilate these compounds as the temperature is increased.

Chilling temperatures reduced the ability of both CS and CR plants to synthesize proteins,[5,47] although the low-temperature inhibition of protein biosynthesis by CR plants gradually diminished.[47] Mohapatra et al.[47] observed that during the initial stages of alfalfa (*Medicago sativa*) cold acclimation, when the ability to snythesize proteins is low, deacclimation by growth at 20°C for 2 d could not reverse the low-temperature inhibition of protein biosynthesis. Even though specific "cold-shock" proteins may be synthesized,[1,9,48] this suggests that chilling can initially induce irreversible inactivation of protein biosynthesis in even CR plants.

Yoshida and Tagawa[49] observed a diversion of electron flow to the alternative respiratory pathway in mitochondria from callus prepared from CS *Cornus stolonifera*. Conversely, in callus from CR *Sambucus sieboldiana*, there was no low-temperature-induced alteration in the apportionment of electrons between the cytochrome path and the alternative path. Unless the alternative pathway was blocked with the inhibitor salicylhydroxamic acid (SHAM), there was no "break" in the Arrhenius plot of mitochondrial oxidative activity of the CS callus. However, in the presence of SHAM, an Arrhenius plot "break" was observed at 13°C for the CB callus. This suggests that the diversion of electrons to the alternative pathway at low temperatures in CS callus results from a low-temperature-sensitive reaction which limits the flow of electrons through the primary respiratory chain. Yoshida and Tagawa[49] suggested that alterations in the regulation of mitochondrial electron flow may be one of the primary consequences of chilling stress in sensitive tissues.

III. STRATEGIES FOR THE THERMAL ACCLIMATION OF PLANT ENZYMES

Many poikilothermic animals have the ability to compensate their metabolic rates in response to changes in temperature.[7] That is, the rates of required metabolic processes are adjusted to offset the accelerating or decelerating effects of temperature variations. In general, enzymes involved in primary pathways of energy metabolism (e.g., glycolysis, Krebs cycle, electron transport) tend to display at least partial compensation to temperature variations.[6] Furthermore, there are often changes in the relative activities of different metabolic pathways during thermal acclimation, shifting metabolism towards biosynthetic reactions at low temperatures. Although such "metabolic reorganizations" have not been conclusively demonstrated in plants, it seems likely that similar processes may occur during the phenotypic acclimation of plants to diurnal or seasonal changes in temperature. The low-temperature acclimation or adaptation of poikilothermic organisms should involve the conservation of enzyme structure, rate, and regulation. As noted by Somero,[6] enzyme adaptation to temperature must be considered as "compromises" which maintain an optimal balance among these traits.

Many of the examples given above of differences between CR and CS plant responses to chilling temperatures have involved CR plants which will ultimately cold acclimate. Such cold hardening or chilling-induced resistance to freezing may involve processes which are different from those required to prevent or reduce chilling damage. Since cold hardening usually requires several weeks of exposure to chilling temperatures, rapid changes in the metabolic processes of CR plants are more likely to be related to chilling acclimation. This possibility becomes more probable when one considers the daily or hourly changes in temperature to which temperate-zone plants are exposed. Therefore, chilling acclimation in response to rapid variations in temperature may be significantly different from chilling or cold acclimation to seasonal temperature changes. As noted by Somero and Hochachka,[7] rapid, short-term compensation of enzyme activity to temperature fluctuations probably involves mechanisms different from those involved in the genotypic adjustment of activation parameters.

A. ENZYME CONCENTRATION

Temperature compensation of plant metabolic processes could result from changes in enzyme concentration with the induction of increased levels of essential enzymes by low temperatures. Kacperska-Palacz[39] noted that one of the initial metabolic events during chilling-induced hardening of plants to freezing stress was the accumulation of water-soluble proteins. The increase in soluble protein concentration partially resulted from *de novo* synthesis rather than a reduction in protein degradation at low temperatures. The levels of membrane-bound proteins also increased during cold acclimation.[48] Since the accumulation of the soluble proteins may not be directly involved with frost hardening,[39] some of these proteins may represent quantitative increases in constitutive enzymes required for metabolic acclimation to chilling.

There are few examples in the literature where the increased activity of plant enzymes produced by chilling can be attributed to elevated enzyme concentrations. The amounts of two enzymes involved in the biosynthesis of phenolic compounds, phenylalanine ammonia lyase and hydroxycinnamoyl CoA quinate hydroxycinnamoyl transferase, increased in many plant species with chilling and may be related to the increased levels of phenolic compounds in chilled plant tissues.[12] The increased specific activity of leucine:tRNA-ligase in cold-treated wheat seedlings has been tentatively attributed to changes in the amount of the enzyme.[50] Since amino acid tRNA-ligases may control the rate of protein biosynthesis, any chilling-induced elevation in the levels of these enzymes may represent thermal compensation, with the expected shift towards the maintenance of biosynthetic pathways at low temperatures. Although apparently not correlated to the chilling sensitivities of *Passiflora* species,[51] the thermal responses of ferredoxin-NADP$^+$ reductase may be involved in the altered activation energy of the electron transport of CS plants at low temperatures.[52] During the cold hardening of wheat at chilling temperatures, ferredoxin-NADP$^+$ reductase was preferentially synthesized,[53] suggesting that increased levels of this enzyme may be required for continued energy production at low temperatures. Since isolated chloroplasts from CR plants had abrupt changes in electron transport activation energy at low temperatures,[29] it has been suggested that there is no correlation between chilling sensitivity and changes in activation energy of photosynthetic electron transport.[16] However, the CR plants had not been subjected to chilling temperatures and, therefore, as suggested above, might not differ from CS plants. Since chilling temperatures can initially reduce protein biosynthesis in CR plants,[47] it seems unlikely that significant increases in the concentration of enzymes from *de novo* synthesis would be a mechanism for the rapid responses of CR plants to chilling temperatures.

B. SUBSTRATE AND COFACTOR LEVELS

Intracellular substrate levels are usually much lower than the concentration required to saturate the enzyme activity. Therefore, increases in substrate levels could elevate enzyme activity at low temperatures. Apparently, there are no examples of systematic increases in substrate concentrations which would reduce the effects of low temperature on plant metabolism. However, the maintenance of the levels of certain substrates, products, or cofactors for specific regulatory enzymes might have significant effects on the low-temperature sensitivity of plants. Many key enzymes involved in metabolism are highly regulated and multimeric. Therefore, as noted above, such enzymes can be subject to cold inactivation through the disruption of hydrophobic interactions. However, the cold inactivation of multimeric enzymes can often be retarded by the presence of substrate, product, or allosteric modifier.[19] If the maintenance of enzyme quaternary structure during low-temperature stress allows metabolism to more rapidly return to normal after rewarming, then potential imbalances produced by variations in the low-temperature sensitivities of metabolic processes might be avoided.

C. MODIFICATION OF ENZYME-SUBSTRATE INTERACTIONS

Thermally induced abrupt changes in enzyme K_m are usually attributed to temperature-dependent alterations in enzyme conformation which modify the configuration of the catalytic site.[28] Since significant temperature-dependent increases in enzyme K_m usually occur near the lethal thermal limits of poikilothermic organisms,[6] variations in the low-temperature stability of enzyme configuration might determine the chilling sensitivities of plants. Based on studies with poikilothermic animals, Somero and Hochachka[7] noted that temperature-dependent changes in apparent K_m are of primary importance in short-term or immediate temperature compensation. It the K_m is directly proportional to temperature, then catalytic rates are less temperature sensitive than when the K_m is temperature independent or varies inversely with temperature changes.

With nonacclimated CR barley, the K_m of the high-affinity root plasma membrane-bound ATPase significantly increased at temperatures below about 12°C and above 32°C, which corresponded to temperature-induced changes in membrane lipid structure.[28] Therefore, these lipid-mediated changes in ATPase protein tertiary or quaternary structure may determine the optimal thermal limits of barley root function. Compared to many poikilothermic animals,[7] this is a rather wide temperature range for optimal function of a membrane-associated enzyme. The apparent insensitivity of the ATPase K_m to temperatures between 12 and 32°C suggests that plants adapted to growth under rapidly varying thermal conditions have evolved enzymes which may not require immediate temperature compensation. The obvious benefit of this situation is the prevention of unnecessary expenditures of energy and resources in the continual "fine tuning" of enzyme structure in response to small temperature fluctuations.

The temperature responses of barley root plasma membrane ATPases also suggests that the maintenance of ATPase activity at temperatures below 12°C requires modification of the lipid-mediated change in enzyme structure. Since plasma membrane ATPases are probably involved in nutrient transport and cellular ionic homeostasis,[6,24] a rapid response to chilling might be required to reduce the electrolyte leakage often associated with chilled tissues.[8,38] Although changes in the types of intra- and intermolecular chemical bonds involved in ATPase structure could reduce the effects of low temperature on enzyme activity, such a process would probably require changes in protein primary structure[54] and, therefore, the biosynthesis of new ATPase proteins. Rather than changing the protein, a rapid modification of the membrane lipid composition could shift the temperature which induces the lipid-mediated alteration in ATPase structure. Temperature-induced changes in plant membrane lipid composition have been examined for a variety of plant and algal species.[27,55,56] Organisms with the ability to acclimate to chilling temperatures may be able to modify their lipid compositions so that the lipid "fluidity" required for normal membrane enzyme function is maintained at low temperatures. This process could be rapid. As discussed by Thompson,[55,57] both alteration of the phospholipid molecular species and desaturation of lipid acyl chains could quickly modify the thermal characteristics of membrane lipids. The rapid rearrangement of existing lipid components presumably as a result of the activation of constitutive enzymes has been termed a "tactical" response[58] and may be essential for the maintenance of membrane-bound enzyme activities during the initial stages of thermal stress. Since the rearrangement of the lipid components has a limited ability to change lipid "fluidity"[55] this process probably represents the means to temporarily reduce the effects of low temperature on lipid-protein interactions until either the temperature increases or lipid metabolism can be changed to produce lipids which would maintain optimal membrane structure under the new temperature conditions.

Although possibly related to cold hardening, the enzyme-substrate affinities of a variety of enzymes are reduced when CR plants are exposed to chilling temperatures. Acclimation to low temperatures reduces the K_m of malate dehydrogenase of both *Lathyrus japonicus*[59] and *Potentilla glandulosa*.[60] Low temperatures significantly reduced the K_m for oxidized glutathione of glutathione reductase from hardened spinach *(Spinacia oleracea)* relative to enzyme from unhardened plants,[61,62] allowing the enzyme from hardened plants to function better at lower temperatures. An increased affinity for CO_2 of RuBP carboxylase accompanied the cold hardening of rye *(Secale cereale)*[63] and potato *(Solanum commersonii)*[64] In contrast to tomato, PEP-carboxylase isolated from wheat or the alpine plant *Caltha intraloba* L. retained a relatively low K_m at chilling temperatures.[23] Since the plants had not been chilled, this may represent an adaptation to permit temperate and alpine plants to rapidly respond to low temperatures, reducing the initial metabolic dysfunctions induced by chilling.

Brouillet and Simon[65] investigated the thermal and kinetic properties of NAD malate dehydrogenase from *Aster* species from contrasting thermal habitats. The K_m of the malate dehydrogenase increased as a positive function in temperature in a fashion similar to the K_m-temperature responses of poikilothermic animals which were defined as positive thermal modulation.[6,7] Positive thermal modulation allows the rapid compensation of enzyme activities in response to sudden temperature variations.[6]

D. MULTIPLE ENZYME FORMS

The structure and function of a single enzyme could develop the required temperature insensitivity to allow satisfactory catalysis and regulation over the entire thermal range of an organism.[7,23] However, temperature insensitivity of an enzymatic reaction could also result from the presence of multiple forms of the enzyme with differing thermal optima.[7] Multiple functional variants of an enzyme could result from either the formation of multiple allelic forms coded at a single gene locus, allozymes, or multiple enzymes coded at different gene loci, isozymes.[7] It should be noted that in the context of this review, the ability of plants to compensate for thermal variations by the production of multiple enzyme forms may represent a genotypic adaptation to temperature fluctuations. That is, if the presence of multiple enzyme forms permits plants to rapidly respond to chilling, then the required enzyme variants may already be present prior to the imposition of the stress conditions. Conversely, new enzyme variants could be biosynthesized or posttranslationally modified in response to inductive low temperatures and, therefore, could be involved in the acclimation to chilling.

Chilling temperatures induce changes in the isozyme patterns of a number of enzymes.[66] The relative proportions of two forms of wheat leaf invertase,[67] the peroxidase isozymes of wheat,[68] alfalfa dehydrogenase isozyme patterns,[69] and forms of spinach leaf glutathione reductase[62] were modified by chilling temperatures. In most cases, the low-temperature-induced changes in the isozyme patterns of these enzymes from hardened tissues may result in the maintenance of enzyme activity at low temperatures. However, it is not clear whether such low-temperature-induced changes in enzyme isozyme patterns represent processes required for the cold hardening of CR plants or depict changes in enzyme structure necessary for metabolic reorganizations which reduce the effects of chilling on the biosynthetic pathways of CR plants.

E. REGULATION OF COMPETING METABOLIC PATHWAYS

As a result of both the consequences of low temperature on enzyme structure and function and the various strategies for the thermal compensation of metabolic processes, temperature variations can significantly modify the relative activities of different metabolic pathways. In poikilothermic animals, a shift in the catabolism of glucose-6-phosphate from glycolysis to the hexosemonophosphate shunt has been observed during cold acclimation.[7] Somero and Hochachka[7] suggested that this change in glucose-6-phosphate metabolism represents an important metabolic reorganization to facilitate biosynthetic processes at low temperatures which may be essential for the synthesis of increased amounts of enzymes required for thermal compensation.

The low-temperature-induced disruption in the regulation between different metabolic pathways, such as the flow of electrons between the normal and alternative pathways in plant mitochondria,[49] could result in chilling injury. Conversely, as discussed by Rychter et al.,[70] the operation of the alternative pathway at low temperatures could be benefical, allowing the rapid oxidation of accumulating sugars and regeneration of NAD^+. Therefore, chilling-induced variations in the relative importance of different metabolic pathways might increase resistance to low temperatures. For example, carbohydrates (e.g., glycerol, ethylene glycol, inositol, glucose, sucrose) have been shown to protect isolated enzymes from low-temperature inactivation.[71] The resistance of plants to low temperatures has been associated with cellular carbohydrate content. The treatment of rice seedlings with fructose or glucose[72] and cotton cotyledons with sucrose[73] reduced their sensitivities to chilling. Potato phosphofructokinase is cold labile,[74] probably resulting from the low-temperature-induced dissociation of enzyme subunits.[75] Dixon and ap Rees[76] suggested that this situation might cause the diversion of hexose phosphates into sucrose biosynthesis rather than into the glycolytic pathway at low temperatures. Although this process may be commercially undesirable for potato tubers, the increased levels of sucrose could reduce the effects of chilling on enzyme stability. This concept is supported by the observed immediate increases in sucrose and fructan levels and sucrose synthetase activity in wheat plants after chilling shock.[77] The results of King et al.[78] indicate that the increased chilling sensitivity of tomato seedlings at the end of a dark period may result from carbohydrate depletion, and the increased tolerance to low temperature following light exposure was a consequence of carbohydrate accumulation. Therefore, the diurnal sensitivity of tomato seedlings to chilling could result from the formation of photosynthetic products. Rapid alterations in the balance between carbon fixation and utilization could result in the accumulation of cryoprotectant carbohydrates. Considering the involvement of fructose 2,6-bisphosphate (F-2,6-P_2) in the regulation of the glycolytic and gluconeogenic pathways for carbohydrate metabolism in plants, Black et al.[79] indicated that ''a fundamental molecular mechanism in plants for coping with physical and biological changes in their environment is to change the intracellular level of F-2,6-P_2.'' Reduced levels of F-2,6-P_2 would channel carbohydrate metabolism toward the production of sucrose and, as a result, may reduce the effects of chilling temperatures. Therefore, it would be of considerable interest to determine the levels of F-2,6-P_2 in both CS and CR plants before and after the imposition of chilling stress.

RuBP carboxylase catalyzes reactions in both photosynthesis and photorespiration.[80] Chilling increased the affinity of RuBP carboxylase for CO_2.[63,64] Since O_2 competes with CO_2 for the active site of RuBP carboxylase,[80] the increased CO_2 affinity at low temperatures should minimize the rate of the apparently wasteful photorespiration, facilitating photosynthesis. The reduced levels of H_2O_2, the byproduct of photorespiration, induced by chilling of both CS and CR plants[81] is consistent with this situation. Considering the regulation of photosynthetic carbon fixation and sucrose synthesis,[82] modification of RuBP carboxylase structure and function during acclimation to chilling could reduce chilling injury by increasing the efficiency of carbon fixation, producing carbohydrates necessary for biosynthesis and possibly cryoprotection. Chilling of cabbage resulted in the rapid appearance of a new form of RuBP carboxylase with lower average hydrophobicity and higher hydrophility.[83] Since the cold lability of RuBP carboxylase apparently involves the disruption of hydrophobic interactions between the enzyme subunits,[18] the replacement of certain hydrophobic amino acids with more hydrophilic residues could increase the low-temperature stability of the enzyme.

IV. CONCLUSIONS

It is apparent that plant adaptation or acclimation to chilling is a complex process, potentially involving essentially every metabolic reaction. As discussed by Raison and Orr,[10] the manifestation of chilling injury may involve various steps with some primary lesion initiating alterations in metabolic processes which ultimately result in tissue damage (see also Chapter 9). Similarly, variations in the resistance of plants to chilling may require a series of metabolic reorganizations which could utilize a variety of mechanisms for reducing the cold lability of enzyme structure and function. The efficacy of any given mechanism to modify the temperature sensitivities of enzymes may depend upon the speed and severity of the imposed temperature stress. Therefore, it is important to attempt to distinguish between processes which would permit rapid, reversible temperature compensation of metabolic rates and those which may be involved in the long-term adaptation of plants to growth under new temperature regimes.

Considering the diversity of animal responses to temperature, there is no reason to expect that all plants have evolved the same mechanism for reducing their sensitivities to temperature variations. Furthermore, CR plants which are able to alter certain metabolic pathways in response to chilling may also have other reactions which are relatively temperature insensitive when compared to CS plants. That is, the identification of a specific phenotypic response to temperature does not preclude the possibility that genotypic adaptations of other reactions can occur within the same plant. For example, using nonacclimated plants, Carey and Berry[84] examined the effects of temperature on ion uptake and respiration of barley and maize. Both CR barley and CS maize had significant changes in activation energy for rubidium uptake at about 10°C. These changes can be attributed to alterations in the physical structure of barley[28,34] and, presumably, maize[85] root plasma membranes. As discussed previously, in the absence of inductive low temperatures, there is no reason to expect differences between CR and CS plants in their responses to chilling. If the temperature-dependence of rubidium uptake reflects the thermal response of plasma membrane-associated processes, then CR barley probably must alter the properties of the plasma membrane in response to chilling. However, unlike maize respiration, which had a change in activation energy at 10°C, barley had a no apparent change in activation energy for respiration between 5 and 35°C.[84] Assuming that significant, low-temperature-dependent changes in the activation energy of a process suggests chilling sensitivity, then a series of metabolic reactions associated with barley mitochondrial membranes are already relatively low-temperature insensitive prior to the imposition of chilling stress.

The similarities between the temperature responses of maize and barley ion uptake[84] are the type of data that have been used by others to refute the Lyons and Raison[3] hypothesis that plant chilling sensitivity is dependent upon temperature-dependent changes in the structure of membrane lipids. It has been argued that if membrane-associated reactions of both CS and CR plants have the same temperature sensitivities, then any low-temperature-dependent changes in membrane structure might have little significance in the ultimate chilling sensitivities of these plants. However, there may be an alternative explanation. A recent study with the blue-green algae, *Anacystis nidulans*, has suggested that lipid phase separations in the plasma membrane may be the site for the initiation of thermal acclimation of algal cells.[86] Therefore, rather than considering temperature-dependent changes in membrane lipid physical structure as primary lesions in plant responses to chilling, it may be equally relevant to consider such changes as primary sensors of chilling conditions. CR plants may be able to respond to low-temperature-induced alterations in lipid structure, modifying metabolic processes to reduce the effects of chilling. Although the same chilling-induced changes in lipid structure may also occur in the membranes of CS plants, injury results from the inability of these plants to react. Therefore, for CS plants, this suggestion is consistent with the proposal of Raison and Orr[10] that temperature-induced lipid structural transitions are primary lesions in plant chilling injury.

REFERENCES

1. **Guy, C. L., Niemi, K. J., and Brambl, R.,** Altered gene expression during cold acclimation of spinach, *Proc. Natl. Acad. Sci. U.S.A.,* 82, 3673, 1985.
2. **Raison, J. K., Chapman, E. A., and Wright, L. C.,** Membrane lipid transitions: their correlation with the climatic distribution of plants, in *Low Temperature Stress in Crop Plants: The Role of the Membrane,* Lyons, J. M., Graham, D., and Raison, J. K., Eds., Academic Press, New York, 1979, 177.
3. **Lyons, J. M. and Raison, J. K.,** Oxidative activity of mitochondria isolated from plant tissues sensitive and resistant to chilling injury, *Plant Physiol.,* 45, 386, 1970.
4. **Berry, J. and Björkman, O.,** Photosynthetic response and adaptation to temperature in higher plants, *Annu. Rev. Plant Physiol.,* 31, 491, 1980.
5. **Brown, G. N.,** Protein synthesis mechanisms relative to cold hardiness, in *Plant Cold Hardiness and Freezing Stress. Mechanisms and Crop Implications,* Li, P. H. and Sakai, A., Eds., Academic Press, New York, 1978, 153.
6. **Somero, G. N.,** Temperature adaptation of enzymes: biological optimization through structure-function compromises, *Annu. Rev. Ecol. Syst.,* 9, 1, 1978.
7. **Somero, G. N. and Hochachka, P. W.,** Biochemical adaptations to temperature, in *Adaptation to Environment: Essays on the Physiology of Marine Animals,* Newell, R. C., Ed., Butterworths, London, 1976, 125.
8. **Lyons, J. M., Raison, J. K., and Steponkus, P. L.,** The plant membrane in response to low temperature, in *Low Temperature Stress in Crop Plants: The Role of the Membrane,* Lyons, J. M., Graham, D., and Raison, J. K., Eds., Academic Press, New York, 1979, 1.
9. **Sarahan, F. and Perras, M.,** Accumulation of high molecular weight protein during cold hardening of wheat (*Triticum aestivum* L.), *Plant Cell Physiol.,* 28, 1173, 1987.
10. **Raison, J. K. and Orr, G. R.,** Proposals for a better understanding of the molecular basis of chilling injury, in *Chilling Injury of Horticultural Crops,* Wang, C. Y., Ed., CRC Press, Boca Raton, FL, 1990, chap. 9.
11. **Raison, J. K. and Lyons, J. M.,** Chilling injury: a plea for uniform terminology, *Plant Cell Environ.,* 9, 685, 1986.
12. **Graham, D. and Patterson, B. D.,** Responses of plants to low, nonfreezing temperatures: proteins, metabolism, and acclimation, *Annu. Rev. Plant Physiol.,* 33, 347, 1982.
13. **Wang, C. Y.,** Physiological and biochemical responses of plants to chilling stress, *HortScience,* 17, 173, 1982.
14. **Christiansen, M. N. and St. John, J. B.,** The nature of chilling injury and its resistance in plants, in *Analysis and Improvement of Plant Cold Hardiness,* Olien, C. R. and Smith, M. N., Eds., CRC Press, Boca Raton, FL, 1981, 1.
15. **Levitt, J.,** *Responses of Plants to Environmental Stresses,* Vol. 1, Academic Press, New York, 1980.
16. **Öquist, G.,** Effects of low temperature on photosynthesis, *Plant Cell Environ.,* 6, 281, 1983.
17. **Alexandrov, V. Ya.,** Conformational flexibility of proteins, their resistance to proteinases and temperature conditions of life, *Curr. Mod. Biol.,* 3, 9, 1969.
18. **Chollet, R. and Anderson, L. L.,** Conformational changes associated with the reversible cold inactivation of ribulose-1,5-bisphosphate carboxylase-oxygenase, *Biochim. Biophys. Acta,* 482, 228, 1977.
19. **Jaenicke, R.,** Enzymes under extremes of physical conditions, *Annu. Rev. Biophys. Bioeng.,* 10, 1, 1981.
20. **Feldberg, R. S. and Datta, P.,** Cold inactivation of l-threonine deaminase from *Rhodospirillum rubrum.* Involvement of hydrophobic interactions, *Eur. J. Biochem.,* 21, 447, 1971.
21. **Shirahashi, K., Hayakawa, S., and Sugiyama, T.,** Cold lability of pyruvate, orthophosphate dikinase in maize leaf, *Plant Physiol.,* 62, 826, 1978.
22. **Taylor, A. D., Slack, C. R., and McPherson, H. G.,** Plants under climatic stress. VI. Chilling and light effects on photosynthetic enzymes of sorghum and maize, *Plant Physiol.,* 54, 696, 1974.
23. **Graham, D., Hockley, D. G., and Patterson, B. D.,** Temperature effects on phosphoenolpyruvate carboxylase from chilling sensitive and chilling resistant plants, in *Low Temperature Stress in Crop Plants: The Role of the Membrane,* Lyons, J. M., Graham, D., and Raison, J. K., Eds., Academic Press, New York, 1979, 453.
24. **McMurchie, E. J.,** Temperature sensitivity of ion-stimulated ATPase associated with some plant membranes, in *Low Temperature Stress in Crop Plants: The Role of the Membrane,* Lyons, J. M., Graham, D., and Raison, J. K., Eds., Academic Press, New York, 1979, 163.
25. **McMurdo, A. C. and Wilson, J. M.,** Chilling injury and Arrhenius plots, *CryoLetters,* 1, 231, 1980.
26. **Minorsky, P. V.,** An heuristic hypothesis of chilling injury in plants: a role for calcium as the primary physiological transducer of injury, *Plant Cell Environ.,* 8, 75, 1985.
27. **Quinn, P. J. and Williams, W. P.,** Plant lipids and their role in membrane function, *Prog. Biophys. Mol. Biol.,* 34, 109, 1978.

28. **Caldwell, C. R. and Haug, A.,** Temperature dependence of barley root plasma membrane-bound Ca^{2+} - and Mg^{2+}-dependent adenosine triphosphatase, *Physiol. Plant.,* 53, 117, 1981.
29. **Nolan, W. G. and Smillie, R. M.,** Multi-temperature effects on Hill reaction activity of barley chloroplasts, *Biochim. Biophys. Acta,* 440, 461, 1976.
30. **Nolan, W. G. and Smillie, R. M.,** Temperature-induced changes in Hill activity of chloroplasts isolated from chilling-sensitive and chilling-resistant plants, *Plant Physiol.,* 59, 1141, 1977.
31. **Silvius, J. R., Read, B. D., and McElhaney, R. N.,** Membrane enzymes: artifacts in Arrhenius plots due to temperature dependence of substrate-binding affinity, *Science,* 199, 902, 1978.
32. **Silvius, J. R. and McElhaney, R. N.,** Non-linear Arrhenius plots and the analysis of reaction and motional rates in biological membranes, *J. Theor. Biol.,* 88, 135, 1981.
33. **Caldwell, C. R.,** Temperature-induced protein conformational changes in barley root plasma membrane-enriched microsomes. II. Intrinsic protein fluorescence, *Plant Physiol.,* 84, 924, 1987.
34. **Caldwell, C. R. and Whitman, C. E.,** Temperature-induced protein conformational changes in barley root plasma membrane-enriched microsomes. I. Effect of temperature on membrane protein and lipid mobility, *Plant Physiol.,* 84, 918, 1987.
35. **Yoshida, S., Kawata, T., Uemura, M., and Niki, T.,** Properties of plasma membrane isolated from chilling-sensitive etiolated seedlings of *Vigna radiata* l., *Plant Physiol.,* 80, 152, 1986.
36. **Tanaka, Y. and Uritani, I.,** Purification and properties of phenylalanine ammonia lyase in cut-injured sweet potato, *J. Biochem.,* 81, 963, 1977.
37. **Jones, R., Wilkins, M. B., Coggins, J. R., Fewson, C. A., and Malcolm, A. D. B.,** Phosphoenolpyruvate carboxylase from the crassulacean plant *Bryophyllum fedtschenkoi, Biochem. J.,* 175, 391, 1978.
38. **Lyons, J. M.,** Chilling injury in plants, *Annu. Rev. Plant Physiol.,* 24, 445, 1973.
39. **Kacperska-Palacz, A.,** Mechanism of cold acclimation in herbaceous plants, in *Plant Cold Hardiness and Freezing Stress. Mechanisms and Crop Implications,* Li, P. H. and Sakai, A., Eds., Academic Press, New York, 1978, 139.
40. **Sakai, A. and Larcher, W.,** *Frost Survival of Plants. Responses and Adaptation to Freezing Stress,* Springer-Verlag, Berlin, 1987.
41. **Powles, S. B.,** Photoinhibition of photosynthesis induced by visible light, *Annu. Rev. Plant Physiol.,* 35, 15, 1984.
42. **Garber, M. P.,** Effect of light and chilling temperature on chilling-sensitive and chilling-resistant plants, *Plant Physiol.,* 59, 981, 1977.
43. **Michalski, W. P. and Kaniuga, Z.,** Photosynthetic apparatus of chilling-sensitive plants. X. Relationship between superoxide dismutase activity and photoperoxidation of chloroplast lipid, *Biochim. Biophys. Acta,* 637, 159, 1981.
44. **Omran, R. G.,** Peroxide levels and the activities of catalase, peroxidase and indoleacetic acid oxidase during and after chilling cucumber seedlings, *Plant Physiol.,* 65, 407, 1980.
45. **Raison, J. K.,** Effect of low temperature on respiration, in *The Biochemistry of Plants. A Comprehensive Treatise,* Davies, D. D., Ed., Academic Press, New York, 1980, 613.
46. **Wade, N. L.,** Physiology of cool-storage disorders of fruit and vegetables, in *Low Temperature Stress in Crop Plants: The Role of the Membrane,* Lyons, J. M., Graham, D., and Raison, J. K., Eds., Academic Press, New York, 1979, 81.
47. **Mohapatra, S. S., Poole, R. J., and Dhindsa, R. S.,** Changes in protein patterns and translatable messenger RNA populations during cold acclimation of alfalfa, *Plant Physiol.,* 84, 1171, 1987.
48. **Mohapatra, S. S., Poole, R. J., and Dhindsa, R. S.,** Alterations in membrane protein-profile during cold treatment of alfalfa, *Plant Physiol.,* 86, 1005, 1988.
49. **Yoshida, S. and Tagawa, F.,** Alteration of respiratory function in chill-sensitive callus due to low temperature stress. I. Involvement of the alternative pathway, *Plant Cell Physiol.,* 20, 1243, 1979.
50. **Weidner, M., Mathee, C., and Schmitz, F.-K.,** Phenotypical temperature adaptation of protein synthesis in wheat seedlings. Qualitative aspects. Involvement of aminoacid:t-RNA ligases, *Plant Physiol.,* 69, 1281, 1982.
51. **Critchley, C., Smillie, R. M., and Patterson, B. D.,** Effect of temperature on photoreproductive activity of chloroplasts from passion fruit species of different chilling sensitivity, *Aust. J. Plant Physiol.,* 5, 443, 1978.
52. **Shneyour, A., Raison, J. K., and Smillie, R. M.,** The effect of temperature on the rate of photosynthetic electron transfer in chloroplasts of chilling-sensitive and chilling-resistant plants, *Biochim. Biophys. Acta,* 292, 152, 1973.
53. **Riov, J. and Brown, G. N.,** Comparative studies on activity and properties of ferredoxin-$NADP^+$ reductase during cold hardening of wheat, *Can. J. Bot.,* 54, 1896, 1976.
54. **Yutani, K., Ogasahara, K., and Sugino, Y.,** Effect of amino acid substitutions on conformational stability of a protein, *Adv. Biophys.,* 20, 13, 1985.
55. **Thompson, G. A., Jr.,** Mechanisms of membrane response to environmental stress, in *Frontiers of Membrane Research in Agriculture,* St. John, J. B., Berlin, E., and Jackson, P. C., Eds., Rowman & Allanheld, Totowa, NJ, 1985, 347.

56. **Kuiper, P. J. C.**, Environmental changes and lipid metabolism of higher plants, *Physiol. Plant.*, 64, 118, 1985.

57. **Ramesha, C. S. and Thompson, G. A., Jr.**, The mechanism of membrane response to chilling. Effect of temperature on phospholipid deacylation and reacylation reactions in the cell surface membrane, *J. Biol. Chem.*, 259, 8706, 1984.

58. **Lynch, D. V. and Thompson, G. A., Jr.**, Retailored lipid molecular species; a tactical mechanism for modulating membrane properties, *Trends Biochem. Sci.*, 9, 442, 1984.

59. **Simon, J. P.**, Adaptation and acclimation of higher plants at the enzyme level: temperature-dependent substrate binding ability of NAD malate dehydrogenase in four populations of *Lathyrus japonicus* Willd (Leguminosae), *Plant Sci. Lett.*, 14, 113, 1979.

60. **Teeri, J. A. and Peet, M. M.**, Adaptation of malate dehydrogenase to environmental temperature variation in two populations of *Potentilla glandulosa* Lindl., *Oecologia*, 34, 133, 1978.

61. **de Kok, L. J. and Oosterhuis, F. A.**, Effects of frost-hardening and salinity on glutathione and sulfhydryl levels and on glutathione reductase activity in spinach leaves, *Physiol. Plant.*, 58, 47, 1983.

62. **Guy, C. L. and Carter, J. V.**, Characterization of partially purified glutathione reductase from cold-hardened and nonhardened spinach leaf tissue, *Cryobiology*, 21, 454, 1984.

63. **Huner, N. P. A. and Macdowall, F. D. H.**, The effects of low temperature acclimation of winter rye on catalytic properties of its ribulose bisphosphate carboxylase-oxygenase, *Can. J. Biochem.*, 57, 1036, 1979.

64. **Huner, N. P. A., Palta, J. P., Li, P. H., and Carter, J. V.**, Comparison of the structure and function of ribulose bisphosphate carboxylase-oxygenase from cold-hardy and nonhardy potato species, *Can. J. Biochem.*, 59, 280, 1981.

65. **Brouillet, L. and Simon, J.-P.**, Adaptation and acclimation of higher plants at the enzyme level: thermal properties of NAD malate dehydrogenase of two species of *Aster* (Asteraceae) and their hybrid adapted to contrasting habitats, *Can. J. Bot.*, 58, 1474, 1980.

66. **Roberts, D. W. A.**, Some possible roles for isozyme substitutions during cold hardening in plants, *Int. Rev. Cytol.*, 26, 303, 1969.

67. **Roberts, D. W. A.**, Changes in the forms of invertase during the development of wheat leaves growing under cold-hardening and nonhardening conditions, *Can. J. Bot.*, 60, 1, 1982.

68. **Roberts, D. W. A.**, A comparison of peroxidase isozymes of wheat plants grown at 6°C and 20°C, *Can. J. Bot.*, 47, 263, 1969.

69. **Krasnuk, M., Jung, G. A., and Witham, F. H.**, Electrophoretic studies of several dehydrogenases in relation to the cold tolerance of alfalfa, *Cryobiology*, 13, 375, 1976.

70. **Rychter, A. M., Ciesla, E., and Kacperska, A.**, Participation of the cyanide-resistant pathway of respiration of winter rape leaves as affected by plant cold acclimation, *Physiol. Plant.*, 73, 299, 1988.

71. **Carpenter, J. F. and Crowe, J. H.**, The mechanism of cryoprotection of proteins by solutes, *Cryobiology*, 25, 244, 1988.

72. **Tajima, K. and Kabaki, N.**, Effects of sugars and several growth regulators on the chilling injury of rice seedlings, *Jpn. J. Crop Sci.*, 50, 411, 1981.

73. **Rikin, A., Gitler, C., and Atsmon, D.**, Chilling injury in cotton (*Gossypium hirsutum* L.): light requirement for the reduction of injury and for the protective effects of abscisic acid, *Plant Cell Physiol.*, 22, 453, 1981.

74. **Pollock, C. J. and ap Rees, T.**, Activities of enzymes of sugar metabolism in cold-stored tubers of *Solanum tuberosum*, *Phytochemistry*, 14, 613, 1975.

75. **Dixon, W. L., Franks, F., and ap Rees, T.**, Cold-liability of phosphofructokinase from potato tubers, *Phytochemistry*, 20, 969, 1981.

76. **Dixon, W. L. and ap Rees, T.**, Carbohydrate metabolism during cold-induced sweetening of potato tubers, *Phytochemistry*, 19, 1653, 1980.

77. **Calderón, P. and Pontis, H. G.**, Increase of sucrose synthetase activity in wheat plants after chilling shock, *Plant Sci.*, 42, 173, 1985.

78. **King, A. I., Joyce, D. C., and Reid, M. S.**, Role of carbohydrates in diurnal chilling sensitivity of tomato seedlings, *Plant Physiol.*, 86, 764, 1988.

79. **Black, C. C., Mustardy, L., Sung, S. S., Kormanik, P. P., Xu, D.-P., and Paz, N.**, Regulation and roles for alternative pathways of hexose metabolism in plants, *Physiol. Plant.*, 69, 387, 1987.

80. **Pierce, J.**, Prospects for manipulating the substrate specificity of ribulose bisphosphate carboxylase/oxygenase, *Physiol. Plant.*, 72, 690, 1988.

81. **MacRae, E. A. and Ferguson, I. B.**, Changes in catalase activity and hydrogen peroxide concentration in plants in response to low temperature, *Physiol. Plant.*, 65, 51, 1985.

82. **Stitt, M., Gerhardt, R., Wilke, I., and Heldt, H. W.**, The contribution of fructose 2,6-bisphosphate to the regulation of sucrose synthesis during photosynthesis, *Physiol. Plant.*, 69, 377, 1987.

83. **Shomer-Ilan, A. and Waisel, Y.**, Cold hardiness of plants: correlation with changes in electrophoretic mobility, composition of amino acids and average hydrophobicity of fraction-1-protein, *Physiol. Plant.*, 34, 90, 1975.

84. **Carey, R. W. and Berry, J. A.,** Effects of low temperature on respiration and uptake of rubidium ions by excised barley and corn roots, *Plant Physiol.,* 61, 858, 1978.

85. **Raison, J. K.,** A biochemical explanation of low-temperature stress in tropical and subtropical plants, in *Mechanisms of Regulation of Plant Growth,* Bull. No. 10, Bielski, R. L., Ferguson, A. R., and Cresswell, M. M., Eds., Royal Society of New Zealand, Wellington, 1974.

86. **Gombos, Z. and Vigh, L.,** Primary role of the cytoplasmic membrane in thermal acclimation evidenced in nitrate-starved cells of the blue-green alga, *Anacystis nidulans, Plant Physiol.,* 80, 415, 1986.

Chapter 11

LIPIDS IN RELATION TO CHILLING SENSITIVITY OF PLANTS

Norio Murata and Ikuo Nishida

TABLE OF CONTENTS

I. INTRODUCTION

Temperature limits the geographical distribution and productivity of plants. Below a particular temperature in the chilling range (e.g., 5 to 15°C), most tropical and subtropical plants are restricted in germination, growth, development, and postharvest longevity. These plants are classified as chilling sensitive and include many economically important crops such as rice, maize, and tropical fruits. In contrast, most plants of temperate origin are endowed with the ability to survive at the chilling temperature and are classified as chilling resistant.

Particular stages in the life cycle of a plant, such as pollen maturation,[1,2] are more sensitive to chilling than other stages of development. Maturity of the tissue may also be important. For example, seedlings appear to be more susceptible to chilling temperature than mature plants.[3]

In the mechanism proposed by Lyons[3] and Raison[4] for the chilling sensitivity of plants, the primary event in chilling injury is the formation of a lipid gel phase in cellular membranes at low temperature, which is followed by a series of processes leading to the death of the cells. When a model membrane goes into the phase-separated state in which both gel and liquid crystalline phases coexist, the membrane becomes leaky to small electrolytes.[5] This phenomenon may happen in plant membranes; thus, the phase-separated state diminishes ion gradients across the membranes that are essential for the maintenance of physiological activities of plant cells.[3] We have shown that this mechanism of chilling injury is operative in the blue-green alga *Anacystis nidulans*, in which the electrolytes leak out from the cytoplasm to the outer medium when the cytoplasmic membranes enter the phase-separated state at low temperatures.[6-10]

In this article, we will review the studies which have been aimed to correlate the composition of membrane lipids with chilling sensitivity in higher plants.

II. MEMBRANE LIPIDS AND THEIR BIOSYNTHESIS

A. LIPID COMPOSITION OF PLANT MEMBRANES

The lipid composition of cellular membranes from higher plant cells is presented in Table 1. The chloroplast membranes are characterized by a high content of glycolipids in addition to phospholipids, whereas the mitochondrial and endoplasmic reticulum membranes contain only phospholipids. On the other hand, the plasma and tonoplast membranes are rich in sterols and sphingolipids in addition to phospholipids. These findings suggest that each kind of cellular membrane is unique in its lipid-lipid and lipid-protein interactions.

Table 2 summarizes the glycerolipid composition of plant-cellular membranes. The thylakoid and inner envelope membranes of chloroplasts are rich in glycolipids such as MGDG, DGDG, and SQDG, whereas the outer envelope membrane contains about equal amounts of glycolipids and phospholipids. In the mitochondrial membranes, endoplasmic reticulum membranes, plasma membranes, and tonoplasts, PC and PE are the major glycerolipid constituents. PI constitutes a substantial fraction of the total phospholipid in endoplasmic reticulum > tonoplasts > plasma membranes. The inner mitochondrial membrane contains a unique phospholipid, DPG.

Abbreviations: ACP, acyl-carrier protein; DG, *sn*-1,2-diacylglycerol; DGDG, digalactosyldiacylglycerol; DPG, diphosphatidylglycerol; MGDG, monogalactosyldiacylglycerol; PA, phosphatidic acid; PC, phosphatidylcholine; PE, phosphatidylethanolamine; PG, phosphatidylglycerol; PI, phosphatidylinositol; PS, phosphatidylserine; SQDG, sulfoquinovosyldiacylglycerol; 16:0, palmitic acid; 16:1t, 3-*trans*-hexadecenoic acid; 16:3, 7,10,13-*cis*-hexadecatrienoic acid 18:1, oleic acid (9-*cis*-octadecenoic acid); 18:2, linoleic acid (9,12-*cis*-octadecadienoic acid); 18:3, α-linolenic acid (9,12,15-*cis*-octadecatrienoic acid); the molecular species are represented as, e.g., 16:0/16:1t for *sn*-1-palmitoyl-2-(3-*trans*)-hexadecenoylglycerol moiety.

TABLE 1
Lipid Composition of Plant Cellular Membranes

Membrane	Plant tissue	Lipid (mol%)				Ref.
		Phospholipid	Glycolipid	Sterol	Sphingolipid	
Chloroplasts	Spinach leaf					11
Thylakoid		16	84	0	0	
Inner envelope		15	85	0	0	
Outer envelope		47	53	tr	tr	
Mitochondria	Castor bean endosperm	100	0	tr	—	12
Endoplasmic reticulum	Castor bean endosperm	99	0	1	—	12
Plasma	Mung bean hypocotyl	49	1	43	7	13
Tonoplast	Mung bean hypocotyl	51	4	28	17	13

TABLE 2
Glycerolipid Composition of Plant Cellular Membranes

Membrane	Lipid (mol %)									Ref.
	MGDG	DGDG	SQDG	PG	PC	PE	PI	DPG	PS	
Chloroplasts from spinach leaves										11
Thylakoid	51	26	7	10	5	0	1	0	0	
Inner envelope	50	30	5	8	6	0	1	0	0	
Outer envelope	17	29	6	10	33	0	5	0	—	
Mitochondria from cauliflower buds										14
Inner	0	0	0	1	29	51	2	17	0	
Outer	0	0	0	2	69	24	5	0	0	
Endoplasmic reticulum from castor bean endosperm	0	0	0	3	43	34	17	0	3	12
Plasma membranes from mung bean hypocotyl	tr	2	0	5	39	44	6	0	4	13
Tonoplasts from mung bean hypocotyl	2	6	0	4	44	30	10	0	4	13

Fatty acids of glycerolipids in plant lipids are unique in a high degree of unsaturation represented by 18:2 and 18:3. Other fatty acid components are 16:0 in most phospholipids and SQDG, 16:1t in PG,[15] and 16:3 in MGDG from 16:3 plants.[16] The physical properties of membrane lipids depend on the constituent molecular species which are defined by a combination of two fatty acid residues at the *sn*-1 and *sn*-2 positions and a polar head group at the *sn*-3 position of glycerol moiety. Table 3 lists the major molecular species of leaf lipids. These molecular species can be categorized into three groups: the first is composed of two polyunsaturated fatty acids, e.g., 18:3/18:3; the second is of a saturated acid and a polyunsaturated acid, e.g., 16:0/18:3; and the third is characterized by two saturated fatty acids, e.g., 16:0/16:0. In addition, PG in chloroplasts, however, contains a unique *trans*-monounsaturated molecular species such as 16:0/16:1t. Molecular species containing 18:0 and 18:1 are minor components in most tissues, but occur at high levels in particular tissues such as anthers.[21]

The major sphingolipid in higher-plant cells is ceramide monoglucoside.[22,23] Its basic

TABLE 3
Major Molecular Species of Leaf Lipids

Lipid		Major molecular species		
MGDG	18:3/18:3	18:3/16:3[a]		
DGDG	18:3/18:3	16:0/18:3	18:3/16:0[a]	
SQDG	16:0/18:3	16:0/18:2	18:3/16:0[a]	16:0/16:0[b]
PG	18:3/16:0	18:3/16:1t	16:0/16:0	16:0/16:1t
PC	16:0/18:2	16:0/18:3	18:2/18:2	18:3/18:3
PE	16:0/18:2	16:0/18:3	18:2/18:2	18:3/18:3
PI	16:0/18:2	16:0/18:3		

[a]   In 16:3 plants.[18,19]
[b]   In some plant species.[20]

structure consists of a molecule each of sphingoid (previously termed sphingosine or long-chain base), fatty acid, and glucose. The structure of plant sphingoid is presented in Figure 1. There are nine kinds of sphingoids which can be classified into three groups: dihydroxy 4-saturated, dihydroxy 4-*trans*-unsaturated, and trihydroxy sphingoids. Each group includes three kinds of sphingoids having either a *cis*-unsaturated, a *trans*-unsaturated, or a saturated bond at the C-8 position (Figure 1). The fatty acids in plant sphingolipids are mostly 2-hydroxy, and saturated or *cis*-monounsaturated.[22-24] Their chain length ranges from C14 to C26.[22] More than ten molecular species of the sphingolipids, which are defined by the combination of sphingoid and fatty acid, have been identified.[25]

Sterols are found in plant tissues as free sterols and sterol derivatives such as steryl glycosides, acyl steryl glycosides, and steryl esters.[26] The sterol composition varies among plant tissues[26] and with environmental conditions.[24,27,28] However, the meaning of such variation is not well understood, partly because little is known about the intracellular distribution of the sterols as well as the regulatory mechanism of their biosynthesis. In most higher-plant tissues Δ^5-24-alkylsterols, consisting mainly of β-sitosterol (24R-ethylcholest-5-en-3β-ol), stigmasterol (24S-ethylcholesta-5,22-dien-3β-ol), and campesterol (24S-methylcholest-5-en-3β-ol), are major components in free sterols as well as sterol derivatives. However, Δ^7-24-alkylsterols predominate in some plant species such as alfalfa, squash, and spinach.[29-32] Although the proportion of β-sitosterol to stigmasterol varies, the sum of these two sterols often amounts to more than 70% of the total sterols.[33] Cholesterol, which is most abundant in animal tissues, is a minor component in plant tissues, but becomes a major component in surface lipids of leaves and fruits.[34]

B. BIOSYNTHESIS OF PLANT LIPIDS

1. Fatty Acids

The fatty acid synthesis in plant cells takes place exclusively within plastids.[35,36] The fatty acid synthetase in the plastids requires ACP as a cofactor and is composed of six dissociable enzymes which catalyze each step of fatty acid synthesis: acetyl-CoA:ACP acetyltransferase, malonyl-CoA:ACP malonyltransferase, 3-oxoacyl-ACP synthetase, 3-oxoacyl-ACP reductase, 3-hydroxyacyl-ACP dehydratase, and enoyl-ACP reductase.[36] The final product of the fatty acid synthetase is 16:0-ACP, a part of which is elongated to 18:0ACP[36] and subsequently desaturated to 18:1-ACP.[37] Part of these fatty acids are transferred from acyl-ACPs to glycerol 3-phosphate and lysophosphatidic acid to yield PA, which is incorporated into glycerolipids; all these reactions take place in plastids. The remaining fatty acids are released from acyl-ACPs in the stroma and, after conversion to acyl-CoA on the outer envelope membrane, exported to the cytoplasm.[38] These fatty acids are incorporated into phospholipids in the endoplasmic reticulum.[39] Although 18:0 is desaturated to 18:1 in

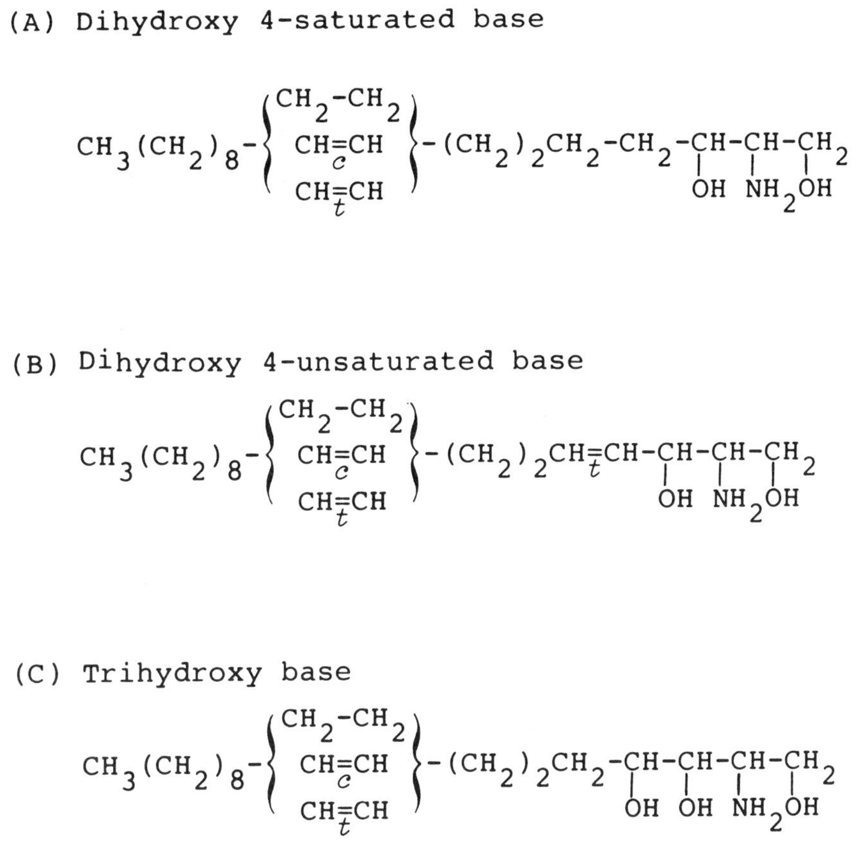

(A) Dihydroxy 4-saturated base

$$CH_3(CH_2)_8- \left\{ \begin{array}{c} CH_2-CH_2 \\ CH\underset{c}{=}CH \\ CH\underset{t}{=}CH \end{array} \right\} -(CH_2)_2CH_2-CH_2-\underset{OH}{\underset{|}{C}}H-\underset{NH_2}{\underset{|}{C}}H-\underset{OH}{\underset{|}{C}}H_2$$

(B) Dihydroxy 4-unsaturated base

$$CH_3(CH_2)_8- \left\{ \begin{array}{c} CH_2-CH_2 \\ CH\underset{c}{=}CH \\ CH\underset{t}{=}CH \end{array} \right\} -(CH_2)_2CH\underset{t}{=}CH-\underset{OH}{\underset{|}{C}}H-\underset{NH_2}{\underset{|}{C}}H-\underset{OH}{\underset{|}{C}}H_2$$

(C) Trihydroxy base

$$CH_3(CH_2)_8- \left\{ \begin{array}{c} CH_2-CH_2 \\ CH\underset{c}{=}CH \\ CH\underset{t}{=}CH \end{array} \right\} -(CH_2)_2CH_2-\underset{OH}{\underset{|}{C}}H-\underset{OH}{\underset{|}{C}}H-\underset{NH_2}{\underset{|}{C}}H-\underset{OH}{\underset{|}{C}}H_2$$

FIGURE 1.   Structure of plant sphingoids. In the major plant sphingolipids, i.e., ceramide mon-oglucoside, glucose is bound to the hydroxy residue at the 1 position with 1-O-β configuration, and fatty acid is bound to the amino group at the 2 position by an amide linkage. The *cis*- and *trans*-configurations of the double bonds are presented by *c* and *t*, respectively.

the ACP-bound form within plastids, all the other desaturation steps take place in a lipid-bound form.[37] The 18:1 is desaturated to 18:2 and further to 18:3 in all the lipid classes in both plastids and cytoplasm. The 16:0, on the other hand, is desaturated only in plastids in which that bound to the *sn*-2 position of PG is desaturated to 16:1t, and that bound to the *sn*-2 position of MGDG, to 16:3.

2. Glycerolipids

The first step of glycerolipid synthesis in higher-plant cells is formation of PA via a stepwise esterification of glycerol 3-phosphate with fatty acids. Two pathways are operative in synthesizing PA in plant tissues,[39-41] i.e., the cytoplasmic (or eukaryotic) pathway and the plastidial (prokaryotic) pathway. The former activity is associated mainly with endo-plasmic reticulum and partly with Golgi bodies and mitochondria,[42-44] and the latter activity is located on the plastid envelopes.[11,45, 46] These two pathways are clearly distinguished by their specificity of fatty acid esterification of the *sn*-2 position of the glycerol moiety.[39] In both the cytoplasmic and plastidial pathways, either 18:1, 16:0, or, in some cases, 18:0 is esterified at the *sn*-1 position of PA. However, at the *sn*-2 position of PA, only 18:1 is esterified in the cytoplasmic pathway, whereas only 16:0 is esterified in the plastidial path-way. As a result, the PA molecular species synthesized in the cytoplasmic pathway are 18:1/18:1, 16:0/18:1, and 18:0/18:1, and those synthesized in the plastidial pathway are 18:1/

Cytoplasmic   Plastidial
pathway   pathway
————————————————————   ————————————————————

┌C16   ┌C18   ┌C18   ┌C16
├C18   ├C18   ├C16   ├C16
└X   └X   └X   └X

FIGURE 2.   The C16 and C18 fatty acid distribution at the *sn*-1 and *sn*-2 positions of *sn*-glycerol in plant glycerolipids, which are synthesized through the cytoplasmic (eukaryotic) pathway and plastidial (prokaryotic) pathways.

16:0, 16:0/16:0, and 18:0/16:0 (Figure 2). Although desaturation of fatty acids on phospholipids and glycolipids produces a complex mixture of lipid molecular species, as shown in Table 3, it is possible to estimate the relative contribution of the cytoplasmic and plastidial pathways in the biosynthesis of cellular glycerolipids in higher plants by the proportions of C18 and C16 fatty acids at the *sn*-2 position of the glycerol moiety.

The endoplasmic reticulum is autonomous for biosynthesis of its own phospholipids.[42,47,48] The PA synthesized in the cytoplasmic pathway is converted into metabolic intermediates, DG and CDP-DG. PC and PE are synthesized from the DG, and PI and PG are made from the CDP-DG. In developing tissues, Golgi membranes are also capable of synthesizing PC and PE.[43,49,50] The contribution of Golgi membranes to PC synthesis in developing maize roots is estimated to be comparable to that of endoplasmic reticulum.[43] The mitochondrion can synthesize PG and DPG[48] and a portion of its PC and PE.[51] However, the contribution of mitochondria to total cellular glycerolipid synthesis is estimated to be only 1 to 2% in the castor bean endosperm.[51]

The plastid is autonomous for biosynthesis of its own PG.[52] PG synthesis from acetate and glycerol 3-phosphate in isolated intact chloroplasts has been demonstrated.[53-56] In the 16:3 plants, which contain both 16:3 and 18:3 in MGDG, certain proportions of MGDG, SQDG, and DGDG are synthesized within plastids. The synthesis of SQDG[57,58] and MGDG[59] in intact chloroplasts isolated from spinach, a 16:3 plant, has been demonstrated. Thus, these lipids are synthesized by the plastidial pathway, and their fatty acid distribution is characterized by C16 at the *sn*-2 position of the glycerol moiety. In the 18:3 plants, which contain 18:3, but no 16:3, the MGDG, SQDG, and DGDG are synthesized by the cytoplasmic pathway and therefore characterized by C18 at the *sn*-2 position of glycerol moiety. The DG moiety of these lipids is synthesized in the form of PA in the endoplasmic reticulum, and transported to the plastids. The lipid form of this transportation is either PC[60] or DG.[59] MGDG, DGDG, and SQDG in 16:3-plants are partly synthesized also through the cytoplasmic pathway.[19] In both 16:3-plants and 18:3-plants, however, only the plastid is the site for assembling the polar head group of MGDG, DGDG, and SQDG onto the DG moiety.[38]

Lipids of the tonoplast and plasma membranes are supplied from other organelles, such as the endoplasmic reticulum and Golgi bodies,[61,62] although direct evidence for this lipid transport has not yet been obtained.

3. Sphingolipids and Sterols

Little is known about the pathway and cellular compartmentation of sphingolipid biosynthesis in higher plant cells. Only Roughan et al.[63] have reported that exogenously added 16:0, but not 18:0 and 18:1, is converted into sphingoid of sphingolipids in spinach leaves. The mechanisms for biosynthesis of unique very long-chain fatty acids of C20-C26 and for their hydroxylation at the C-2 position are a subject of future study.

In contrast, the biosynthesis of plant sterols has been extensively studied.[27,33,64-66] A pathway from mevalonic acid to squalene-2,3-oxide is common in both plants and animals.

TABLE 4
Phase Transition Temperatures of Lipid Molecular
Species in Plant Membranes

Lipid	Molecular species	Phase transition temperature (°C)	Ref.
PG	16:0/16:0-PG	42	70
		56 ($+Mg^{2+}$)	
	16:0/16:1t-PG	32	71
		42 ($+Mg^{2+}$)	
	18:1/16:0-PG	−3	72
		8 ($+Mg^{2+}$)	
SQDG	16:0/16:0-SQDG	43	73
		38 ($+Mg^{2+}$)	
PC	18:1/18:1-PC	−20	68
	16:0/18:1-PC	−5	67
	18:0/18:1-PC[a]	3	68

[a] This species may exist in anthers.[21]

In higher-plants, the latter compound is specifically cyclized to yield the first sterol compound, cycloartenol. Subsequent modification of cycloartenol yields a complex mixture of plant sterols. Interconversions of sterols and sterol derivatives have been reviewed elsewhere.[26,28]

III. PHASE TRANSITION OF MEMBRANE LIPIDS

A. GLYCEROLIPIDS

The effect of unsaturation and chain length of constituent fatty acids on the thermotropic phase behavior of glycerolipids has been intensively studied in bilayer membranes of artificially synthesized lipids.[67] Molecular species containing only saturated fatty acids, such as 16:0 and 18:0, undergo the gel to liquid-crystalline phase transition above 40°C. The phase transition of PC molecular species containing one *cis*-double bond occurs near 0°C, and introduction of another *cis*-double bond into the second fatty acid decreases the phase transition temperature further to about −20°C.[68] The *trans*-double bond, however, is much less effective in decreasing phase transition temperature. The transition temperature of 9-*trans*-18:1/9-*trans*-18:1-PC is 9°C,[69] as compared with that of 9-*cis*-18:1/9-*cis*-18:1-PC at −20°C.[67] The phase transition temperature of 16:0/16:0-PC is 41°C, and that of 18:0/18:0-PC is 56°C, indicating that elongation of the fatty acid moieties increases the phase transition temperature by about 15°C. These results of the model membrane system can be applied to the phase transition of plant cellular membranes in relation to the chilling sensitivity of plants.

Among lipid molecular species in plant membranes, as listed in Table 3, only a very limited number have been subjected to the study of thermotropic phase behavior (Table 4). The PG molecular species, 16:0/16:0 and 16:0/16:1t, undergo the phase transition at 42 and 32°C, respectively, in the absence of Mg^{2+} (Table 4). The presence of Mg^{2+}, which exists in the chloroplast stroma at 18 ± 5 mM,[74] increases the phase transition temperature by 14°C in the 16:0/16:0 species and 10°C in the 16:0/16:1t species. This observation is consistent with the reported effect of divalent cations on the phase transition of acidic phospholipids.[75] The 18:0/16:0 and 18:0/16:1t species of PG, which are present at low levels in plant membranes,[76] are very likely to change phase at temperatures higher than the corresponding 16:0/16:0 and 16:0/16:1t species, respectively, since elongation of the fatty acyl chain increases the transition temperature. The phase transition of the *cis*-monounsaturated

PG species, 18:1/16:0, which exists in leaves of evergreen trees,[77] occurs at 8°C in the presence of Mg^{2+} and below $-3°C$ in its absence (Table 4). Therefore, this molecular species may contribute to the phase transition of chloroplast membranes. The phase behavior of other major molecular species of PG, 18:2/16:0, 18:3/16:0, 18:2/16:1t and 18:3/16:1t, has not been studied. However, their transition temperatures are estimated to be below 0°C, since they are more unsaturated than the 18:1/16:0 species.

The saturated species of SQDG, 16:0/16:0, which exists in some plant species,[20] undergoes the phase transition at 43°C in the absence of Mg^{2+} and at 38°C in its presence.[73] It is interesting that a divalent cation decreases the phase transition temperature of this acidic glycolipid, in contrast to its effect on the acidic phospholipids. Although the phase transition temperatures of the major SQDG species 16:0/18:3, 18:3/16:0 and 18:3/18:3 have not been determined, they are pressumed to be below 0°C because of the high degrees of unsaturation.

The remaining chloroplast lipids, MGDG and DGDG, comprise only highly unsaturated molecular species such as 18:3/18:3, 18:3/16:3 (MGDG), and 16:0/18:3 (DGDG). Although the transition temperatures of these molecular species have not been determined, Shipley et al.[78] reported that the total MGDG isolated from spinach leaves undergoes the phase transition at $-30°C$

Recently, Orr et al.[79] have studied the molecular species composition and phase behavior of DPG from mitochondria. They found that the major DPG molecular species contain 18:2 and 18:3 as fatty acid constituents, regardless of whether the lipid comes from chilling-sensitive or chilling-resistant plants. Therefore, their phase transition temperatures are expected to be far below 0°C. This result was confirmed directly by calorimetric and fluorescent probe measurements.

Among the molecular species of PC present in plant cellular membranes, the phase transition temperature has been determined only in minor components such as 16:0/18:1 ($-5°C$) and 18:1/18:1 ($-20°C$) (Table 4). Since the major PC molecular species present in the plant membranes (e.g., 16:0/18:2, 16:0/18:3, 18:2/18:2, and 18:3/18:3) are more unsaturated than the above-mentioned species, the phase transition temperature of any of the PC molecular species in plant membranes should be lower than 0:°C. The 18:0/18:1-PC, which exists in membranes of anthers and may be related to the chilling sensitivity of this organ,[21] changes its phase at 3°C (Table 4).

These findings suggest that if there are glycerolipid molecular species which induce a membrane phase transition well above 0:°C, they should be fully saturated or *trans*-monounsaturated ones. Of all the glycerolipids from leaf cells, only PG contains high levels of these molecular species.[55,76,80,81] This conclusion is supported by the observation of Murata and Yamaya[70] that PG, but no other lipid class, from chilling-sensitive plants undergoes the gel to liquid-crystalline phase transition at room temperature or above. In some chilling-sensitive plants, low levels of saturated molecular species of SQDG are also found.[20,82,83]

Although the saturated or *trans*-monounsaturated lipid molecular species are suspected to initiate the phase transition of biological membranes at chilling temperatures, the temperature critical for injury of the chilling-sensitive plants, 5 to 15°C, is far below the phase transition temperature of the saturated and *trans*-monounsaturated species. This may be attributable to the lipid-lipid and lipid-protein interactions in the membrane which decrease the phase transition temperature of the high-melting-point molecular species.

The phase behavior of binary lipid mixtures consisting of saturated and *cis*-unsaturated lipid molecular species has been extensively studied,[68,84-88] although these studies use lipid molecular species which are absent, or present only at low levels, in plant membranes. It is clearly demonstrated, however, that the phase transition temperature of a saturated lipid molecular species is reduced by the presence of a low-melting-point mono- or diunsaturated molecular species. A similar phase behavior is observed in a binary mixture of different lipid classes.[89] These observations suggest that the abundant polyunsaturated lipid molecular species decrease the phase transition temperature of 16:0/16:0-PG and 16:0/16:1t-PG in the cellular membrane system. This is confirmed by the phase behavior of lipids extracted from plant membranes.[81,90]

The effect of proteins on thermotropic phase behavior of phospholipid membranes have been investigated with reconstituted proteoliposomes.[87,91-94] Papahadjopoulos et al.[92] have classified the lipid-protein interaction into three groups. In the first group, an extrinsic protein such as ribonuclease and polylysine binds onto the surface of the lipid bilayer through electrostatic force, and this interaction increases the enthalpy of the phase transition of the lipid bilayer with either an increase or no change in the phase transition temperature. In the second group, an extrinsic protein such as cytochrome c and A-1 basic myelin protein binds onto the surface of lipid bilayer by partial penetration and/or deformation of the bilayer, and this interaction significantly decreases both temperature and enthalpy of the lipid phase transition. The above two interactions require the net negative charge on the lipid bilayer for the binding of protein.

On the other hand, the third group of interaction caused by the intrinsic protein, such as the N-2 apoprotein of myelin proteolipid and gramicidin A, is primarily nonpolar in character. This type of interaction does not significantly affect the phase transition temperature of the lipid bilayer, but decreases the enthalpy of the phase transition. The intrinsic proteins in the lipid bilayer appear to interact with a limited number of lipid molecules, namely, boundary lipids[95] or annular lipids,[96] in which the rotational freedom is greatly restricted.[87] When the ratio of intrinsic protein to lipid is high such that the membrane consists primarily of protein and its annular lipid, the lipid phase transition is significantly broadened.[97] A theoretical prediction suggests that the intrinsic protein has antagonistic effects on the lipid phase transition temperature of such lipid bilayers, depending on the strength of the Van der Waals interaction between lipid and protein molecules.[98] When the interaction is weak, the protein decreases the lipid phase transition temperature. When the interaction is strong, the protein increases the transition temperature. The annular lipid is regarded to regulate enzyme activities of intrinsic proteins.[96] In the Na/K-ATPase from rabbit kidney outer medulla, for example, the fluidity of annular lipids, but not the bulk lipids, appears to control the activity.[87,99]

B. SPHINGOLIPIDS AND STEROLS

Recently, Ohnishi et al.[25] have determined phase transition temperatures of major molecular species of plant sphingolipid, i.e., cerebroside. The molecular species having only *trans*-double bond(s) show the phase transition above 45°C. The transition temperature of molecular species having one *cis*-double bond ranges between 20 and 35°C, regardless of the presence or absence of a *trans*-double bond. A species having two *cis*-double bonds has a transition temperature of 10°C. These findings suggest that almost all the molecular species of plant sphingolipids undergo a phase transition at room temperature or above. As will be discussed later, chilling-resistant plants, as well as chilling-sensitive plants, contain a large proportion of sphingolipid molecular species with a phase transition temperature above room temperature.[25] The sphingolipids extracted from the plasma membranes and tonoplasts of mung bean hypocotyls undergo the phase transition at 38 and 35°C, respectively.[100] The demonstration that sphingolipids exist in tonoplast and plasma membranes suggests that this class of lipids may induce a phase transition in these membranes at chilling temperatures.

Cholesterol does not form bilayers when dispersed in water at physiological temperature,[101] but can be incorporated into bilayers of other lipids such as PC, PE, and sphingolipid. The characteristic of these binary-mixture membranes has been extensively studied in order to understand the physiological role of cholesterol in biological membranes.[89,102,103] As the ratio of cholesterol to glycerolipid is increased, both temperature and enthalpy of the gel to liquid-crystalline phase transition of the glycerolipid decrease. When the ratio is increased to 1:2, the endothermic signal of differential scanning calorimetry completely disappears. This observation is explained as follows: below the phase transition temperature of the glycerolipid, cholesterol molecules intercalate with glycerolipid molecules so that the lipid bilayer is "liquefied", whereas above the phase transition temperature, cholesterol molecules restrict mobility of the acyl chains of glycerolipid molecules in the liquid-crystalline phase so that the lipid bilayer becomes more "condensed". Therefore, it is suggested that a lipid bilayer containing an appropriate amount of cholesterol maintains an intermediate fluidity over a wide range of temperature.[89] Since the effect of plant sterols on the phase behavior of 16:0/16:0-PC and 14:0/14:0-PC is very similar to that of cholesterol,[104,105] they probably also act as modulators of membrane fluidity.[106,107]

When cholesterol is mixed with a mixture of two molecular species of the same glycerolipid class, it preferentially interacts with the molecular species having the lower transition temperature.[103] This result indicates that cholesterol prefers to be located in the liquid-crystalline rather than the gel phase. On the other hand, when cholesterol is mixed with a mixture of different lipid classes, it preferentially interacts with the lipids in the order of sphingomyeline > PC > PE, regardless of their phase transition temperatures.[108] The last observation suggests that sterols exert their greatest effect over the phase behavior of sphingolipids in plasma membranes and tonoplasts, which contain high proportions of sphingolipids and sterols.

IV. LIPIDS POSSIBLY RELATED TO CHILLING SENSITIVITY

A. PHOSPHATIDYLGLYCEROL (PG)
1. Molecular Species

Chilling-sensitive plants usually contain much higher proportions of 16:0/16:0 plus 16:0/16:1t species of PG than chilling-resistant plants. In about 20 plant species examined in our laboratory, the sum of the contents of these molecular species ranges from 3 to 19% of the total PG in the chilling-resistant plants, and from 26 to 65% in the chilling-sensitive plants.[76,80] This may indicate that these two molecular species in chloroplast membranes are closely associated with the chilling sensitivity of plants. Roughan[55] has surveyed the fatty acid composition of PG in 74 plant species to confirm the correlation between chilling sensitivity and the percentage of saturated plus *trans*-monounsaturated PG molecular species. In a recent work by Li *et al.*,[109] the correlation was shown to hold for rice varieties having different sensitivities to chilling.

In addition to 16:0/16:0-PG and 16:0/16:1t-PG, 18:0/16:0-PG and 18:0/16:1t-PG species exist at low proportions. As described before, the transition temperatures of the latter two species are expected to be higher than those of the former two species. These four molecular species are called "high-melting-point PG molecular species" according to Kenrick and Bishop,[83] or "disaturated PG molecular species" according to Roughan.[55] In the latter case, the *trans*-monounsaturated species are considered as saturated ones because the *trans*-double bond does not cause much decrease in the phase transition temperature as the *cis*-double bond.[67]

The content of saturated plus *trans*-monounsaturated PG species relative to the total PG in etioplast and amyloplast membranes is almost identical with that in chloroplasts of the same plant.[110] This suggests that the mechanism of PG synthesis and the temperature-dependent behavior of the membranes are essentially the same in these types of plastids.

Sparace and Mudd,[54] Andrews and Mudd,[53] and Roughan[56] have demonstrated that PG is synthesized in chloroplasts from glycerol 3-phosphate and acetate. Based on this finding, and on the positional distribution of fatty acids in PG molecular species, we have proposed biosynthetic pathways for the various PG molecular species,[76] presented in Figure 3. In pathway A, 18:1 is esterified to the *sn*-1 position of glycerol 3-phosphate, and 16:0 to the

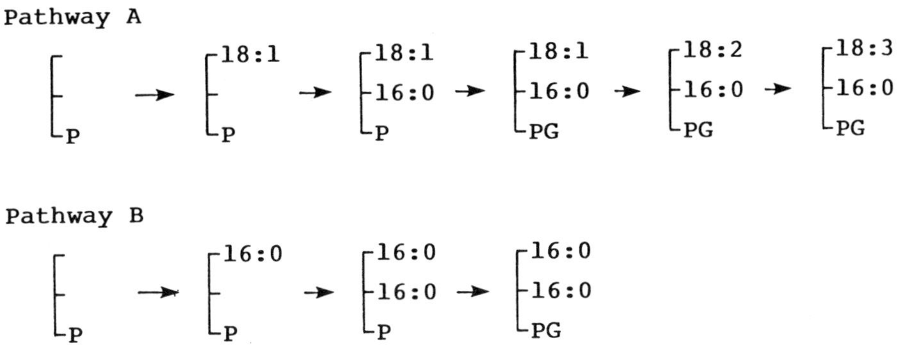

Pathway A

Pathway B

FIGURE 3. Two pathways proposed for biosynthesis of the PG molecules in the plastids of chilling-resistant and chilling-sensitive plants. Pathway A is dominant in the chilling-resistant plants, whereas both pathways A and B are of comparable activity in the chilling-sensitive plants. The *trans*-desaturation to form 16:1t at the *sn*-2 position of the glycerol moiety is not presented in this scheme. (P) phosphate; (PG) glycerophosphate.

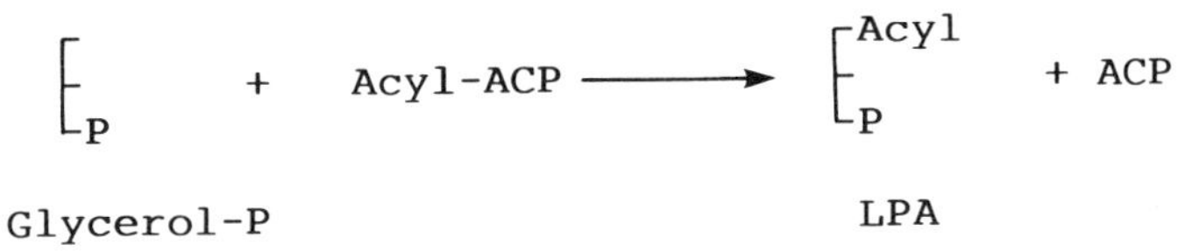

FIGURE 4. The reaction catalyzed by plastidial acyl-ACP:glycerol-3-phosphate acyltransferase. Glycerol-P, glycerol 3-phosphate; LPA, Lysophosphatidic acid.

sn-2 position. After the PA thus produced is converted to PG, most of the 18:1 at the *sn*-1 position is desaturated to 18:2 and subsequently to 18:3. A part of 16:0 at the *sn*-2 position is converted to 16:1t (this desaturation is not described in Figure 3). The combination of the fatty acids thus produced forms a variety of *cis*-unsaturated molecular species which undergo phase transition at about 0°C or below. In pathway B, 16:0, and a small proportion of 18:0, are esterified to the *sn*-1 positions of glycerol 3-phosphate and then 16:0 to the *sn*-2 position. After the conversion of PA into PG, a part of 16:0 at the *sn*-2 position is *trans*-desaturated to 16:1t (this is not described in Figure 3), whereas 16:0 and 18:0 at the *sn*-1 position are not desaturated at all, resulting in the formation of four molecular species of high melting points: 16:0/16:0, 16:0/16:1t, 18:0/16:0, and 18:0/16:1t.

In chilling-resistant plants, in which the sum of the 16:0/16:0 plus 16:0/16:1t content is low, pathway A should be favored over pathway B. In chilling-sensitive plants, on the other hand, pathway A and pathway B should be comparably active, resulting in production of saturated and *trans*-monounsaturated PG species in addition to *cis*-unsaturated molecular species.

If this scheme is valid, chilling sensitivity or resistance in higher plants is a consequence of variation in the proportion of 16:0/16:0 plus 16:0/16:1t PG molecular species, which, in turn, must result from the preferential transfer of 16:0 or 18:1 to the *sn*-1 position of glycerol 3-phosphate. This reaction is catalyzed by a stromal enzyme, acyl-ACP:glycerol-3-phosphate acyltransferase (EC 2.3.1.15, abbreviated as glycerol-P acyltransferase). This enzyme transfers the acyl group from acyl-ACP to the *sn*-1 position of glycerol 3-phosphate to yield lysophosphatidic acid (Figure 4). The preferential acyl transfer becomes possible if one of the following mechanisms operates in the reaction in Figure 4. In the first possible mechanism, glycerol-P acyltransferases from chilling-sensitive and chilling-resistant plants have different selectivities toward 18:1-ACP and 16:0-ACP; i.e., the enzyme in the chilling-resistant plants has a rather strict selectivity for 18:1-ACP over 16:0-ACP, whereas that in the chilling-sensitive plants is unspecific for either of the substrates and transfers both 18:1 and 16:0 at comparable rates. This mechanism, which is most likely to explain the correlation between the 16:0/16:0 plus 16:0/16:1t PG species and chilling sensitivity of plants, will be discussed in detail in the next section.

In the second possible mechanism, it is assumed that no great difference exists in the substrate selectivity of glycerol-P acyltransferase between the chilling-sensitive and chilling-resistant plants, but the pool sizes of 18:1-ACP and 16:0-ACP are different; i.e., 18:1-ACP is more abundant than 16:0-ACP in the chilling-resistant plants, whereas both of donors of acyl-ACP are at about the same levels in the chilling-sensitive plants. Soll and Roughan,[111] and Roughan[112] have determined the acyl donor pool sizes in chilling-resistant spinach and chilling-sensitive *Amaranthus* and found that 18:1-ACP and 16:0-ACP are at about the same levels in both plants. Therefore, this possibility can be excluded as an explanation for the correlation of the 16:0/16:0 plus 16:0/16:1t PG species with the chilling sensitivity of plants.

In the third possible mechanism, the substrate selectivity of glycerol-P acyltransferase and the acyl donor pool sizes are both assumed to be about the same between chilling-sensitive and chilling-resistant plants, but the relative contents of two isomeric forms of ACP[113] are different; an isomeric form, ACP I, which is a more efficient acyl donor for glycerol-P acyltransferase than ACP II,[113] binds a greater amount of 18:1 than 16:0 in chilling-resistant plants, whereas it binds comparable amounts of 18:1 and 16:0 in chilling-sensitive plants. This possibility has not been systematically tested.

Although the high-melting-point PG molecular species are very likely to be a main determinant to initiate the phase transition in the chloroplast membranes, other parameters can be considered which may modulate the thermotropic phase behavior of PG molecular species: (1) The relative content of *trans*-unsaturated species to the total high-melting-point species may affect the phase behavior of PG, since the introduction of a *trans*-unsaturated bond decreases the phase transition temperature by about 10°C (Table 4). It is noted that the 16:1t to 16:0 ratio in PG is lower in the inner envelope membrane than the thylakoid membrane.[114] The mechanism for *trans*-desaturation is not known at all, but a specific enzyme is supposed to catalyze this reaction.[115] (2) The degree of *cis*-unsaturation among the *cis*-unsaturated PG species may affect the phase behavior of the saturated and *trans*-monounsaturated species. As described in the previous section, in a mixture of molecular species having different transition temperatures, the transition temperature of the high-transition-temperature species is decreased by the presence of the low-transition-temperature species. Moreover, the decrease in the transition temperature of the high-transition-temperature species depends on the transition temperature of the low-transition-temperature species. Therefore, an increase in the degree of *cis*-unsaturation of the *cis*-unsaturated (and, therefore, low-transition-temperature) species is expected to decrease the phase transition temperature of saturated and *trans*-monounsaturated PG species in the chloroplast membranes. (3) Strong interaction of the intrinsic proteins with membrane lipids increases the phase transition temperature of the lipids. This is well demonstrated in the cyanobacterium *Anacystis nidulans*, in which the plasma membrane and thylakoid membrane have very similar lipid and fatty acid compositions, but different protein contents. The relative content of protein to the total membrane constituents on a weight basis is 70% in the thylakoid membrane and 40% in the plasma membrane.[116] This difference in the protein content explains why the phase transition temperature of the thylakoid membrane is higher by 10°C than that of the plasma membrane.[9] It is noted that the protein to lipid ratio in the thylakoid membrane is higher than that in envelope membrane in higher plant chloroplasts.[117]

2. Acyl-ACP:glycerol-3-phosphate Acyltransferase

The substrate selectivity of glycerol-P acyltransferase in plastids toward the acyl group in the acyl-ACP is very likely to be a determinant of chilling sensitivity in higher plants.

This enzyme was partially purified by Bertram and Heinz[118] in a soluble form from leaves of two chilling-resistant plants (pea and spinach). Recently, we have purified the enzyme from cotyledons of a chilling-sensitive plant (squash).[119] Squash cotyledons contain three isozymes, designated as AT1, AT2, and AT3, and the latter two have been purified to single components after 40,000- and 32,000-fold purification, respectively. Molecular masses estimated for AT1, AT2, and AT3 are about 30 kDa, 40 kDa, and 40 kDa, respectively, and their isoelectric points are pI = 6.6, 5.6, and 5.5, respectively. We have also found at least two isozymes of glycerol-P acyltransferase, each having different isoelectric points and molecular masses, in all the plants examined (rice, barley, spinach, pea, sunflower, and maize).[120]

To determine whether the glycerol-P acyltransferase from chilling-sensitive plants exhibits fatty acid selectivities different from those of resistant plants, the activities of the enzymes from squash, spinach, and pea were compared.[121,122] The enzymes from spinach and pea preferentially acylated 18:1 to the *sn*-1 position of glycerol 3-phosphate. Of the three isozymes from squash, AT1 showed a preference for 18:1-ACP, as did spinach and pea enzymes. AT2 and AT3, on the other hand, hardly discriminated between 18:1-ACP and 16:0-ACP. In addition, the observed selectivity of AT1 for 18:1-ACP over 16:0-ACP was substantially reduced when the pH of the reaction mixture was increased from 7.4 to 8.0, the latter being the stromal pH of illuminated chloroplasts. The fatty acid selectivity of glycerol-P acyltransferase can therefore explain the low content of 16:0/16:0 plus 16:0/16:1t PG molecular species in chilling-resistant plants and the high content of these molecular species in chilling-sensitive plants.[122]

Direct verification of the proposed role of glycerol-P acyltransferase in the chilling sensitivity of higher plants could be obtained by cloning the gene for glycerol-P acyltransferase and subsequently transforming plants with the cloned gene. For this purpose, a cDNA clone for glycerol-P acyltransferase has been isolated from a cDNA library which is constructed from the poly(A)$^+$RNA obtained from greening squash cotyledons.[123] When the amino acid sequence deduced from the cDNA sequence was compared with the amino terminal sequence determined for AT2, it became apparent that glycerol-P acyltransferase is synthesized as a precursor protein with a leader sequence of 28 amino acid residues and processed to become a mature protein of 368 amino acid residues. Experiments for transforming plants in terms of the lipid molecular species composition and chilling sensitivity is now in progress in the authors' laboratory.

B. SULFOQUINOVOSYLDIACYLGLYCEROL (SQDG)

Another glycerolipid which contains a saturated molecular species is SQDG.[20,82,83] The content of saturated SQDG molecular species (16:0/16:0 plus 18:0/16:0) relative to the total SQDG ranges from 0 to 20%,[20] which is much lower than in the case of PG. Although there is little correlation between the saturated SQDG and chilling sensitivity,[20] the saturated SQDG, if present, may interact with the saturated and *trans*-monounsaturated PG molecular species by stimulating the formation of gel phase domains in the chloroplast membranes. The SQDG is synthesized by both plastidial and cytoplasmic pathways.[19] The SQDG molecular species synthesized by the cytoplasmic pathway are 16:0/C18 and C18/C18, in which the C18 fatty acids are *cis*-unsaturated. The SQDG molecular species synthesized by the plastidial pathway are C18(*cis*-unsaturated)/16:0 and 16:0/16:0. Therefore, saturated molecular species of SQDG originate with the transfer of 16:0 from 16:0-ACP to the *sn*-1 position of glycerol 3-phosphate which is catalyzed by glycerol-P acyltransferase in the chloroplasts. Thus, the enzyme responsible for synthesis of the 16:0/16:0 and 16:0/16:1t species of PG is also responsible for production of saturated SQDG species. This is supported by the finding that plants which contain saturated molecular species of SQDG always contain high levels of the saturated plus *trans*-monounsaturated molecular species of PG.[20,82,83]

C. SPHINGOLIPIDS AND STEROLS

Recently, Ohnishi *et al.*[25] have studied the molecular species composition of sphingo-lipids from chilling-sensitive and chilling-resistant plants. Of the 13 plants studied, only wheat and rye contained a *cis*-diunsaturated species, which undergoes the phase transition at about 10°C. The other 11 plants, which included both chilling-sensitive and chilling-resistant species, contained sphingolipid molecular species with no or only one *cis*-double bond per molecule, and therefore had a phase transition temperature ranging from 20 to 60°C. These findings suggest that sphingolipids may induce a phase transition in tonoplasts and/or plasma membranes at chilling temperature. However, in contrast to the data for PG, no clear correlation exists between the content of *cis*-mono- or *cis*-diunsaturated sphingolipid species and the chilling sensitivity of plants.[25]

The enzymes responsible for *cis*-desaturation of the sphingoid and fatty acid moieties of plant sphingolipids have not as yet been studied or isolated. Until the enzymes involved in the biosynthesis of various sphingolipid molecular species are identified and characterized, no attempt to alter the phase behavior of sphingolipids can be made.

As descirbed in a previous section, sterols and sphingolipids most strongly interact with each other, and sterols diminish the phase transition of sphingolipids. This may occur in plasma and tonoplast membranes in plant cells. Therefore, it is possible that the sterol to sphingolipid ratio in plasma and tonoplast membranes determines the phase behavior of these membranes and, thus, the chilling sensitivity of plants. However, the mechanism to regulate the sterol and sphingolipid contents is not known at present.

V. CONCLUSIONS

Higher plant membranes contain seven major glycerolipids, sphingolipids, and sterols. Among the glycerolipids PG and SQDG, localized in the plastid membranes, are contained molecular species with phase transition temperatures above 30°C, namely: 16:0/16:0-PG, 16:0/16:1t-PG, 18:0/16:0-PG, 18:0/16:1t-PG, 16:0/16:0-SQDG, and 18:0/16:0-SQDG. The relative content of the sum of these molecular species is well correlated with chilling sensitivity of higher plants. All the other molecular species of PG, SQDG, and other glycerolipids undergo phase transitions below 5°C. Therefore, it is hypothesized that the PG and SQDG molecular species with high phase transition temperatures are a determinant of chilling sensitivity of plants (PG hypothesis). Molecular species of sphingolipid undergo phase transition in the temperature range between 10 and 60°C, but its molecular species composition is not well correlated with chilling sensitivity.

The PG and SQDG molecular species with high phase transition temperatures are synthesized within plastids. Nuclear-coded, plastidial enzymes, acyl-ACP:glycerol 3-phosphate acyltransferase, from chilling-sensitive and chilling-resistant plants reveal different substrate selectivities which can explain the correlation of lipid molecular species with chilling sensitivity. This enzyme has been isolated from a chilling-sensitive plant, and its cDNA has been cloned.

In the near future, genes for this enzyme will be isolated from several species of chilling-sensitive and chilling-resistant plants and be introduced to other plants having different chilling sensitivities. This will provide a direct answer to the PG hypothesis. If the hypothesis is proven to be true, it will open new perspectives for transforming crop plants from chilling-sensitive to chilling-resistant. This result would promise higher productivity of crops in the temperate and subarctic area and also improvement of storage of horticultural products.

ACKNOWLEDGMENT

This work was supported in part by a grant from the NIBB Program for Biomembrane

Research and Grants-in-Aid for Scientific Research (61440002) and Cooperative Research (62304004) from the Ministry of Education, Science and Culture, Japan.

REFERENCES

1. **Satake, T. and Koike, S.,** Sterility caused by cooling treatment at the flowering stage in rice plants, *Jpn. J. Crop Sci.,* 52, 207, 1983.
2. **Patterson, B. D., Mutton, L., Paull, R. E., and Nguyen, V. Q.,** Tomato pollen development: stages sensitive to chilling and a natural environment for the selection of resistant genotypes, *Plant Cell Environ.,* 10, 363, 1987.
3. **Lyons, J. M.,** Chilling injury in plants, *Annu. Rev. Plant Physiol.,* 24, 445, 1973.
4. **Raison, J. K.,** The influence of temperature-induced phase changes on kinetics of respiratory and other membrane-associated enzymes, *J. Bioenerg.,* 4, 258, 1973.
5. **De Gier, J., Block, M. C., van Dijck, P. W. M., Mombers, C., Verkley, A. J., van der Neut-Kok, E. C. M., and van Deenen, L. L. M.,** Relations between liposomes and biomembranes, *Ann. N.Y. Acad. Sci.,* 308, 85, 1978.
6. **Ono, T. and Murata, N.,** Chilling susceptibility of the blue-green alga *Anacystis nidulans.* I. Effect of growth temperature, *Plant Physiol.,* 67, 176, 1981.
7. **Ono, T. and Murata, N.,** Chilling susceptibility of the blue-green alga *Anacystis nidulans.* II. Stimulation of the passive permeability of cytoplasmic membrane at the chilling temperatures, *Plant Physiol.,* 67, 182, 1981.
8. **Ono, T. and Murata, N.,** Chilling susceptibility of the blue-green alga, *Anacystis nidulans.* III. Lipid phase of cytoplasmic membrane, *Plant Physiol.,* 69, 125, 1982.
9. **Murata, N., Wada, H., and Hirasawa, R.,** Reversible and irreversible interaction of photosynthesis in relation to the lipid phases of membranes in the blue-green algae (cyanobacteria) *Anacystis nidulans* and *Anabaena variabilis, Plant Cell Physiol.,* 25, 1027, 1984.
10. **Murata, N. and Nishida, I.,** Lipids of blue-green algae (cyanobacteria), in *The Biochemistry of Plants,* Vol. 9, Stumpf, P. K., Ed., Academic Press, New York, 1987, 315.
11. **Block, M. A., Dorne, A.-J., Joyard, J., and Douce, R.,** Preparation and characterization of membrane fractions enriched in outer and inner envelope membranes from spinach chloroplasts. II. Biochemical characterization, *J. Biol. Chem.,* 258, 13281, 1983.
12. **Donaldson, R. P. and Beevers, H.,** Lipid composition of organelles from germinating castor bean endosperm, *Plant Physiol.,* 59, 259, 1977.
13. **Yoshida, S. and Uemura, M.,** Lipid composition of plasma membranes and tonoplasts isolated from etiolated seedlings of mung bean (*Vigna radiata* L.), *Plant Physiol.,* 82, 807, 1986.
14. **Douce, R., Alban, C., Bligny, R., Block, M. A., Coves, J., Dorne, A.-J., Journet, E.-P., Joyard, J., Neuburger, M., and Rebeille, F.,** Lipid distribution and synthesis within the plant cell, in *The Metabolism, Structure, and Function of Plant Lipids,* Stumpf, P. K., Mudd, J. B., and Nes, W. D., Eds., Plenum Press, New York, 1987, 255.
15. **Haverkate, F. and van Deenen, L. L. M.,** Isolation and chemical characterization of phosphatidyl glycerol from spinach leaves, *Biochim. Biophys. Acta,* 106, 78, 1965.
16. **Jamieson, G. R. and Reid, E. H.,** The occurrence of hexadeca-7,10,13-trienoic acid in the leaf lipids of angiosperms, *Phytochemistry,* 10, 1837, 1971.
17. **Nishihara, M., Yokota, K., and Kito, M.,** Lipid molecular species composition of thylakoid membranes, *Biochim. Biophys. Acta,* 617, 12, 1980.
18. **Heinz, E.,** Enzymatic reaction in galactolipid biosynthesis, in *Lipids and Lipid Polymers in Higher Plants,* Tevini, M. and Lichtenthaler, H. K., Eds., Springer-Verlag, Berlin, 1977, 102.
19. **Bishop, D. G., Sparace, S. A., and Mudd, J. B.,** Biosynthesis of sulfoquinovosyldiacylglycerol in higher plants: the origin of diacylglycerol moiety, *Arch. Biochem. Biophys.,* 240, 851, 1985.
20. **Murata, N. and Hoshi, H.,** Sulfoquinovosyl diacylglycerols in chilling-sensitive and chilling-resistant plants, *Plant Cell Physiol.,* 25, 1241, 1984.
21. **Toriyama, S., Hinata, K., Nishida, I., and Murata, N.,** Prominent difference of glycerolipids among anther walls, pollen grains and leaves of rice and maize, *Plant Cell Physiol.,* 29, 615, 1988.
22. **Ohnishi, M., Ito, S., and Fujino, Y.,** Characterization of sphingolipids in spinach leaves, *Biochim. Biophys. Acta,* 752, 416, 1983.
23. **Ohnishi, M., Ito, S., and Fujino, Y.,** Structural characterization of sphingolipids in leafy stems of rice, *Agric. Biol. Chem.,* 49, 3327, 1985.

24. **Lynch, D. V. and Steponkus, P. L.,** Plasma membrane lipid alterations associated with cold acclimation of winter rye seedlings (*Secale cereale* L. cv. Puma), *Plant Physiol.,* 83, 761, 1987.

25. **Ohnishi, M., Imai, H., Kojima, M., Yoshida, S., Fujino, Y., and Ito, S.,** Separation of cerebroside species in plants by reversed-phase HPLC and their phase transition temperature, Abstr. 19th World Cong. Int. Soc. Fat Research, 1988.

26. **Mudd, J. B.,** Sterol interconversions, in *The Biochemistry of Plants,* Vol. 4, Stumpf, P. K., Ed., Academic Press, New York, 1980, 509.

27. **Grunwald, C.,** Plant sterols, *Annu. Rev. Plant Physiol.,* 26, 209, 1975.

28. **Goad, L. J., Zimowski, J., Evershed, R. P., and Male, V. L.,** The steryl esters of higher plants, in *The Metabolism, Structure, and Function of Plant Lipids,* Stumpf, P. K., Mudd, J. B., and Nes, W. D., Eds, Plenum Press, New York, 1987, 95.

29. **Fernholz, E. and Moore, M. L.,** The isolation of α-spinasterol from alfalfa, *J. Am. Chem. Soc.,* 61, 2467, 1939.

30. **Armarego, W. L. F., Goad, L. J., and Goodwin, T. W.,** Biosynthesis of α-spinasterol from (2-^{14}C, (4R)-4-^{3}H$_1$) mevalonic acid by *Spinacea oleracea* and *Medicago sativa, Phytochemistry,* 12, 2181, 1973.

31. **Artaud, J., Iatrides, M.-C., and Gaydou, E. M.,** Co-occurrence of Δ^5- and Δ^7-sterols in two *Gleditsia* species. A reassessment of the sterol composition in oils rich in Δ^7-sterols, *Phytochemistry,* 23, 2303, 1984.

32. **Garg, V. K. and Nes, W. R.,** Studies on the C-24 configurations of Δ^7-sterols in the seeds of *Cucurbita maxima, Phytochemistry,* 23, 2919, 1984.

33. **Benveniste, P.,** Sterol biosynthesis, *Annu. Rev. Plant Physiol.,* 37, 275, 1986.

34. **Noda, M., Tanaka, M., Seto, Y., Aiba, T., and Oku, C.,** Occurrence of cholesterol as a major sterol component in leaf surface lipids, *Lipids,* 23, 439, 1988.

35. **Stumpf, P. K.,** Biosynthesis of saturated and unsaturated fatty acids, in *The Biochemistry of Plants,* Vol. 4, Stumpf, P. K., Ed., Academic Press, New York, 1980, 177.

36. **Stumpf, P. K.,** The biosynthesis of saturated fatty acids, in *The Biochemistry of Plants,* Vol. 9, Stumpf, P. K., Ed., Academic Press, New York, 1987, 121.

37. **Jaworski, J. G.,** Biosynthesis of monoenoic and polyenoic fatty acids, in *The Biochemistry of Plants,* Vol. 9, Stumpf, P. K., Ed., Academic Press, New York, 1987, 159.

38. **Joyard, J. and Douce, R.,** Galactolipid synthesis, in *The Biochemistry of Plants,* Vol. 9, Stumpf, P. K., Ed., Academic Press, New York, 1987, 215.

39. **Roughan, G.,** On the control of fatty acid compositions of plant glycerolipids, in *The Metabolism, Structure, and Function of Plant Lipids,* Stumpf, P. K., Mudd, J. B., and Nes, W. D., Eds., Plenum Press, New York, 1987, 247.

40. **Roughan, P. G. and Slack, C. R.,** Cellular organization of glycerolipid metabolism, *Annu. Rev. Plant Physiol.,* 33, 97, 1982.

41. **Roughan, P. G. and Slack, C. R.,** Glycerolipid synthesis in leaves, *Trends Biochem. Sci.,* 9, 383, 1984.

42. **Moore, T. S., Jr.,** Biochemistry and biosynthesis of plant acyl lipids, in *Structure, Function and Metabolism of Plant Lipids,* Siegenthaler, P.-A. and Eichenberger, W., Eds., Elsevier, Amsterdam, 1984, 83.

43. **Sauer, A. and Robinson, D. G.,** Subcellular localization of enzymes involved in lecithin biosynthesis in maize roots, *J. Exp. Bot.,* 36, 1257, 1985.

44. **Sparace, S. A. and Moore, T. S., Jr.,** Phospholipid metabolism in plant mitochondria. Submitochondrial sites of synthesis, *Plant Physiol.,* 63, 963, 1979.

45. **Dorne, A.-J., Block, M. A., Joyard, J., and Douce, R.,** The galactolipid:galactolipid galactosyltransferase is located on the outer surface of the outer membrane of the chloroplast envelope, *FEBS Lett.,* 145, 30, 1982.

46. **Andrews, J., Ohlrogge, J. B., and Keegstra, K.,** Final step of phosphatidic acid synthesis in pea chloroplasts occurs in the inner envelope membrane, *Plant Physiol.,* 78, 459, 1985.

47. **Moore, T. S., Jr.,** Phospholipid biosynthesis, *Annu. Rev. Plant Physiol.,* 33, 235, 1982.

48. **Moore, T. S., Jr.,** Regulation of phospholipid headgroup composition in castor bean endosperm, in *The Metabolism, Structure, and Function of Plant Lipids,* Stumpf, P. K., Mudd, J. B., and Nes, D. W., Eds., Plenum Press, New York, 1987, 265.

49. **Morre, D. J., Nyquist, S., and Rivera, E.,** Lecithin biosynthetic enzymes of onion stem and the distribution of phosphorylcholine-cytidyltransferase among cell fractions, *Plant Physiol.,* 45, 800, 1970.

50. **Montague, M. J. and Ray, P. M.,** Phospholipid-synthesizing enzymes associated with Golgi dictyosomes from pea tissue, *Plant Physiol.,* 59, 225, 1977.

51. **Sparace, S. A. and Moore, T. S., Jr.,** Phospholipid metabolism in plant mitochondria. II. Submitochondrial site of synthesis of phosphatidylcholine and phosphatidylethanolamine, *Plant Physiol.,* 67, 261, 1981.

52. **Mudd, J. B. and Dezacks, R.,** Synthesis of phosphatidylglycerol by chloroplasts from leaves of *Spinacia oleracea* L. (spinach), *Arch. Biochem. Biophys.,* 209, 584, 1981.

53. **Andrews, J. and Mudd, J. B.,** Phosphatidylglycerol synthesis in pea chloroplasts. Pathway and localization, *Plant Physiol.,* 79, 259, 1985.

54. **Sparace, S. A. and Mudd, J. B.,** Phosphatidylglycerol synthesis in spinach chloroplasts: characterization of the newly synthesized molecule, *Plant Physiol.,* 70, 1260, 1982.
55. **Roughan, P. G.,** Phosphatidylglycerol and chilling sensitivity in plants, *Plant Physiol.,* 77, 740, 1985.
56. **Roughan, P. G.,** Cytidine triphosphate-dependent, acyl-CoA-independent synthesis of phosphatidylglycerol by chloroplasts isolated from spinach and pea, *Biochim. Biophys. Acta,* 835, 527, 1985.
57. **Joyard, J., Blée, E., and Douce, R.,** Sulfolipid synthesis from $^{35}SO_4^{-2}$ and [1-^{14}C]acetate in isolated intact spinach chloroplasts, *Biochim. Biophys. Acta,* 879, 78, 1986.
58. **Kleppinger-Sparace, K. F. and Mudd, J. B.,** Biosynthesis of sulfoquinovosyldiacylglycerol in higher plants. The incorporation of $^{35}SO_4^{2-}$ by intact chloroplasts in darkness, *Plant Physiol.,* 84, 682, 1987.
59. **Roughan, P. G., Holland, R., and Slack, R.,** The role of chloroplasts and microsomal fractions in polar-lipid synthesis from [1-^{14}C] acetate by cell-free preparations from spinach *(Spinacia oleracea)* leaves, *Biochem. J.,* 188, 17, 1980.
60. **Ohnishi, J. and Yamada, M.,** Glycerolipid synthesis in *Avena* leaves during greening of etiolated seedlings. II. α-Linolenic acid synthesis, *Plant Cell Physiol.,* 21, 1607, 1980.
61. **Mollenhauer, H. H. and Morre, D. J.,** The Golgi apparatus, in *The Biochemistry of Plants,* Vol. 1, Tolbert, N. E., Ed., Academic Press, New York, 1980, 437.
62. **Marty, F., Branton, D., and Leigh, R. A.,** Plant vacuoles, in *The Biochemistry of Plants,* Vol. 1, Tolbert, N. E., Ed., Academic Press, New York, 1980, 625.
63. **Roughan, P. G., Thompson, G. A., Jr., and Cho, S. H.,** Metabolism of exogenous long-chain fatty acids by spinach leaves, *Arch. Biochem. Biophys.,* 259, 481, 1987.
64. **Nes, W. R.,** The biochemistry of plant sterols, *Adv. Lipid Res.,* 15, 233, 1977.
65. **Goodwin, T. W.,** Biosynthesis of sterols, in *The Biochemistry of Plants,* Vol. 4, Stumpf, P. K., Ed., Academic Press, New York, 1980, 485.
66. **Goodwin, T. W.,** Biosynthesis of terpenoids, *Annu. Rev. Plant Physiol.,* 30, 369, 1979.
67. **Silvius, J. R.,** Thermotropic phase transitions of pure lipids in model membranes and their modification by membrane proteins, in *Lipid Protein Interactions,* Jost, P. C. and Griffith, O. H., Eds., John Wiley & Sons, New York, 1982, 239.
68. **Phillips, M. C., Hauser, H., and Paltauf, F.,** The inter- and intra-molecular mixing of hydrocarbon chains in lecithin/water systems, *Chem. Phys. Lipids,* 8, 127, 1972.
69. **Van Dijck, P. W. M., de Kruijff, B., van Deenen, L. L. M., de Gier, J., and Demel, R. A.,** The presence of cholesterol for phosphatidylcholine in mixed phosphatidylcholine-phosphatidylethanolamine bilayers, *Biochim. Biophys. Acta,* 455, 576, 1976.
70. **Murata, N. and Yamaya, J.,** Temperature-dependent phase behavior of phosphatidylglycerols from chilling-sensitive and chilling-resistant plants, *Plant Physiol.,* 74, 1016, 1984.
71. **Bishop, D. G. and Kenrick, J. R.,** Thermal properties of 1-hexadecanoyl-2-*trans*-3-hexadecenoyl phosphatidylglycerol, *Photochemistry,* 26, 3065, 1987.
72. **Higashi, S., Okuyama, H., and Murata, N.,** Unpublished data.
73. **Bishop, D. G., Kenrick, J. R., Kondo, T., and Murata, N.,** Thermal properties of membrane lipids from two cyanobacteria, *Anacystis nidulans* and *Synechococcus* sp., *Plant Cell Physiol.,* 27, 1593, 1986.
74. **Schröppel-Meier, G. and Kaiser, W. M.,** Ion homeostasis in chloroplasts under salinity and mineral deficiency. I. Solute concentrations in leaves and chloroplasts from spinach plants under NaCl or $NaNO_3$ salinity, *Plant Physiol.,* 87, 822, 1988.
75. **Jacobson, K. and Papahadjopoulos, D.,** Phase transitions and phase separations in phospholipid membranes induced by changes in temperature, pH, and concentration of bivalent cations, *Biochemistry,* 14, 152, 1975.
76. **Murata, N.,** Molecular species composition of phosphatidylglycerols from chilling-sensitive and chilling-resistant plants, *Plant Cell Physiol.,* 24, 81, 1983.
77. **Sekiya, J., Koiso, H., Morita, A., and Hatanaka, A.,** Molecular species composition of phosphatidylglycerol in leaves of Camellia species and chilling sensitivity, in *The Metabolism, Structure, and Function of Plant Lipids,* Stumpf, P. K., Mudd, J. B., and Nes, W. D., Eds., Plenum Press, New York, 1987, 377.
78. **Shipley, G. G., Green, J. P., and Nichols, B. W.,** The phase behavior of monogalactosyl, digalactosyl, and sulphoquinovosyl diglycerides, *Biochim. Biophys. Acta,* 311, 531, 1973.
79. **Orr, G., Raison, J. K., and Murata, N.,** Unpublished data, 1988.
80. **Murata, N., Sato, N., Takahashi, N., and Hamazaki, Y.,** Compositions and positional distributions of fatty acids in phospholipids from leaves of chilling-sensitive and chilling-resistant plants, *Plant Cell Physiol.,* 23, 1071, 1982.
81. **Raison, J. K. and Wright, L. C.,** Thermal phase transitions in the polar lipids of plant membranes. Their induction by disaturated phospholipids and their possible relation to chilling injury, *Biochim. Biophys. Acta,* 731, 69, 1983.
82. **Kenrick, J. R. and Bishop, D. G.,** Phosphatidylglycerol and sulphoquinovosyldiacylglycerol in leaves and fruits of chilling-sensitive plants, *Phytochemistry,* 25, 1293, 1986.

83. **Kenrick, J. R. and Bishop, D. G.,** The fatty acid composition of phosphatidylglycerol and sulfoquino-vosyldiacylglycerol of higher plants in relation to chilling sensitivity, *Plant Physiol.,* 81, 946, 1986.

84. **Phillips, M. C., Ladbrooke, B. D., and Chapman, D.,** Molecular interactions in mixed lecithin systems, *Biochim. Biophys. Acta,* 196, 35, 1970.

85. **Blume, A. and Ackermann, T.,** A calorimetric study of the lipid phase transitions in aqueous dispersions of phosphorylcholine-phosphorylethanolamine mixtures, *FEBS Lett.,* 43, 71, 1974.

86. **Marbrey, S. and Sturtevant, J. M.,** Investigation of phase transitions of lipids and lipid mixtures by high sensitivity differential scanning calorimetry, *Proc. Natl. Acad. Sci. U.S.A.,* 73, 3862, 1976.

87. **Lee, A. G.,** Lipid phase transitions and phase diagrams. II. Mixtures involving lipids, *Biochim. Biophys. Acta,* 472, 285, 1977.

88. **Curatolo, W., Sears, B., and Neuringer, L. J.,** A calorimetry and deuterium NMR study of mixed model membranes of 1-palmitoyl-2-oleylphosphatidylcholine and saturated phosphatidylcholines, *Biochim. Biophys. Acta,* 817, 261, 1985.

89. **Oldfield, E. and Chapman, D.,** Dynamics of lipids in membranes: heterogeneity and the role of cholesterol, *FEBS Lett.,* 23, 285, 1972.

90. **Orr, G. R. and Raison, J. K.,** Compositional and thermal properties of thylakoid polar lipids of *Nerium oleander* L. in relation to chilling sensitivity, *Plant Physiol.,* 84, 88, 1987.

91. **Chapman, D., Urbina, J., and Keough, K. M.,** Biomembrane phase transitions. Studies of lipid-water systems using differential scanning calorimetry, *J. Biol. Chem.,* 249, 2512, 1974.

92. **Papahadjopoulos, D., Moscarello, M., Eylar, E. H., and Isac, T.,** Effects of proteins on thermotropic phase transitions of phospholipid membranes, *Biochim. Biophys. Acta,* 401, 317, 1975.

93. **Gomez-Fernandez, J. C., Goni, F.M., Bach, D., Restall, C. J., and Chapman, D.,** Protein-lipid interaction. Biophysical studies of $(Ca^{2+} + Mg^{2+})$-ATPase reconstituted systems, *Biochim. Biophys. Acta,* 598, 502, 1980.

94. **Grant, C. W. M. and McConnell, H. M.,** Glycophorin in lipid bilayers, *Proc. Natl. Acad. Sci. U.S.A.,* 71, 4653, 1974.

95. **Jost, P. C., Griffith, O. H., Capaldi, R. A., and Vanderkooi, G.,** Evidence for boundary lipid in membranes, *Proc. Natl. Acad. Sci., U.S.A.,* 70, 480, 1973.

96. **Warren, G. B., Houslay, M. D., Metcalfe, J. C., and Birdsall, N. J. M.,** Cholesterol is excluded from the phospholipid annulus surrounding and active calcium transport protein, *Nature (London),* 255, 684, 1975.

97. **Curatolo, W., Sakura, J. D., Small, D. M., and Shipley, G. G.,** Protein-lipid interaction: recombinants of the proteolipid apoprotein of myelin with dimyristoyllecithin, *Biochemistry,* 16, 2313, 1977.

98. **Marčelja, S.,** Lipid-mediated protein interaction in membranes, *Biochim. Biophys. Acta,* 455, 1, 1976.

99. **Kimelberg, H. K. and Papahadjopoulos, D.,** Effects of phospholipid acyl chain fluidity, phase transitions, and cholesterol on $(Na^+ + K^+)$-stimulated adenosine triphosphatase, *J. Biol. Chem.,* 249, 1071, 1974.

100. **Yoshida, S., Washio, K., Kenrick, J., and Orr, G.,** Thermotropic properties of lipids extracted form plasma membrane and tonoplast isolated from chilling-sensitive mung bean (*Vigna radiata* [L] Wilczek), *Plant Cell Physiol.,* 29, 1411, 1988.

101. **Israelachvili, J. N., Marcelja, S., and Horn, R. G.,** Physical principles of membrane organization, *Q. Rev. Biophys.,* 13, 121, 1980.

102. **Oldfield, E. and Chapman, D.,** Molecular dynamics of cerebroside-cholesterol and sphingomyelin-cho-lesterol interactions: implications for myelin membrane structure, *FEBS Lett.,* 23, 303, 1972.

103. **Demel, R. A. and de Kruyff, B.,** The function of sterols in membranes, *Biochim. Biophys. Acta,* 457, 109, 1976.

104. **Ghosh, D. and Tinoco, J.,** Monolayer interactions of individual lecithins with natural sterols, *Biochim. Biophys. Acta,* 266, 41, 1972.

105. **McKersie, B. D. and Thompson, J. E.,** Influence of plant sterols on the phase properties of phospholipid bilayers, *Plant Physiol.,* 63, 802, 1979.

106. **McElhaney, R. N.,** The effect of membrane-lipid phase transitions on membrane structure and on the growth of *Acholeplasma laidlawii* B, *J. Supramol. Struct.,* 2, 617, 1974.

107. **Lyons, J. M., Raison, J. K., and Steponkus, P. L.,** The plant membrane in response to low temperature: an overview, in *Low Temperature Stress in Crop Plants,* Lyons, J. M., Graham, D., and Raison, J. K., Eds., Academic Press, New York, 1979, 1.

108. **Demel, R. A., Jansen, J. W. C. M., van Dijck, P. W. M., and van Deenen, L. L. M.,** The preferential interaction of cholesterol with different classes of phospholipids, *Biochim. Biophys. Acta,* 465, 1, 1977.

109. **Li, T., Lynch, D. V., and Steponkus, P. L.,** Molecular species composition of phosphatidylglycerols from rice varieties differing in chilling sensitivity, *Cryo-Letters,* 8, 314, 1987.

110. **Murata, N. and Kurisu, K.,** Fatty acid compositions of phosphatidylglycerols from plastids in chilling-sensitive and chilling-resistant plants, in *Structure, Function and Metabolism of Plant Lipids,* Siegenthaler, P.-A. and Eichenberger, W., Eds., Elsevier, Amsterdam, 1984, 551.

111. **Soll, J. and Roughan, G.,** Acyl-acyl carrier protein pool sizes during steady-state fatty acid synthesis by isolated spinach chloroplasts, *FEBS Lett.,* 146, 189, 1982.

112. **Roughan, P. G.,** Acyl lipid synthesis by chloroplasts isolated from the chilling-sensitive plant *Amaranthus lividus* L., *Biochim. Biophys. Acta,* 878, 371, 1986.

113. **Guerra, D. J., Ohlrogge, J. B., and Frentzen, M.,** Activity of acyl carrier protein isoforms in relation of plant fatty acid metabolism, *Plant Physiol.,* 82, 448, 1986.

114. **Bahl, J., Francke, B., and Monéger, R.,** Lipid composition of envelopes, prolamellar bodies and other plastid membranes in etiolated, green and greening wheat leaves, *Planta,* 129, 193, 1976.

115. **Browse, J., McCourt, P., and Somerville, C. R.,** A mutant of *Arabidopsis* lacking a chloroplast-specific lipid, *Science,* 227, 763, 1985.

116. **Omata, T. and Murata, N.,** Isolation and characterization of the cytoplasmic membranes from the blue-green alga (cyanobacterium) *Anacystis nidulans, Plant Cell Physiol.,* 24, 1101, 1983.

117. **Douce R., Holtz, R. B., and Benson, A. A.,** Isolation and properties of the envelope of spinach chloroplasts, *J. Biol. Chem.,* 248, 7215, 1973.

118. **Bertrams, M. and Heinz, E.,** Positional specificity and fatty acid selectivity of purified *sn*-glycerol 3-phosphate acyltransferases from chloroplasts, *Plant Physiol.,* 68, 653, 1981.

119. **Nishida, I., Frentzen, M., Ishizaki, O., and Murata, N.,** Purification of isomeric forms of acyl-[acyl-carrier-protein]:glycerol-3-phosphate acyltransferase from greening squash cotyledons, *Plant Cell Physiol.,* 28, 1071, 1987.

120. **Dubacq, J.-P., Douady, D., Nishida, I., and Murata, N.,** Unpublished data.

121. **Frentzen, M., Heinz, E., McKeon, T. A., and Stumpf, P. K.,** Specificities and selectivities of glycerol-3-phosphate acyltransferase and monoacylglycerol-3-phosphate acyltransferase from pea and spinach chloroplasts, *Eur. J. Biochem.,* 129, 629, 1983.

122. **Frentzen, M., Nishida, I., and Murata, N.,** Properties of the plastidial acyl- (acyl-carrier-protein):glycerol-3-phoshate acyltransferase form the chilling-sensitive plant, squash *(Cucurbita moschata), Plant Cell Physiol.,* 28, 1195, 1987.

123. **Ishizaki, O., Nishida, I., Agata, K., Eguchi, G., and Murata, N.,** Cloning and nucleotide sequence of cDNA for a glycerol-3-phosphate acyltransferase from squash, *FEBS Lett.,* 238, 424, 1988.

Chapter 12

RELATION OF CHILLING STRESS TO MEMBRANE PERMEABILITY

Takao Murata

TABLE OF CONTENTS

I. SYMPTOMS OF CHILLING INJURY AND MEMBRANE PERMEABILITY

There are many symptoms of chilling injury of horticultural crops associated with the changes in membrane permeability. Pitting, sheet pitting, shriveling, wilting, scald, discoloration, browning, watery breakdown, abnormal respiration, and ethylene production are typical symptoms of chilling injury in the horticultural crops originating in subtropical and tropical regions.[1,2] Among them, pitting, sheet pitting, shriveling, and wilting may be considered as a result of increased permeability of vapor from the cells to ambient atmosphere. In the case of pitting, increased permeability may be limited to special cells in the parenchymal tissues. Increased permeability of membranes may cause the promotion of an enzyme-substrate interaction, resulting in the occurrence of scald, discoloration, browning, and an increase in respiratory rate. There may be changes in membrane permeability of phenol substances in the course of browning of tissues. Changes in membrane properties may also induce abnormal ethylene production by tissues suffering from chilling stress.

The effects of chilling stress on the structure and function of the membrane of plant cells, plasma, mitochondria, and chloroplast have received increased attention for the past 20 years. Much attention has been focused upon phase separation (transition) in membrane lipids, lipid compositions, fatty acids, membrane protein, and membrane-associated enzymes (see Chapters 9 to 11). The results of electron spin resonance (ESR) spectra of spin labels and fluorescence exhibit dramatic changes at critical temperatures. These results suggest the direct effect of chilling stress on the structure and function of the membranes.

Arrhenius plots of the repiratory rate of tissues,[3,4] oxygen uptake of mitochondria isolated from chilling-sensitive plant tissues,[5-8] and ethylene production from the sensitive plants[9] demonstrated discontinuity at the critical temperature. Thus, there is much evidence which indicates the direct effect of chilling stress on the membrane fluidity of chilling-sensitive plant cells.

II. MEMBRANE PERMEABILITY AND LEAKAGE

Membrane permeability is an expression of the freedom with which water and solutes can pass through the membrane. Methods for direct measurement of membrane permeability in intact crops have not been completely established. Therefore, in most cases, membranes in the excised tissues, callus, or isolated organelles have to be used for the measurement of permeability.

Permeability can be assessed by the measurement of the rate of leakage of solutes, including ions, amino acids, sugars, and pigments,[10] from the tissues to the medium or by the measurement of the rate of uptake of ions, ^{3}H-amino acids,[11,12] and carbohydrates into the tissues from the incubation medium. Leakage of solutes through the membranes includes active and passive transport, which is affected by ionic strength, pH, respiratory inhibitors, and certain chemicals.[10,13] Membrane permeability also depends upon ripening[14-18] and senescense[17] of fruit tissues. In some cases, tissues tend to burst during incubation in water, resulting in an abnormal tissue permeability.[19] As mentioned above, measurement of leakage from leafy vegetables is difficult because of the covering with epidermal tissues. Thus, the measurement of leakage of solutes from the tissue and uptake of solutes into the tissue from incubation medium is always accompanied by structural and physiological problems besides chilling stress.

III. ARRHENIUS PLOTS OF RATE OF SOLUTE LEAKAGE

Arrhenius plots of the rate of solute leakage can be obtained from the rate of solute leakage at different temperatures. Figure 1 shows the log of the rate of potassium ion leakage

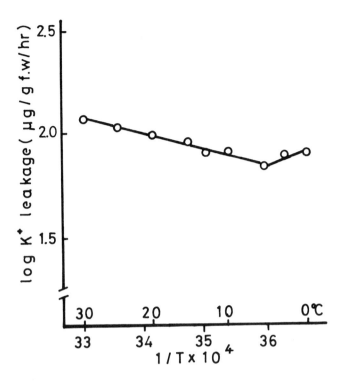

FIGURE 1.   Arrhenius plots of the rate of potassium ion leakage from discs of parenchymal tissues of cucumber fruit cv. Horai.

form the discs of parenchymal tissues of freshly harvested cucumber fruit (*Cucumis sativus* L.) vs. 1/T for the temperature range 0 to 30°C. Therefore, this figure gives an expression of the Arrhenius plot of the rate of potassium ion leakage. This figure exhibits a straight line with an upward bend at 5.5°C which is near the critical temperature for the occurrence of chilling injury of cucumber fruit. This tendency in the Arrhenius plots has been observed in other fruit tissues of the Cucurbitaceae (*Cucumis melo* L, *Cucurbita moshata* Duch., *Cucurbita pepo* L., *Momordica charantine* L., *Sechium edule* Swartz),[20-22] rind tissues of citrus fruit (*Citrus unshiu* Marc., *C. shaerocarpa* hort ex Tanaka, *C. hassaku* hort ex Tanaka, *C. glaberrima* hort ex Tanaka, *C. iyo* hort ex Tanaka, *C. natsudaidai* Hayata),[23] and sweet pepper (*Capsicum annum* L.).[24] In all cases, the temperature of break points generally coincided with the critical temperature for the occurrence of chilling injury of each of the fruits and vegetables (Table 1).

Hirata et al.[25] have also reported that a break point of an Arrhenius plot for potassium ion leakage in young and mature leaves of water convolvulus (*Ipomoea aquatica* Fork; chilling-sensitive vegetable) was found between 10 and 16°C, and a marked increase of the rate of potassium ion leakage occurred in young leaves (more sensitive) at 10°C.

Paull and co-workers[11,12] have investigated the uptake and leakage of [³H]-leucine and ⁸⁶Rb by leaf fragments and root of *Lycopersicon esculentum* (domestic) and *L. hirsutum* (wild tomato) at different temperatures and found that the plots of the log of ³H-amino acid uptake against temperature exhibited a line with a downward bend at about 11°C for domestic and 6 to 10°C for wild tomatoes originating from different altitudes. On the other hand, the plots of the log of ³H-amino acid leakage against temperature exhibited a line with an upward bend showing a marked increase below the critical temperature range of 5 to 10°C and a gradual increase above these temperatures. Many researchers have found abrupt breaks in the Arrhenius plots of respiratory rate, rate of ethylene production, velocity of protoplasmic

TABLE 1

**Temperature for Break Points in the Arrhenius Plots of
Rate of Potassium Ion Leakage from Discs of Parenchyma
and Rind Tissues**

Cucurbitaceae	°C	Citrus	°C
Oriental pickling melon	5.0	Kabosu	4.0
Cucumber			
cv. Pixie	5.0	Ponkan	4.5
cv. Horai	5.5	Satsuma mandarin	5.0
Chayote	5.5	Hassaku	6.5
Winter squash	6.0	Kinukawa	7.5
Summer squash (pepo)	9.0	Natsudaidai	11.5
Bitter melon	11.5	Iyo	12.5

streaming, velocity of IAA transport, growth rate, activity of mitochondrial oxidation, and motion of spin labels with membrane lipids of chilling-sensitive plants. Some of these researchers have suggested curves rather than straight lines with breaks. There are arguments on the validity of the break in the Arrhenius plots of biological and biophysical measurement.[26]

The reasons for the increased rate of leakage at a temperature lower than the break point before the appearance of chilling injury symptoms are unclear. However, it is reasonable to expect that membrane permeability would increase at chilling temperatures due to the effect of phase separation of membrane lipids, since solute permeability is maximal where the bilayers undergo gel to liquid-crystalline phase transition. Some investigators[27,28] have, in fact, shown the existence of a maximum of the permeability for electrolyte in the vicinity of the transition temperature of membrane lipids. In addition, there may be decreases in active uptake of solutes into the tissue on account of low respiratory rate at chilling temperature.

On the other hand, Nobel[29] has reported that the curves relating reflection coefficients and temperatures (10 to 30°C) for permeability of the chloroplast membrane of chilling-sensitive *Phaseolus vulgaris* L. and *Lycopersicon esculentum* Mill. exhibited a line with a downward bend at near 11°C, whereas the curve for chilling-resistant *Pisum sativum* L. and *Spinacia oleracea* L. were smooth. Jensen and Taylor[30] have determined the rate of water movement through the various tissues with respect to time and found there was a straight line relationship in the temperature range from 10 to 40°C.

This controversy concerning the trends of the bends in the Arrhenius plots remains to be solved, but these abrupt breaks in the Arrhenius plots of the rate of leakage or uptake of solutes may suggest a direct effect of chilling stress on the membrane properties of chilling-sensitive crops.

Besides chilling treatment, direct effects on membrane permeability may also result by the treatment with certain detergents. In fact, the occurrence of chilling injury of cucumber fruit during storage at chilling temperatures was found to be increased slightly by the treatment with 0.1% Triton X-100.[20] However, there was no significant difference in the shapes of Arrhenius plots of the rate of potassium ion leakage from the fruit tissues between the nontreated control and the detergent-treated sample.[20]

IV. CHANGES IN THE RATE OF SOLUTE LEAKAGE DURING STORAGE AT CHILLING TEMPERATURES

Since Lieberman et al.[31] reported a continuous increase of potassium ion leakage from the tissues of sweet potato root stored at chilling temperatures, comparable results have been

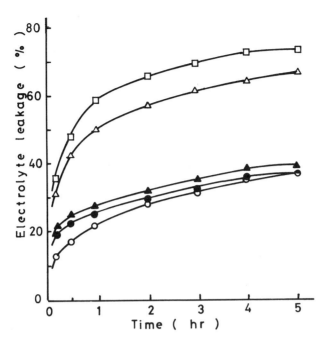

FIGURE 2. Time course of the rate of electrolyte leakage from the discs of cucumber fruit, cv. Natsuakihushinari, previously stored at 0 to 20°C for 12 d. (—○—) before storage; (—●—) 20°C, 12 d; (—▲—) 10°C, 12 d; (—△—) 5°C, 12 d; (—□—) 0°C, 12 d.

reported by many researchers on this subject which showed a rapid increase in the permeability of membranes of chilling-sensitive plant tissues after the occurrence of chilling symptoms including leaves,[32-38] fruit,[20-24,39-48] and root tissue.[49] Among these investigators, Lieberman et al.[31] and Yamawaki et al.[44] showed that there was a gradually continuous increase of ion leakage before the occurrence of chilling injury.

Figure 2 shows the time course of electrolyte leakage into deionized water from the discs of cucumber fruit prepared after being stored at 0, 5, 10, and 20°C for 12 d. An increased rate of leakage was observed in the discs from cucumber fruit stored at temperatures of 0 and 5°C, which were below the critical temperature for chilling injury. The shapes of curves for the time course of electrolyte leakage into deionized water and into isotonic solution were slightly different between two incubation media. However, the trends in the curves were almost the same in both cases (Figure 3). This result means that the increased leakage from the fruit tissues stored at chilling temperatures was due to increased permeability of the membrane as a result of chilling injury and did not depend on the bursting of the cells of fruit tissues in the incubation medium. There was no substantial increase in the ion leakage before the occurrence of chilling injury. A sudden increase in the ion leakage almost coincided with the time of occurrence of chilling injury (Figure 4). The same tendency was also found in the case of potassium, sodium, and magnesium ion leakage from the tissues of chilled sweet potato.[20] Many investigators have reported comparable results for the changes in membrane permeability including natsudaidai,[40] peach,[41] eggplant,[42] tomato,[39,45,46] papaya,[48] and yam.[49] Thus, the sudden rise in membrane permeability after the occurrence of chilling symptoms is a secondary effect of chilling injury.

King and Ludford[45] have found that mature green tomatoes analyzed immediately after chilling showed higher electrolyte leakage in chilling-sensitive lines (New Yorker, Early Cherry) than in chilling-tolerant lines (Line 79-546, Small Cherry, and Line 281) when leakage from different lines of tomato fruit was compared. As previously discribed, mem-

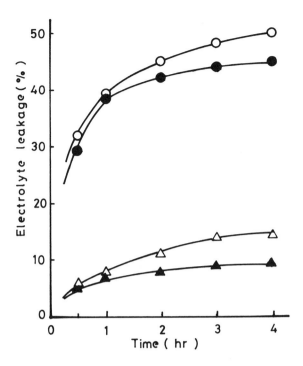

FIGURE 3.   Time course of the rate of electrolyte leakage from
the discs of cucumber fruit incubated with deionized water and 0.4
M mannitol solution at 20°C. (—△—) before storage, in water:
(—▲—) before storage, in mannitol solution; (—○—) 5°C, 15 d,
in water; (—●—) 5°C, 15 d, in mannitol solution.

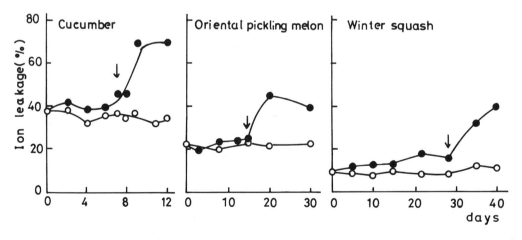

FIGURE 4.   Changes in the rate of ion leakage from the discs of Cucurbitaceae fruit stored at 0°C (—●—) and
20°C (—○—) into deionized water (% of leaked ions for 4 h/ total ions. Arrows show the time for the occurrence
of chilling symptoms).

brane permeability of the fruit, especially climacteric type, exhibit a trend of increasing
permeability during ripening[14-17] and senescence.[17] Membrane permeability of ripe fruit
tissues may be affected by both senescence and chilling factors. Therefore, comparison of
leakage from young and ripe fruit tissues is always accompanied by some confusion. Autio
and Bramlage[46] have suggested that chilling enhances ion leakage 32% above the controls
for mature green and red tomatoes by using fruit treated with silverthiosulfate (STS) to

prevent ripening. Yam is one of the most chilling sensitive crops. Olorunda and MacKlon[49] have indicated that incipient chilling injury could be rapidly detected by a change in ion absorption and salt retention capacity of tuber disks. They suggested that the changes in membrane permeability were considered to be among the causative factors.

It is well known that the temperature affects the relative humidity of the ambient atmosphere. A change of temperature is accompanied by a change in relative humidity (RH). Therefore, it is necessary to consider the simultaneous effect of humidity when the temperature has been lowered. Wright and Simon[33] have reported that RH in the growing conditions affects both the rate of ion leakage from the leaf tissues and the severity of chilling injury of cucumber seedlings. Simon[50] has mentioned in his review that about half of the phospholipids from the leaf cells disappeared after 3 d of chilling at 85% RH, whereas leaves chilled at 100% RH lost no phospholipids. Changes in phospholipid content in the membrane may affect membrane permeability.

Most of the published investigations on membrane permeability have demonstrated an increased rate of leakage of solutes from the chilling-sensitive horticultural crops as a result of chilling stress. However, there are some exceptions which show no increase of membrane permeability during chilling stress. For example, the tissues from sweet pepper, which is a chilling-sensitive crop, exhibit no increase of electrolyte leakage after storage at chilling temperatures. The problem of this inconsistency concerning membrane permeability has not been solved. However, with regard to the tissues of sweet pepper, it is reasonable to consider that the structural differences in other fruit tissues may affect the rate of electrolyte leakage,[51] because the discs of sweet pepper are covered by the epidermal tissues on both sides which prevents leakage of ions.

V. CONCLUSION

Enough comprehensive data are not available for the relationship between membrane permeability and phase separation of membrane lipids to be fully understood. Therefore, it is better to avoid a hasty conclusion. However, it may be reasonable to infer that a slight increase in membrane permeability of sensitive horticultural crops before the occurrence of chilling symptoms is due to the phase separation (transition) of polar lipids in the membranes and/or to the lowering of active transport of solutes into the cells. A sudden increase in membrane permeability at the time of onset of chilling injury may be the result of denaturation of membranes such as tonoplast and plasmalemma.

There are several exceptions for the trends of membrane permeability in the crops at chilling temperatures. Some chilling-sensitive crops do not show a sudden rise of membrane permeability, even after the occurrence of chilling injury. Some chilling-resistant crops (e.g., carrot and onion) exhibit the discontinuity in the Arrhenius plots of the rate of ion leakage from sliced tissues. These exceptions lead to complications concerning the relationship between membrane permeability and chilling stress. Further studies are needed to understand these exceptions.

REFERENCES

1. **Couey, H. M.,** Chilling injury of crops of tropical and subtropical origin, *HortScience,* 17, 162, 1982.
2. **Bramlage, W. J.,** Chilling injury of crops of temperate origin, *HortScience,* 17, 165, 1982.
3. **Purvis, A. C.,** Relation of grapfruit and orange flavedo tissue in relation to chilling and non-chilling temperatures and respiratory inhibitors, *J. Am. Soc. Hortic. Sci.,* 105, 209, 1980.
4. **Murata, T., Tatsumi, Y., Iwamoto, M., and Nishimoto, T.,** On the chilling sensitivity of citrus fruit in relation to respiration, *Bull. Fac. Agric. Shizuoka Univ.,* 30, 23, 1980.

5. **Lyons, J. M. and Raison, J. K.,** Oxidative activity of mitochondria isolated from plant tissues sensitive and resistant to chilling injury, *Plant Physiol.,* 45, 386, 1970.
6. **Raison, J. K.,** Temperature-induced phase changes in membrane lipids and their influence on metabolic regulation, *Symp. Soc. Exp. Biol.,* XXVII, 485, 1973.
7. **McGlasson, W. B. and Raison, J. K.,** Occurrence of a temperature-induced phase transition in mitochondria isolated apple fruit, *Plant Physiol.,* 52, 390, 1973.
8. **Kane, O. and Marcellin, P.,** Incidence of ripening and chilling on the oxidative activity and fatty acid composiiton of the mitochondria from mango fruit, *Plant Physiol.,* 61, 634, 1978.
9. **Mattoo, A. K., Baker, J. E., Chalutz, E., and Lieberman, M.,** Effect of temperature on the ethylene-synthesizing system in apple, tomato and *Penicillium digitatum, Plant Cell Physiol.,* 18, 715, 1977.
10. **Matsuo, T., Yoneda, T., and Ito, S.,** The properties of betacyanin leakage in red beet discs exposed to tert-butylhydroperoxide, *J. Jpn. Soc. Hortic. Sci.,* 55, 332, 1986.
11. **Paull, R. E., Patterson, B. D., and Graham, D.,** Amino acid uptake by tomato leaf tissue under chilling stress: uptake conditions and a comparison of *Lycopersicon* species at different chilling resistance, *Aust. J. Plant Physiol.,* 6, 475, 1979.
12. **Paull, R. E.,** Temperature induced leakage from chilling-sensitive and chilling-resistant plants, *Plant Physiol.,* 68, 149, 1981.
13. **McCullugh, C. H. R. and Simon, E. W.,** The effect of indoacetate on phospholipid levels and membrane permeability, *J. Exp. Bot.,* 24, 841, 1973.
14. **Sacher, J. A.,** Relation between changes in membrane permeability and the climacteric in banana and avocado, *Nature (London),* 195, 577, 1962.
15. **Sacher, J. A.,** Permeability characteristics and amino acid incorporation during senescence (ripening) of banana tissue, *Plant Physiol.,* 41, 701, 1966.
16. **Lewis, T. L. and Martin, D.,** Changes in rate of leakage of potassium from excised disks of apple fruit held at 20°C after harvest, *Aust. J. Biol. Sci.,* 22, 1577, 1969.
17. **Anderson, W. P.,** Ion transport in the cells of higher plant tissues, *Annu. Rev. Plant Physiol.,* 23, 55, 1972.
18. **Brady, C. J., O'Connell, P. B. H., Smydzuk, J., and Wade, N. L.,** Permeability, sugar accumulation and respiration rate in ripening banana fruits, *Aust. J. Biol. Sci.,* 23, 1143, 1970.
19. **Simon, E. W.,** Leakage from fruit cells in water, *J. Exp. Bot.,* 28, 1147, 1977.
20. **Murata, T. and Tatsumi, Y.,** Ion leakage in chilled plant tissues, in *Low Temperature Stress in Crop Plants: The Role of the Membrane,* Lyons, J. M., Graham, D., and Raison, J. K., Eds., Academic Press, 1979, 141.
21. **Murata, T. and Tatsumi, Y.,** Chilling injury in *Cucurbitaceae* fruits, in *Recent Advances In Food Science and Technology,* Hua Shiang Yuan Publishing, Taipei, 1981, 476.
22. **Tatsumi, Y. and Murata, T.,** Relation between chilling sensitivity of *Cucurbitaceae* fruit and membrane permeability, *J. Jpn. Soc. Hortic. Sci.,* 50, 108, 1981.
23. **Murata, T.,** Physiological disorders of citrus fruits in Japan, in *Proc. Int. Soc. Citricult.,* 2, 776, 1981.
24. **Tatsumi, Y. and Murata, T.,** Studies on chilling injury of fruit and vegetables. I. Chilling injury of cucumber fruit with special reference to permeability of membrane of tissues, *J. Jpn. Soc. Hortic. Sci.,* 47, 105, 1978.
25. **Hirata, K., Chachin, K., and Iwata, T.,** Changes of K^+ leakage, free amino acid contents and phenylpropanoid metabolism in water convolvulus (*Ipomoea aquatica* Fork) with reference to the occurrence of chilling injury, *J. Jpn. Soc. Hortic. Sci.,* 55, 516, 1987.
26. **McMurdo, A. C. and Wilson, J. M.,** Chilling injury and Arrhenius plots, *CryoLetters,* 1, 231, 1980.
27. **Papahadjopoulos, D., Jacobson, K., Nir, S., and Isac, T.,** Phase transition in phospholipid vesicles: fluorescence polarization and permeability measurement concerning the effect of temperature and cholesterol, *Biochim. Biophys. Acta,* 311, 330, 1973.
28. **Blok, M. C., Van der Neut-Kok, E. C. M., Van Deenen, L. L. M., and DeGier, J.,** The effect of chain length and lipid phase transitions on the selective permeability properties of liposomes, *Biochim. Biophys. Acta,* 406, 187, 1975.
29. **Nobel, P. S.,** Temperature dependence of the permeability of chloroplasts from chilling-sensitive and chilling-resistant plants, *Planta,* 115, 369, 1974.
30. **Jensen, R. D. and Taylor, S. A.,** Effect of temperature on water transport through plants, *Plant Physiol.,* 36, 639, 1961.
31. **Lieberman, M., Craft, C. C., Audia, W. V., and Wilcox, M. S.,** Biochemical studies of chilling injury in sweet potatoes, *Plant Physiol.,* 33, 307, 1958.
32. **Minchin, A. and Simon, E. W.,** Chilling injury in cucumber leaves in relation to temperature, *J. Exp. Bot.,* 24, 1231, 1973.
33. **Wright, M. and Simon, E. W.,** Chilling injury in cucumber leaves, *J. Exp. Bot.,* 24, 400, 1973.
34. **Patterson, B. D., Murata, T., and Graham, D.,** Electrolyte leakage induced by chilling in *Passiflora* species tollerant to different climates, *Aust. J. Plant Physiol.,* 3, 435, 1976.

35. **Tanczos, O. Gy.,** Influence of chilling on electrolyte permeability, oxygen uptake and 2,4-dinitrophenol stimulated oxygen uptake in leaf discs of the thermophilic *Cucumis sativus, Physiol. Plant.,* 41, 289, 1977.
36. **Wright, M.,** The effect of chilling on ethylene production, membrane permeability and water loss of leaves of *Phaseolus vulgaris, Planta,* 120, 63, 1974.
37. **Lindstrom, O. M. and Carter, J. V.,** Injury to potato leaves exposed to subzero temperatures in the absence of freezing, *Planta,* 164, 512, 1985.
38. **MacRae, E. A., Hardacre, A. K., and Ferguson, I. B.,** Comparison of chlorophyll fluorescence with several other techniques used to assess chilling sensitivity in plants, *Physiol. Plant.,* 67, 659, 1986.
39. **Lewis, T. L. and Workman, M.,** The effect of low temperature on phosphate esterification and cell membrane permeability in tomato and cabbage leaf tissue, *Aust. J. Biol. Sci.,* 17, 147, 1964.
40. **Iwata, T., Nakagawa, K., and Ogata, K.,** Physiological studies of chilling injury in natsudaidai (Citrus natsudaidai Hayata) fruit. I, *J. Jpn. Soc. Hortic. Sci.,* 37, 383, 1965.
41. **Furmanski, R. J. and Buescher, R. M.,** Influence of chilling on electrolyte leakage and internal conductivity of peach fruits, *HortScience,* 14, 167, 1979.
42. **Abe, K. and Ogata, K.,** Chilling injury in eggplant fruit. V. Changes of K ion leakage and content of phospholipids during storage and effect of phenolic compound on K ion leakage, phospholipid content and ultrastructural change of eggplant fruit sections, *J. Jpn. Soc. Hortic. Sci.,* 47, 111, 1978.
43. **Tatsumi, Y., Iwamoto, M., and Murata, T.,** Electrolyte leakage from the discs of *Cucurbitaceae* fruits associated with chilling injury, *J. Jpn. Soc. Hortic. Sci.,* 50, 114, 1981.
44. **Yamawaki, K., Yamauchi, N., Chachin, K., and Iwata, T.,** Relationship of mitochondrial enzyme activity to chilling injury of cucumber fruit, *J. Jpn. Soc. Hortic. Sci.,* 52, 93, 1983.
45. **King, M. M. and Ludford, P. M.,** Chilling injury and electrolyte leakage in fruit of different tomato cultivars, *J. Am. Soc. Hortic. Sci.,* 108, 74, 1983.
46. **Autio, W. R. and Bramlage, W. J.,** Chilling sensitivity of tomato fruit in relation to ripening and senescence, *J. Am. Soc. Hortic. Sci.,* 111, 201, 1986.
47. **Goto, M., Minamide, T., Fujii, M., and Iwata, T.,** Preventive effect of cold-shock on chilling injury of mume (*Prunus mume* SIEB et ZUCC) fruit in relation to changes of permeability and fatty acid composition of membrane, *J. Jpn. Soc. Hortic. Sci.,* 53, 210, 1984.
48. **Chan, H. T., Jr., Sanxter, S., and Couey, H. M.,** Electrolyte leakage and ethylene production induced by chilling injury of papayas, *HortScience,* 20, 1070, 1985.
49. **Olorunda, A. O. and MacKlon, A. E. S.,** Effect of storage at chilling temperature on ion absorption, salt retention capacity and respiratory pattern in yam tubers, *J. Sci. Food Agric.,* 27, 405, 1976.
50. **Simon, E. W.,** Phospholipids and plant membrane permeability, *New Phytol.,* 73, 377, 1974.
51. **Morrod, R. S.,** A new method for measuring the permeability of plant cell membranes using epidermis-free leaf discs, *J. Exp. Bot.,* 25, 521, 1974.

Chapter 13

RELATION OF CHILLING STRESS TO CARBOHYDRATE COMPOSITION

Albert C. Purvis

TABLE OF CONTENTS

I. INTRODUCTION

Vegetative and fruit tissues of most tropical and subtropical plant species, as well as many temperate species, are sensitive to low but above-freezing temperatures and are injured after prolonged exposure to such temperatures. Sensitivity varies, however, with stage of development or the particular physiological processes and biochemical reactions occurring in the tissues at the time the plant material is exposed to low temperatures.[1-6]

Actively growing seedlings are generally more sensitive to low temperatures than are seedlings which have arrested growth. Variability in chilling sensitivity also exists among cultivars within a species.[7-12]

Grapefruit harvested at midseason are also less sensitive to low temperatures during storage than are fruit harvested early or late in the season,[3,4] and fruit harvested from the interior tree canopy are less sensitive than the fruit harvested from the exterior tree canopy.[13] Grapefruit stored at low temperatures immediately after harvest develop visible symptoms of chilling injury more rapidly than do fruit which are held for several days at nonchilling temperatures before storage at low temperatures.[4,14]

A complete understanding of what constitutes the variable resistance to chilling injury in the tissues of sensitive species is not known. Although largely circumstantial, there is increasing evidence that carbohydrate level and/or composition influence the sensitivity of plant tissues to low temperatures. Environmental conditions which lead to the accumulation of soluble carbohydrates in plant tissues also reduce the sensitivity of the plant material to low temperatures. This chapter will deal with the environmental conditions which lead to carbohydrate accumulation in plant tissues and reduced sensitivity of plant tissues to low temperatures and the possible roles that carbohydrates may play in reducing the sensitivity of plant tissues to low temperatures. The terminology proposed by Raison and Lyons[15] will be used where possible.

II. INFLUENCE OF ENVIRONMENTAL FACTORS ON SUGAR ACCUMULATION AND CHILLING SENSITIVITY

In contrast to the numerous studies showing strong correlations between the soluble carbohydrate levels and composition in plant tissues and frost hardiness,[16-22] few studies have shown a positive correlation between either carbohydrate level or composition and resistance to chilling injury. Diurnal changes in chilling sensitivity of tomato seedlings were correlated with endogenous carbohydrate content.[23] High levels of reducing sugars in grapefruit peel also correlated positively with resistance to chilling injury.[3,4,24] However, since it was subsequently found that low field temperatures, as well as low storage temperatures, induce invertase activity in grapefruit peel, it is not certain whether reducing sugar accumulation is an essential part of the chilling-resistance mechanism or is simply a consequence of chilling stress.[25] Nevertheless, environmental factors which affect photosynthesis directly, as well as translocation and respiration, i.e., the utilization of carbohydrates, also altered the sensitivity of plant tissues to low temperatures.

A. LOW TEMPERATURE

It has been widely reported that a reduction in temperature results in an increase in soluble carbohydrate accumulation in plant tissues.[3,4,17-28] Cotton seedlings exposed to 15°C day and 10°C night temperatures accumulated sugars and starch and were not injured during subsequent exposure to 5°C, compared to similar seedlings exposed to 35°C day and 30°C night temperatures which had no increased sugar and starch accumulation.[29]

The temperature during the night before chilling markedly affected the amount of chilling injury that developed in tomato seedlings.[23] Seedlings held at 10°C night temperatures before chilling were much less sensitive to chilling exposure at the end of the dark period than were the seedlings which had been held at 20 and 26°C night temperatures. Although carbohydrate levels were not reported, other studies show that low night temperatures reduce the utilization of carbohydrates.[30,31]

Low temperatures also have an effect on carbohydrate composition. Starch, which is the major storage carbohydrate in many plant tissues, is hydrolyzed to soluble carbohydrates when the tissues are exposed to or stored at low temperatures.[32-34] In some plant tissues, sucrose is hydrolyzed to reducing sugars by the activity of a low-temperature-induced invertase.[25]

B. LIGHT

Seedlings of several chilling-sensitive genera exhibited greater tolerance to low temperatures in the dark when the exposure to low temperature was initiated during the diurnal light period than when the exposure was initiated at the end of the dark period.[23,35-37] Chilling sensitivity was highest during the last hours of the diurnal dark period, and chilling tolerance was regained quickly during the diurnal light period.[23] Since increased light intensity resulted in less chilling injury, the light response is probably related to photosynthesis, rather than being a photomorphogenic response.[23,36] Furthermore, application of the photosynthetic inhibitor, dichlorophenyl dimethyl urea (DCMU), negated the light effect in cotton cotyledon discs.[36] Chilling injury is greater, however, when the plants are exposed to low temperatures in the light than in the dark.[37-39] Photosynthetic rates are also lower at chilling than at nonchilling temperatures.

C. CARBON DIOXIDE

Carbon dioxide enrichment of the atmosphere during chilling ameliorated chilling injury in okra.[40] In experiments at ambient CO_2 concentrations (350 μl l^{-1}), okra plants failed to survive at temperatures below 26/20°C (day/night), whereas in enriched CO_2 atmospheres (up to 1000 μl l^{-1}), the plants not only grew to maturity, but also produced fruit. Thus, CO_2 enrichment compensated for the adverse effects of cool temperatues on the growth of chilling-sensitive okra plants. Carbohydrate levels were not reported, but in other studies increasing the ambient CO_2 concentration resulted in increased carbohydrates.

Carbon dioxide enrichment of the atmosphere also ameliorated chilling injury of *Echinochloa crus-galli* and *Eleusine indica*.[41] Chilling generates chlorotic bands on the leaf blades of these sensitive C_4 grasses. Less chlorosis was observed in plants which were grown in a CO_2-enriched atmosphere (675 μl l^{-1}) than in a normal atmosphere when they were subsequently chilled at 7°C.

D. EXOGENOUS SUGAR

The most direct evidence, however, that soluble carbohydrates play a role in the resistance of sensitive plants to chilling injury comes from studies involving the exogenous application of carbohydrates before the chilling treatment. Injury to chilling-sensitive cotton cotyledon discs floated on sucrose solutions in darkness before they were exposed to chilling temperatures while they were floating on distilled water was less than for similar discs which were floated on distilled water before and during chilling.[36] Furthermore, the effect of sucrose solutions on chilling sensitivity was dependent upon the concentration. Discs floated on 0.5% sucrose showed less injury than discs floated on 0.1% sucrose. Treatment of rice seedlings with glucose and fructose before chilling in the dark also increased their ability to survive the chilling treatment.[42] In a different experiment, sucrose appeared to be less effective in reducing chilling injury. On the other hand, the exogenous application of sucrose, glucose, and fructose to excised tomato shoots before chilling markedly lessened the severity of chilling injury; and sucrose was more effective than either glucose or fructose.[23] Mannitol treatment hastened chilling injury.[23] How much sugar was taken up from the solutions, or if some sugars were taken up preferentially over others is not known.

III. ROLE OF SUGARS IN CHILLING RESISTANCE

The critical threshold level of soluble carbohydrates and the primary sugar responsible for reduced sensitivity of plant tissues to low temperatures have not been established. Indeed, the critical level and composition of carbohydrate may differ with species and tissue.

A. MECHANISM OF CHILLING INJURY

In order to establish how soluble carbohydrates may alter the chilling sensitivity of plants, an understanding of the mechanism of chilling injury is necessary. Although a specific mechanism for all species has not been unequivocally delineated, it is generally agreed that the initial event is a direct effect of low temperature on cellular constituents which results in changes in the fluidity of the membrane lipids and/or conformational changes in enzyme and structural proteins.[1,43-45] Such direct effects are readily reversible if the plant material is transferred from chilling to nonchilling temperatures before injury occurs. The length of time during which the effects of low temperature are reversible depends upon the species and varies from a few hours for the most sensitive species to several weeks for the least sensitive species.

Several biochemical and physiological processes are altered as a consequence of the direct effect of low temperature on cellular constituents.[1,44] Thus, prolonged chilling leads to metabolic imbalances, enzyme inactivation, membrane damage, and eventual cell death. Solute leakage and loss of specialized membrane functions, which reflect membrane damage, often times immediately precede the appearance of the visible symptoms of chilling injury such as wilting, water soaking, pitting, and necrosis. Cellular reactions which are responsible for enzyme inactivation and membrane damage have not been delineated. The progressive breakdown of cellular proteins and membrane deterioration with advancing senescence have been associated with the inability of tissues to utilize oxygen.[46] A variety of macromolecules in cells are damaged by the various species of activated oxygen. Unsaturated fatty acids are especially prone to attack, and there is a strong temporal correlation between free radical-induced lipid peroxidation and increased permeability with advancing senescence. Vianello et al.[47] reported that lipid peroxidation is inducible in soybean mitochondria and microsomes by various peroxidizing systems and is partially prevented by certain respiratory substrates. The prevention of lipid peroxidation by succinate, malate, and NADH was mainly ascribed to the reduction of Coenzyme Q-10 which, in such a form, acts as a potent antioxidant. Whether free radicals and lipid peroxidation are involved in chilling injury is not known. Free radical scavengers, however, have been found to ameliorate chilling injury in several sensitive species.[48] More recently, a positive relationship was found between the activity of a manganese-containing superoxide dismutase and chilling resistance in strains of *Chlorella ellipsoidea*.[49] Thus, membrane deterioration during chilling could be caused by activated oxygen species.

Soluble carbohydrates can have either a direct or an indirect role in protecting cellular constituents of sensitive plants at chilling temperatures. Furthermore, qualitative as well as quantitative aspects of carbohydrates may be involved in the resistance mechanism. Soluble carbohydrates can influence the chilling resistance mechanism in at least three ways. First, carbohydrates contribute to the osmotic potential of the cell, thereby decreasing the cell water potential and reducing water loss from the tissue. Second, certain carbohydrates can stabilize cell membranes and enzymes by binding directly to the constitutive molecules. Third, carbohydrates serve as an energy source for plant cells.

B. OSMOTIC EFFECT

It is well documented that restricting water loss during exposure to low temperatures greatly reduces chilling injury in vegetative tissues and fruits of several plant species.[50-59] Water loss from plant tissues is a function of both the permeability of the tissue to water vapor and the magnitude of the difference between the water potential of the tissue and the water potential of the atmosphere. Antitranspirants,[57] waxes,[54] oils,[60] polyethylene films,[51-55] and maintenance of high relative humidity[57-59] restrict water loss from plant tissues and have been reported to ameliorate chilling injury. Hydrocooling grapefruit before storage at low temperatures reduced the vapor pressure gradient between the fruit and the storage room atmosphere and resulted in a reduction in initial water loss from the fruit and a delay in the development of pitting.[50]

The accumulation of sugars increases the osmotic potential and decreases the water potential of cells. The increase in sugars during the hardening of 'Red Blush' grapefruit seedlings accounted for 25 to 50% of the measured change in freezing point depression of the leaves.[28] However, if the role of sugars in ameliorating chilling injury is strictly that of an osmotic agent, it could be fulfilled by numerous other molecules. Indeed, the chilling resistance of grapefruit has been correlated positively with elevated proline levels in the peel tissue.[61] In other studies, however, osmotic agents, such as mannitol and inorganic salts, were ineffective in reducing chilling injury in cotton cotyledon discs and tomato seedlings.[23,36] Whether these osmotic agents were taken up by the plant tissues was not reported. Furthermore, although interior and exterior canopy grapefruit differ markedly in their resistance to chilling injury, no differences exist in the levels of total soluble carbohydrates, sucrose, or reducing sugars in their peel tissues.[13] Although soluble carbohydrates contribute to the osmotic potential of plant cells, they apparently have a more direct role in ameliorating chilling injury in sensitive plant tissues.

C. MEMBRANE AND ENZYME STABILIZATION

It was suggested that membrane deterioration is a primary factor in the physiological and visible manifestations of chilling injury. Ben-Yehoshua et al.[62] have postulated that water stress is the primary factor involved in the disintegration of cell membranes. Since, in their opinion, water protects membranes, restricting water loss during low-temperature storage, therefore, would reduce chilling injury of sensitive plant tissues. An increase in ''bound water'' occurred during the hardening of 'Red Blush' grapefruit trees by low temperatures.[28] The increased hydration capacity of the tissue was associated with the ability of the tissue to resist freezing. Various sugars stabilize cell membranes during the freezing and thawing of plant tissues.[63,64] The sugars apparently substitute for water molecules in critical areas of the membranes so that membrane integrity is maintained during desiccation resulting from water loss and freezing and thawing.

Trehalose, a nonreducing disaccharide, has been found to inhibit hydration-dependent phase transitions and fusion between vesicles during the dehydration of the gastrulae of brine shrimp, certain nematodes, spores of various fungi, and the resurrection plant *Selaginella lepidophylla*.[65] Trehalose, at concentrations of up to 20% of the dry weight of these organisms, is synthesized during the induction of anhydrobiosis and is degraded during the resumption of active metabolism. Although trehalose is not a common sugar in most plant tissues, other sugars, such as sucrose and the reducing sugars which accumulate in plant tissues when the plants are exposed to low temperatures, could perform a similar function.

It is also possible that sugars can substitute for water in the hydration of proteins, particularly that small fraction of water which makes up the main hydration of proteins. Thus, the structural conformation of enzymes may be maintained by binding particular sugars at certain critical sites during the chilling of plant tissues. Steponkus[66] suggested that the low-temperature acclimation of *Hedera helix* consisted of two phases. The first phase involves carbohydrate accumulation in the tissue. The second phase involves synthesis of proteins which have a greater affinity or ability to bind sugar molecules. A new pattern of isoenzymes which are more functional at low temperatures are produced in the tissues of some plants when they are exposed to hardening temperatures.[67,68] Cold acclimation of winter rape plants resulted in lower sensitivity of pyruvate kinase to adenosine triphosphate (ATP), loss of sensitivity to alanine, and decreased sensitivity to temperatures below 10°C.[69] It is not known if the enzymes synthesized during acclimation to low temperatures have a greater ability to bind sugar molecules than the enzymes in nonacclimated plant tissues. However, the K_m value for invertase in pea root is lower after pretreatment of the plants at low temperature, which may contribute to the ability of this species to adapt to low temperatures.[70]

D. RESPIRATORY SUBSTRATE

The protective effect of sugar in the resistance of plant tissues to low temperatures is apparently not due to the level of sugars or to a particular sugar per se, since grapefruit from the interior and exterior canopies have similar levels of total soluble carbohydrates, sucrose, and reducing sugars in their peels; yet they differ greatly in their sensitivity to low temperatures. Furthermore, reducing sugars accumulate in the peels of grapefruit from the interior and exterior canopies at similar rates during low-temperature storage, but chilling injury symptoms develop more rapidly and more severely in fruit from the exterior canopy.[71] Hence, the resistance of grapefruit to chilling injury appears to be related more to some aspect of metabolism involving sugars than to the levels of sugars.

The respiratory response of plant tissues to low temperatures has been used as a criterion to distinguish between chilling-sensitive and chilling-resistant species.[1,43,44,72-75] Chilling-sensitive species exhibit a disproportionate decrease in respiratory rates at temperatures below the critical threshold temperature and accelerated respiratory rates immediately after transfer from chilling to warmer temperatures.[1,43,73-76]

Sugars have a central role in metabolism as respiratory substrates. Unlike animal cells, where respiration is coupled tightly to adenylate charge, respiration of plant tissues is regulated, in part, by the availability of respiratory substrate. Several studies have shown a significant positive correlation between respiratory rates and the sugar content actually usable in the tissues.[70,77-83] The differing ability of the roots of chilling-resistant pea and chilling-sensitive maize to grow at low temperatures was associated with the maintenance of adequate supplies of sugars to the root tips.[70] In maize root, low temperatures caused a sharp and continuing fall in sugars to the growing tip, whereas the fall in pea root was temporary. The fall in sugars in maize root was accompanied by a reduction in respiration rate and the cessation of growth. An early study reported higher respiratory rates in grapefruit during the seasonal period when reducing sugar levels in the peel were highest and the fruit were resistant to chilling injury.[84]

Although glycolysis is regulated, in part, by the availability of sugars, electron transport in the mitochondria is coupled tightly to energy conservation, and continued metabolism of sugars through the glycolytic pathway when the energy demands of the cells are low would result in the accumulation of anaerobic metabolites. Few studies, however, have actually reported increased levels of acetaldehyde and ethanol in the tissue during low-temperature storage.[85-88] Instead, acetaldehyde and ethanol are produced after transfer of the fruit to nonchilling temperatures in quantities proportional to the duration and temperature of exposure.[89,90] Furthermore, increasing the oxygen concentration to 100% during low-temperature storage did not reduce chilling injury in cucumber fruits.[91] The onset and development of chilling injury was associated with an excessive rather than a limited use of oxygen.[72] Therefore, plant tissues which are resistant to chilling injury must be able to regulate respiratory metabolism at low temperatures.

Plant mitochondria possess two pathways of respiratory electron transport: the conventional cyanide-sensitive cytochrome pathway and a cyanide- and azide-insensitive alternative pathway which is unique to plant tissues. The capacity of the alternative pathway increases in the tissues of several plant species during their exposure to low temperatures,[10,76,92-97] and cultivars and tissues which are resistant to low temperatures develop a greater potential for respiratory electron flux through the alternative pathway then do cultivars and tissues which are sensitive to low temperatures.[10,76,93,98] The role of sugars in the induction and activity of the alternative pathway in grapefruit and other tissues, however, has not been established. Furthermore, in most studies, it is not clear whether the alternative pathway is actually utilized at low temperatures. The alternative pathway is inefficient in energy conservation, i.e., one ATP is synthesized vs. three ATPs from the cytochrome pathway. It is generally agreed, therefore, that the alternative pathway functions when the cellular energy charge is high or when there is an imbalance between the supply of carbohydrates and the requirement of carbohydrates for structural growth, energy production, storage, and osmoregulation.[92,99-102] Since the growth of sensitive species ceases at the threshold temperatures for chilling injury, the flow of respiratory electrons through the alternative pathway would enhance oxygen utilization when energy demands of the tissues are low and sugar levels are high. If, indeed, irreversible membrane damage results from the peroxidation of membrane lipids resulting from the inability of the tissue to utilize oxygen in normal metabolic processes, then enhanced respiration from accumulated sugars in concert with respiratory electron flux through the alternative pathway could reduce the sensitivity of plant tissues to low temperatures.

IV. CONCLUSIONS

Environmental factors which contribute to the accumulation of sugars in plant tissues also lead to a decrease in their sensitivity to low temperatures. Whether the sugars are directly or indirectly involved in the resistance of plant tissues to chilling injury is unresolved. In fact, there is little published evidence to indicate that sugars are involved in chilling resistance. Although sugars contribute to the osmotic potential of the cell and can stabilize cellular constituents during desiccation by substituting for water molecules at critical sites on the macromolecules in the cell, their role in reducing the sensitivity of plant tissues to low temperatures is more likely to be metabolic. It has been proposed that the inability of plant tissues to utilize oxygen metabolically leads to the peroxidation of membrane lipids and other macromolecules and, as a consequence, membrane deterioration. Since respiration of plant tissues is controlled by sugars and energy charge, accumulated sugars in concert with respiratory electron flux through the alternative pathway can stimulate respiration at low temperatures when the energy demands of the cells are low. Enhanced respiration, therefore, would decrease the oxygen available for nonmetabolic oxidative reactions which lead to cellular destruction.

REFERENCES

1. **Lyons, J. M.,** Chilling injury in plants, *Annu. Rev. Plant Physiol.,* 24, 445, 1973.
2. **Christiansen, M. N.,** Periods of sensitivity to chilling in germinating cotton, *Plant Physiol.,* 42, 431, 1967.
3. **Purvis, A. C., Kawada, K., and Grierson, W.** Relationship between midseason resistance to chilling injury and reducing sugar level in grapefruit peel, *HortScience,* 14, 227, 1979.
4. **Purvis, A. C. and Grierson, W.,** Accumulation of reducing sugar and resistance of grapefruit peel to chilling injury as related to winter temperatures, *J. Am. Soc. Hortic. Sci.,* 107, 139, 1982.

5. **Kosiyachinda, S. and Young, R. E.,** Chilling sensitivity of avocado fruit at different stages of the respiratory climacteric, *J. Am. Soc. Hortic. Sci.,* 101, 665, 1976.

6. **Chen, N.-M. and Paull, R. E.,** Development and prevention of chilling injury in papaya fruit, *J. Am. Soc. Hortic. Sci.,* 111, 639, 1986.

7. **Karnok, K. J. and Beard, J. B.,** Morphological responses of *Cynodon* and *Stenotaphrum* to chilling temperatures as affected by gibberellic acid, *HortScience,* 18, 95, 1983.

8. **Orr, W., de la Roche, A. I., Singh, J., and Voldeng, H.,** Imbibitional chilling injury in cultivars of soybeans differing in temperature sensitivity to pod formation and maturation periods, *Can. J. Bot.,* 61, 2996, 1983.

9. **Bramlage, W. J., Leopold, A. C., and Specht, J. E.,** Imbibitional chilling sensitivity among soybean cultivars, *Crop Sci.,* 19, 811, 1979.

10. **van de Venter, H. A.,** Cyanide-resistant respiration and cold resistance in seedlings of maize *(Zea mays L.),* Ann. Bot., 56, 561, 1985.

11. **Picha, D. H.,** Chilling injury, respiration, and sugar changes in sweet potatoes stored at low temperature, *J. Am. Soc. Hortic. Sci.,* 112, 497, 1987.

12. **King, M. M. and Ludford, P. M.,** Chilling injury and electrolyte leakage in fruit of different tomato cultivars, *J. Am. Soc. Hortic. Sci.,* 108, 74, 1983.

13. **Purvis, A. C.,** Influence of canopy depth on susceptibility of 'Marsh' grapefruit to chilling injury, *HortScience,* 15, 731, 1980.

14. **Hatton, T. T. and Cubbedge, R. H.,** Conditioning Florida grapefruit to reduce chilling injury during low-temperature storage, *J. Am. Soc. Hortic. Sci.,* 107, 57, 1982.

15. **Raison, J. K. and Lyons, J. M.,** Chilling injury: a plea for uniform terminology, *Plant Cell Environ.,* 9, 685, 1986.

16. **Levitt, J.,** *Responses of Plants to Environmental Stresses,* Vol. 1, Academic Press, New York, 1980, 169.

17. **Sieckmann, S. and Boe, A. A.,** Low temperature increases reducing and total sugar concentrations in leaves of boxwood (*Buxus sempervirens* L.) and cranberry (*Vaccinium macrocarpon* Ait.), *HortScience,* 13, 439, 1978.

18. **Raese, J. T., Williams, M. W., and Billingsley, H. D.,** Cold hardiness, sorbitol, and sugar levels of apple shoots as influenced by controlled temperature and season, *J. Am. Soc. Hortic. Sci.,* 103, 796, 1978.

19. **Young, R.,** Cold hardening in citrus seedlings as related to artificial hardening conditions, *J. Am. Soc. Hortic. Sci.,* 94, 612, 1969.

20. **Young, R.,** Cold hardening in 'Redblush' grapefruit as related to sugars and water soluble proteins, *J. Am. Soc. Hortic. Sci.,* 94, 252, 1969.

21. **Sakai, A.,** Relation of sugar content to frost-hardiness in plants, *Nature (London),* 185, 698, 1960.

22. **Steponkus, P. L. and Lanphear, F. O.,** The relationship of carbohydrates to cold acclimation of *Hedera helix* L. cv. Thorndale, *Physiol. Plant.,* 21, 777, 1968.

23. **King, A. I., Joyce, D. C., and Reid, M. S.,** Role of carbohydrates in diurnal chilling sensitivity of tomato seedlings, *Plant Physiol.,* 86, 764, 1988.

24. **Harvey, E. M. and Rygg, G. L.,** Field and storage studies on changes in the composition of the rind of the Marsh grapefruit in California, *J. Agric. Res.,* 52, 747, 1936.

25. **Purvis, A. C. and Rice, J. D.,** Low temperature induction of invertase activity in grapefruit flavedo tissue, *Phytochemistry,* 22, 831, 1983.

26. **Yelenosky, G. and Guy, C. L.,** Carbohydrate accumulation in leaves and stems of 'Valencia' orange at progressively colder temperatures, *Bot. Gaz.,* 138, 13, 1977.

27. **Purvis, A. C. and Yelenosky, G.,** Sugar and proline accumulation in grapefruit flavedo and leaves during cold hardening of young trees, *J. Am. Soc. Hortic. Sci.,* 107, 222, 1982.

28. **Young, R. and Peynado, A.,** Changes in cold hardiness and certain physiological factors of Red Blush grapefruit seedlings as affected by exposure to artificial hardening temperatures, *Proc. Am. Soc. Hortic. Sci.,* 86, 244, 1965.

29. **Guinn, G.,** Changes in sugars, starch, RNA, protein, and lipid-soluble phosphate in leaves of cotton plants at low temperatures, *Crop Sci.,* 11, 262, 1971.

30. **Hilliard, J. H. and West, S. H.,** Starch accumulation associated with growth reduction at low temperatures in a tropical plant, *Science,* 168, 494, 1970.

31. **Taylor, A. O., Jepsen, N. M., and Christeller, J. T.,** Plants under climatic stress. III. Low temperature, high light effects on photosynthetic products, *Plant Physiol.,* 49, 798, 1972.

32. **Dear, J.,** Brief communications: a rapid degradation of starch at hardening temperatures, *Cryobiology,* 10, 78, 1973.

33. **Glier, J. H. and Caruso, J. L.,** Brief communications: low-temperature induction of starch degradation in roots of a biennial weed, *Cryobiology,* 10, 328, 1973.

34. **Amir, J., Kahn, V., and Unterman, M.,** Respiration, ATP level, and sugar accumulation in potato tubers during storage at 4°, *Phytochemistry,* 16, 1495, 1977.

35. **King, A. I., Reid, M. S., and Patterson, B. D.,** Diurnal changes in the chilling sensitivity of seedlings, *Plant Physiol.,* 70, 211, 1982.
36. **Rikin, A., Gitler, C., and Atsmon, D.,** Chilling injury in cotton (*Gossypium hirsutum* L.): light requirement for the reduction of injury and for the protective effect of abscisic acid, *Plant Cell Physiol.,* 22, 453, 1981.
37. **Pomeroy, M. K. and Mudd, J. B.,** Chilling sensitivity of cucumber cotyledon protoplasts and seedlings, *Plant Physiol.,* 84, 677, 1987.
38. **Yelenosky, G.,** Chilling injury in leaves of citrus plants at 1.7°C, *HortScience,* 17, 385, 1982.
39. **Wise, R. R., McWilliam, J. R., and Naylor, A. W.,** A comparative study of low-temperature-induced ultrastructural alterations of three species with differing chilling sensitivities, *Plant Cell Environ.,* 6, 525, 1983.
40. **Sionit, N., Strain, B. R., and Beckford, H. A.,** Environmental controls on the growth and yield of okra. I. Effects of temperature and of CO_2 enrichment at cool temperature, *Crop Sci.,* 21, 885, 1981.
41. **Potvin, C.,** Amelioration of chilling effects by CO_2 enrichment, *Physiol. Veg.,* 23, 345, 1985.
42. **Tajima, K. and Kabaki, N.,** Effects of sugars and several growth regulators on the chilling injury of rice seedlings, *Jpn. J. Crop. Sci.,* 50, 411, 1981.
43. **Lyons, J. M. and Raison, J. K.,** Oxidative activity of mitochondria isolated from plant tissues sensitive and resistant to chilling injury, *Plant Physiol.,* 45, 386, 1970.
44. **Graham, D. and Patterson, B. D.,** Responses of plants to low, nonfreezing temperatures: proteins, metabolism, and acclimation, *Annu. Rev. Plant Physiol.,* 33, 347, 1982.
45. **Raison, J. K. and Orr, G. R.,** Phase transitions in liposomes formed from the polar lipids of mitochondria from chilling-sensitive plants, *Plant Physiol.,* 81, 807, 1986.
46. **Thompson, J. E., Legge, R. L., and Barber, R. F.,** The role of free radicals in senescence and wounding, *New Phytol.,* 105, 317, 1987.
47. **Vianello, A., Macri, F., Cavallini, L., and Bindoli, A.,** Induction of lipid peroxidation in soybean mitochondria and protection by respiratory substrates, *J. Plant Physiol.,* 125, 217, 1986.
48. **Wang, C. Y. and Baker, J. E.,** Effects of two free radical scavengers and intermittent warming on chilling injury and polar lipid composition of cucumber and sweet pepper fruits, *Plant Cell Physiol.,* 20, 243, 1979.
49. **Clare, D. A., Rabinowitch, H. D., and Fridovich, I.,** Superoxide dismutase and chilling injury in *Chlorella ellipsoidea, Arch. Biochem. Biophys.,* 231, 158, 1984.
50. **Purvis, A. C.,** Importance of water loss in the chilling injury of grapefruit stored at low temperature, *Sci. Hortic.,* 23, 261, 1984.
51. **Purvis, A. C.,** Relationship between chilling injury of grapefruit and moisture loss during storage: amelioration by polyethylene shrink film, *J. Am. Soc. Hortic. Sci.,* 110, 385, 1985.
52. **McDonald, R. E.,** Effects of vegetable oils, CO_2, and film wrapping on chilling injury and decay of lemons, *HortScience,* 21, 476, 1986.
53. **Miller, W. R. and Risse, L. A.,** Film wrapping to alleviate chilling injury of bell peppers during cold storage, *HortScience,* 21, 467, 1986.
54. **Grierson, W.,** Chilling injury in tropical and subtropical fruits. IV. The role of packaging and waxing in minimizing chilling injury of grapefruit, *Proc. Trop. Reg. Am. Soc. Hortic. Sci.,* 15, 76, 1971.
55. **Wardowski, W. F., Grierson, W., and Edwards, G. J.,** Chilling injury of stored limes and grapefruit as affected by differentially permeable packaging films, *HortScience,* 8, 173, 1973.
56. **Christiansen, M. N. and Ashworth, E. N.,** Prevention of chilling injury to seedling cotton with antitranspirants, *Crop Sci.,* 18, 907, 1978.
57. **Wilson, J. M.,** The mechanism of chill- and drought-hardening of *Phaseolus vulgaris* leaves, *New Phytol.,* 76, 257, 1976.
58. **Wright, M. and Simon, E. W.,** Chilling injury in cucumber leaves, *J. Exp. Bot.,* 24, 400, 1973.
59. **Morris, L. L. and Platenius, H.,** Low temperature injury to certain vegetables after harvest, *Proc. Am. Soc. Hortic. Sci.,* 36, 609, 1938.
60. **Aljuburi, H. J. and Huff, A.,** Reduction in chilling injury to stored grapefruit (*Citrus paradisi* Macf.) by vegetable oils, *Sci. Hortic.,* 24, 53, 1984.
61. **Purvis, A. C.,** Free proline in peel of grapefruit and resistance to chilling injury during cold storage, *HortScience,* 16, 160, 1981.
62. **Ben-Yehoshua, S., Shapiro, B., Chen, Z. E., and Lurie, S.,** Mode of action of plastic film in extending life of lemon and bell pepper fruits by alleviation of water stress, *Plant Physiol.,* 73, 87, 1983.
63. **Sakai, A. and Yoshida, S.,** The role of sugar and related compounds in variations of freezing resistance, *Cryobiology,* 5, 160, 1968.
64. **Santarius, K. A.,** The protective effect of sugars on chloroplast membranes during temperature and water stress and its relationship to frost, desiccation and heat resistance, *Planta,* 113, 105, 1973.
65. **Crowe, J. H. and Crowe, L. M.,** *Effects of Dehydration on Membranes and Membrane Stabilization at Low Water Activities,* Vol. 5, Chapman, D., Ed., Academic Press, New York, 1984, 57.
66. **Steponkus, P. L.,** Cold acclimation of *Hedera helix, Plant Physiol,* 47, 175, 1971.

67. **Duke, S. H., Schrader, L. E., and Miller, M. G.,** Low temperature effects on soybean (*Glycine max* [L.] Merr. cv. Wells) mitochondrial respiration and several dehydrogenases during imbibition and germination, *Plant Physiol.,* 60, 716, 1977.

68. **Duke, S. H. and Doehlert, D. C.,** Root respiration, nodulation, and enzyme activities in alfalfa during cold acclimation, *Crop Sci.,* 21, 489, 1981.

69. **Sobczyk, E. A., Rybka, Z., and Kacperska, A.,** Modification of pyruvate kinase activity in cold-sensitive and cold-resistant leaf tissues, *Z. Pflanzenphysiol.,* 114, 285, 1984.

70. **Crawford, R. M. M. and Huxter, T. J.,** Root growth and carbohydrate metabolism at low temperatures, *J. Exp. Bot.,* 28, 917, 1977.

71. **Purvis, A. C.,** Soluble sugars and respiration of flavedo tissue of grapefruit stored at low temperatures, *HortScience,* 24, 320, 1989.

72. **Eaks, I. L. and Morris, L. L.,** Respiration of cucumber fruits associated with physiological injury at chilling temperatures, *Plant Physiol.,* 31, 308, 1956.

73. **Eaks, I. L.,** Physiological studies of chilling injury in citrus fruits, *Plant Physiol.,* 35, 632, 1960.

74. **Eaks, I. L.,** Effect of chilling on the respiration of oranges and lemons, *Proc. Am. Soc. Hortic. Sci.,* 87, 181, 1965.

75. **Purvis, A. C.,** Respiration of grapefruit and orange flavedo tissue in relation to chilling and non-chilling temperatures and respiratory inhibitors, *J. Am. Soc. Hortic. Sci.,* 105, 209, 1980.

76. **Purvis, A. C.,** Low temperature induced azide-insensitive oxygen uptake in grapefruit flavedo tissue, *J. Am. Soc. Hortic. Sci.,* 110, 782, 1985.

77. **Day, D. A. and Lambers, H.,** The regulation of glycolysis and electron transport in roots, *Physiol. Plant.,* 58, 155, 1983.

78. **Saglio, P. H. and Pradet, A.,** Soluble sugars, respiration, and energy charge during aging of excised maize root tips, *Plant Physiol.,* 66, 516, 1980.

79. **Azcon-Bieto, J., Day, D. A., and Lambers, H.,** The regulation of respiration in the dark in wheat leaf slices, *Plant Sci. Lett.,* 32, 313, 1983.

80. **Azcon-Bieto, J., Lambers, H., and Day, D. A.,** Effect of photosynthesis and carbohydrate status on respiratory rates and the involvement of the alternative pathway in leaf respiration, *Plant Physiol.,* 72, 598, 1983.

81. **Hrubec, T. C., Robinson, J. M., and Donaldson, R. P.,** Effects of CO_2 enrichment and carbohydrate content on the dark respiration of soybeans, *Plant Physiol.,* 79, 684, 1985.

82. **Coggeshall, B. M. and Hodges, H. F.,** The effect of carbohydrate concentration on the respiration rate of soybean, *Crop Sci.,* 20, 86, 1980.

83. **Bryce, J. H. and ap Rees, T.,** Effects of sucrose on the rate of respiration of the roots of *Pisum sativum,* *J. Plant Physiol.,* 120, 363, 1985.

84. **Rygg, G. L. and Harvey, E. M.,** Behavior of pectic substances and naringin in grapefruit in the field and in storage, *Plant Physiol.,* 13, 571, 1938.

85. **Hulme, A. C., Smith, W. H., and Wooltorton, L. S. C.,** Biochemical changes associated with the development of low-temperature breakdown in apples, *J. Sci. Food Agric.,* 15, 303, 1964.

86. **Murata, T.,** Physiological and biochemical studies of chilling injury in bananas, *Physiol. Plant.,* 22, 401, 1969.

87. **Davis, P. L.,** Further studies of ethanol and acetaldehyde in juice of citrus fruits during the growing season and during storage, *Proc. Fla. State Hortic. Soc.,* 84, 217, 1971.

88. **Davis, P. L., Hofmann, R. C., and Hatton, T. T., Jr.,** Temperature and duration of storage on ethanol content of citrus fruits, *HortScience,* 9, 376, 1974.

89. **Eaks, I. L.,** Effect of chilling on respiration and volatiles of California lemon fruit, *J. Am. Soc. Hortic. Sci.,* 105, 865, 1980.

90. **Pantastico, E. B., Soule, J., and Grierson, W.,** Chilling injury in tropical and subtropical fruits. II. Limes and grapefruit, *Proc. Trop. Reg. Am. Soc. Hortic. Sci.,* 12, 171, 1968.

91. **Eaks, I. L.,** Effect of modified atmospheres on cucumbers at chilling and non-chilling temperatures, *Proc. Am. Soc. Hortic. Sci.,* 67, 473, 1956.

92. **Elthon, T. E., Stewart, C. R., McCoy, C. A., and Bonner, W. D., Jr.,** Alternative respiratory path capacity in plant mitochondria: effect of growth temperature, the electrochemical gradient, and assay pH, *Plant Physiol.,* 80, 378, 1986.

93. **Voinikov, V. K.,** Aftereffect of cooling on functional activity of the mitochrondria of wheat, couch grass, and wheat-couch grass hybrids, *Sov. Plant Physiol.,* 25, 593, 1978.

94. **Leopold, A. C. and Musgrave, M. E.,** Respiratory changes with chilling injury of soybeans, *Plant Physiol.,* 64, 702, 1979.

95. **Kiener, C. M. and Bramlage, W. J.,** Temperature effects on the activity of the alternative respiratory pathway in chill-sensitive *Cucumis sativus,* *Plant Physiol.,* 68, 1474, 1981.

96. **Cole, M. E., Solomos, T., and Faust, M.,** Growth and respiration of dormant flower buds of *Pyrus communis* and *Pyrus calleryana,* *J. Am. Soc. Hortic. Sci.,* 107, 226, 1982.

97. **Janes, H. W., Wulster, G., and Frenkel, C.,** Cyanide-resistant respiration in potato tubers as affected by cold storage, in *Proc. Natl. Symp. Controlled Atmospheres for Storage and Transport of Perishable Agricultural Commodities,* Richardson, D. G. and Mehuriuk, M., Eds., Timber Press, Beaverton, OR, 1982, 181.
98. **McCaig, T. N. and Hill, R. D.,** Cyanide-insensitive respiration in wheat: cultivar differences and effects of temperature, carbon dioxide, and oxygen, *Can. J. Bot.,* 55, 549, 1977.
99. **Lambers, H.,** The physiological significance of cyanide-resistant respiration in higher plants, *Plant Cell Environ.,* 3, 293, 1980.
100. **Steingrover, E.,** The relationship between cyanide-resistant root respiration and the storage of sugars in the taproot in *Daucus carota* L., *J. Exp. Bot.,* 32, 911, 1981.
101. **van der Plas, L. H. W. and Wagner, M. J.,** Regulation of the activity of the alternative oxidase in callus forming discs from potato tubers, *Physiol. Plant.,* 58, 311, 1983.
102. **Smakman, G. and Hofstra, R. (J. J.),** Energy metabolism of *Plantago lanceolata,* as affected by change in root temperature, *Physiol. Plant.,* 56, 33, 1982.

Chapter 14

RELATION OF CHILLING STRESS TO RESPIRATION

James M. Lyons and R. W. Breidenbach

TABLE OF CONTENTS

I. INTRODUCTION

Aside from describing visual symptoms of the response of sensitive plants to chilling treatments, recording some altered type of respiratory behavior was the focus of early attempts to develop hypotheses to elucidate the mechanism(s) by which low temperatures might induce injury in chilling-sensitive plants.[1,2] Although respiration is a key pathway in intermediary metabolism which would logically be examined, the most likely reason that it received such early and continued focus in the study of chilling injury is the ease in measuring or estimating changes in the *rate* of metabolism by following O_2 consumed or CO_2 produced in respiration. The current body of literature on chilling injury in plants includes a number of reviews and chapters,[3-11] texts,[12-14] and symposia proceedings,[15,16] each of which contain a significant review or discussion in relation to the published literature on the impact of chilling treatments on respiration, and there is a chapter by Raison[17] devoted entirely to this topic.

In any attempt to relate respiratory behavior to chilling injury, one must distinguish between the initial response, or primary event, of a chilling-sensitive plant to low temperatures and that of the subsequent or secondary events which are usually associated with degenerative changes accompanying injury leading to death of the tissues.[18] The primary event is essentially instantaneous, it occurs at some critical or threshold temperature, and, in the short term, is reversible. The primary event might be a change in membrane lipid structure, a conformational change in some regulatory enzyme or structural protein, or an alteration in the cytoskeletal structure of the cytoplasm. Secondary events, on the other hand, are both time and temperature dependent and, in the short term, are reversible if the chilling stress is removed and the plant material warmed at nonchilling temperatures. However, if the stress is maintained beyond some time limit for that particular plant material, the cascade of degenerative events becomes irreversible, and warming exacerbates the symptoms of injury. These secondary events can include disruptions such as metabolic or ionic imbalances, increased levels of cytosolic calcium, general loss of cellular integrity, or similar events that lead to visible symptoms of injury. The impact of chilling temperatures on the respiratory metabolism of sensitive plant species should be interpreted as a manifestation of one of these possible secondary events, not as a primary event.

This chapter will discuss the various hypotheses put forth in relation to respiration and chilling and attempt to place in perspective the role of altered respiration as a *reporter* of the primary event — a manifestation of secondary events reflecting metabolic changes ultimately leading to the visible symptoms of injury.

II. AN OVERVIEW OF RESPIRATORY METABOLISM

Respiration is a highly integrated metabolism in which many enzyme systems participate for the purpose of exchanging substrates and energy between the cell and its environment. The metabolic pathways associated with respiration have three main roles in the life of a plant:

1. To carry out the extraction of chemical energy from organic substrates and to provide the plant with a supply of chemical energy in the form of ATP.
2. To perform the conversion of exogenous substrates into the building blocks or precursors of the macromolecular components of cells.
3. To provide for the assembly of these substrates into the intermediates required for biosyntheses — the nucleic acids, proteins, lipids, and other macromolecular cell components.

Aerobic cells obtain most of their energy from respiration, that is, the transfer of electrons from organic fuel molecules to molecular oxygen. About 42% of the total energy produced is useful to the cell, with the remainder dissipated as heat.[19-21] Respiration involves a sequential series of reactions which includes three basic stages:

1.   Glycolysis, the breakdown of glucose into pyruvate, leading to the oxidative formation of acetyl CoA.
2.   The degradation of acetyl residues by the Krebs tricarboxylic acid cycle to yield carbon dioxide and hydrogen atoms.
3.   The transport of electrons equivalent to those hydrogen atoms to molecular oxygen, a process which is accompanied by the coupled phosphorylation of ADP.

The overall equation can be written as follows:

$$C_6H_{12}O_6 + 6O_2 + 38ADP + 38P_i \rightarrow 6CO_2 + 44H_2O + 38ATP$$

Considerable variation exists in the rate of this respiratory metabolism, and it is a function of both internal characteristics of the particular plant material and external factors in the environment. Characteristics of the tissue, such as whether it is meristematic or dormant, whether it is immature, mature, or senescent, and whether it is a storage organ such as a root or tuber, fruit, or flower, are all factors determining respiratory activity. Generally, storage organs have low respiration rates, vegetative tissues moderate, and meristematic tissues such as growing shoots or flower buds, very high. Respiration is generally quite high during early stages of development, decreasing as the plant organs mature. With detached plant parts, the respiration rate generally declines steadily after detachment. An exception to this pattern can be found in climacteric fruits, where the rate declines gradually to a low value at maturity, then increases sharply during ripening, and again declines as the tissue senesces.[22] Temperature is the most influential external factor in determining respiration rate, and, like other chemical reactions, respiration decreases as the temperature is decreased.

Measurement of the respiration rate can be made by determining the amount of carbohydrate consumed, O_2 taken up, CO_2 given off, or heat produced. As indicated in the previous section, most of the literature is based on estimates of respiratory rates based upon gas exchange, i.e., O_2 or CO_2, because of the ease and accuracy of gas analysis methods in comparison to methods for measuring changes in the other components.

III. ALTERED RESPIRATION IN RESPONSE TO CHILLING TEMPERATURES

A. TEMPERATURE COEFFICIENTS (Q_{10})

As a general feature, the respiration rate of plants decreases as the temperature is decreased. Within the normal physiological temperature range, e.g., 0 to 40°C, the velocity of biological reactions decrease two- to three-fold with every 10°C decrease in temperature, and the temperature coefficient for a 10°C interval is called the Q_{10}, which can be calculated as follows:

$$Q_{10} = \left(\frac{R_2}{R_1}\right)^{10/(T_2 - T_1)}$$

where R_1 and R_2 are the rates of respiration at T_1 and T_2, and T_1 and T_2 are the respective temperatures in °C. This allows calculation of expected respiration rates at certain temperatures from a known rate at another temperature. Q_{10} values for known respiration rates of plant materials vary among the temperature ranges with higher values at the lower temperature

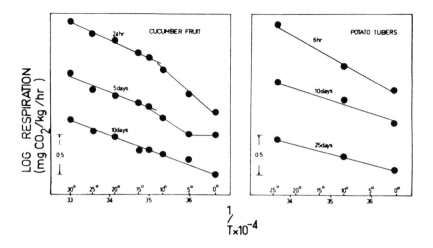

FIGURE 1. Plots of the log respiration rate vs. the reciprocal of the absolute temperature for cucumber fruit and potato tubers. The data for cucumber fruit calculated from that of Eaks and Morris[27] and for potato tubers, from Platenius.[31] The different plots represent different time periods, as indicated, after the plant materials were placed in the temperature treatments.

ranges than at the upper ranges. For example, Q_{10} values of 2.5 to 4.0 occur at 0 to 10°C, and 1.5 to 2.0 at 20 to 30°C.[20]

While the normal pattern of respiration is a gradual decline with time after harvest (and such is the case with chilling-insensitive plant tissues at all temperatures and chilling-sensitive tissues above the chilling threshold), anomalous respiratory behavior has been reported for sensitive plant tissues when held at temperatures below their threshold for chilling. Faster respiration rates (following an initial brief decline) have been observed for chilling-sensitive plant tissues, e.g., sweet potato roots,[23,24] tomato fruits,[25,26] cucumbers,[27] snap beans,[28] and a number of others. Lower Q_{10} values were observed at temperatures in the chilling range for papaya in one report,[29] but others found higher coefficients in this same range,[30] and similar values for both sensitive and insensitive plant materials[31] indicate that high Q_{10} values in the lower temperature ranges are not limited to sensitive plants. In fact, an increase in respiratory Q_{10} at lower temperatures of many plant species has been reported[32] and, hence, should not be considered as necessarily characteristic of chilling sensitivity.

It is again important to examine this reported observed increased rate during storage with the concept of an initial primary response to chilling temperatures followed by a series of subsequent responses associated with an injured tissue. For example, if the data for cucumbers[27] are presented as a plot of the log rate vs. a reciprocal of the absolute temperature (Figure 1) for time periods of 24 h, and 5 and 10 d after being placed at the treatment temperatures, the differences between the initial effect of temperature on respiration and those changes in respiration which occur after damage to the tissue has occurred are readily apparent. At 24 h, the respiration rate decreases with a Q_{10} value of 1.7 between 30 and 13°C. Below 10°C, there is an increase in the rate of reduction of respiration as the temperature is decreased further, and the Q_{10} rose to 2.9 over this range. Visible symptoms of injury occurred between 24 h and 5 d in fruit held at 0 and 5°C. By 10 d in storage, injury and increased respiration had occurred at all temperatures at or below 10°C, causing the slope of the plot in this temperature region to be elevated to the slope exhibited at the higher temperatures. A similar plot for data from the chilling-insensitive potato tuber[31] shows a linear slope over the entire temperature range from 25 to 0.5°C. While the rate of respiration gradually decreases with time in storage (as does the respiration of cucumbers at nonchilling temperatures), there is no change in the course of respiration or linearity in the plot (Figure

1). Thus, in evaluating the impact of chilling treatments on the respiratory process, it is important that the initial reversible response to brief exposures be distinguished from the response observed when the amount of chilling has been sufficient to cause irreversible degenerative changes in the tissue.

B. POSTCHILLING "BURST" OF RESPIRATION

Another respiratory phenomenon observed in relation to chilling injury is the greatly exaggerated rate, or "burst", of CO_2 production (or O_2 uptake) which occurs at warm temperatures after transfer from a chilling treatment. This respiratory stimulation upon transfer has been observed for many plant species,[2] but with chilling-sensitive species it is greatly amplified and can be used as an index of the severity of chilling injury in those tissues. For example, when cucumber fruits are transferred to 25°C after 4 d at 5°C, there is a burst in respiration, but the rate rapidly returns to that of fruits held continuously at 25°C, indicating that the 4-d treatment was not sufficient to cause permanent damage to the tissues.[27] In contrast, chilling for 8 or 12 d induced a similar peak in respiration upon transfer to 25°C, but the rate did not return to a normal level. It is also important to note that symptoms of injury became apparent in the 8-d and 12-d treatments, but not in the 4-d treatment. Similar responses have been reported for zucchini squash,[33] citrus,[34] tomatoes,[25] snap beans,[28] and bananas.[35] Studies by Kiener and Bramlage[36] with cucumber hypocotyls have shown that the alternative respiratory pathway was operating to a much greater extent during this burst than it was either before or after the burst. Involvement of the alternative pathway in chilling will be discussed in a subsequent section.

C. RESPIRATORY QUOTIENTS (RQs)

The respiratory quotient (RQ), i.e., the ratio of the volume of CO_2 produced to the O_2 consumed, has been examined by many workers in an attempt to understand the nature of the altered course in respiration observed in sensitive plant tissues at chilling temperatures. For example, Platenius[31] compared the RQs of a number of plant species, but found no differences which could be related to chilling sensitivity. On the other hand, Eaks and Morris[27] found that the RQ of cucumber fruit held at 0°C was initially below unity, but that it increased with time and rose to above unity after 7 d of chilling, whereas the RQ for fruit held at 15°C was approximately unity throughout the experiment. The increase in RQ with time at 0°C most likely reflected degenerative changes in the tissues since only 2 d at this temperature was sufficient to induce visible symptoms of injury. It was suggested that these RQ values below unity at chilling temperatures indicated an altered metabolism in which protein or lipids were being respired, or some other biochemical dysfunction was occurring. However, based on these types of studies, few conclusions can be drawn which infer that RQs less than unity represent a specific response of chilling-sensitive plants.

IV. PHYSIOLOGICAL SIGNIFICANCE OF CHILLING-INDUCED ALTERED RESPIRATION

A. METABOLIC IMBALANCES

To relate some form of altered respiration induced by chilling to observed symptoms of injury, "imbalances" in metabolism have been proposed repeatedly in the literature. For example, in 1928, Sellschop and Salmon[37] suggested that injury was due to formation of toxic substances resulting from the loss of "coordination" of various cellular processes at low temperatures (and in fact, Christiansen[38] showed that the linear cumulative long-term adverse effects of chilling could be viewed as evidence that toxins quantitatively accumulated with exposure time), and Plank[39] proposed that the temperature for maximum injury was where the temperature coefficient for the "vital" enzymatic process differed the most. Luyet and Gehenio[40] reviewed the theories proposed to explain the mechanism of chilling injury, and imbalances in metabolism, or differential changes in velocities of interrelated chemical reactions, were central in their list. These concepts were fully summarized by Fidler[41] in his review on low-temperature injury and need not be repeated here. Many studies since then have reported that compounds closely associated with intermediary metabolism, such as pyruvate, acetaldehyde, ethanol, and keto acids, accumulate in chilled tissue.[35,42] Furthermore, it is generally recognized that the accumulation of products of fermentation induced by anaerobic conditions, such as ethanol and acetaldehyde, are formed from intermediates of glycolysis and, hence, indicate an "imbalance" in metabolism representing a relative decrease in respiratory activity and an increase in glycolysis.[43]

The occurrence of excessive concentrations of these products of intermediary metabolism in chilling-stressed tissues is unequivocal; however, to interpret the significance of these observations is difficult, as the time sequence of their occurrence in relation to actual tissue injury is not established. Do these compounds themselves lead to injury, or do they simply reflect the injury? This anomalous respiratory behavior may in itself be a symptom of some primary disturbance and not necessarily a contributing factor to the development of visual symptoms. As an example, Smagula and Bramlage[44] concluded that it was not possible from existing data to distinguish whether the acetaldehyde which accumulated as a metabolic product from chilled tissue was a cause or an effect of cellular injury and disorganization.

At the cellular level, it has been clearly demonstrated that the respiration of mitochondria from chilling-sensitive plant tissues is significantly depressed at temperatures below 10°C, indicating that the immediate response of sensitive tissue to chilling is a depression of respiratory activity.[45] Hence, the metabolic imbalances observed following exposure to chilling temperatures could be explained by a disproportionately greater depression of mitochondrial respiration compared with the decrease in glycolytic activity. Raison,[17] in his discussion on the effect of temperature on enzyme reactions, stated that the imbalances in the respiratory metabolism observed in the tissues of plants sensitive to low temperature was the result of the differential change in the E_a (Arrhenius activation energy) of closely associated enzyme systems, and that this change in E_a resulted from a change in the molecular ordering of the membrane lipids in response to chilling. At first glance, it might appear as though there is little consistency between the observation of a depressed respiration of mitochondria derived from chilling-sensitive plant tissues and the increased respiratory activity of those intact tissues in response to chilling treatments; however, this can be reconciled if one keeps in mind the time course involved. That is, the depressed mitochondrial respiration is essentially immediate and reversible, i.e., a quick result of the primary response; within the intact tissue, the imbalances set up would soon lead to the disruption of normal cellular metabolism and increased respiration typically observed with injured tissues, the so-called "wound" or "stress" respiration would occur.[46]

B. ENERGY SUPPLY

Observation of the increased respiratory rates associated with chilling injury has led to questioning of the impact of chilling on cellular respiration and the possible uncoupling of oxidation from phosphorylation. Shichi and Uritani[47] reported that disks prepared from sweet potato roots lost the ability to respond to dinitrophenol at 20°C after 10 to 15 d at 0°C, and this led them to propose that the respiratory mechanism became uncoupled and, hence, adenosine triphosphate, (ATP) formation retarded as a result of chilling. Lewis and Workman[48] demonstrated a marked decline in the incorporation of ^{32}P at 20°C in tissue slices from mature-green tomato fruit following chilling at 0°C, which also suggested a reduced capacity for oxidative phosphorylation. Wheaton,[24] on the other hand, found that root segments from chilling-sensitive corn seedlings were highly responsive to 2,4-dinitrophenol (DNP) after 3 h of chilling at 5°C, but he did not extend his treatments beyond this time period. Tanczos[49] showed that with leaf disks from the chilling-sensitive cucumber (*Cucumis sativus*) that the increase in respiration and the uncoupling by DNP was only evident during the first 2 h of chilling, and it was not observed in leaves chilled longer than 4 d. That is, these physiological responses to chilling were only observed on leaf disks after some damage had occurred. These observations are in agreement with results with mitochondria isolated from chilling-sensitive plants[45,50] where it was shown that the phosphorylative efficiency (as measured by respiratory control or ADP:O ratios) of those from sensitive as well as resistant plant tissues were not affected directly by low temperatures. Stewart and Guinn[51] found there was a continual decrease in ATP concentration with time at chilling for cotton seedlings, but that this response was prevented by hardening the seedlings at 15°C for 2 d prior to chilling. They also found that there was a general shift in phosphorolytic activity associated with chilling injury rather than damage to the phosphorylating mechanism alone.[52] Wilson and McMurdo[53] reported that ATP and ADP levels in *Phaseolus vulgaris* leaves chilled at 5°C and 100% RH actually increased over the first 24 h and remained high for 7 d. Even after 8 or 9 d of chilling in a saturated atmosphere when visible signs of injury appeared, there was only slight reduction in the ATP and ADP levels, and this led them to conclude that a decrease in the ATP supply below that necessary to maintain the metabolic integrity of the cytoplasm could not be considered as the cause of chilling injury in leaves at 5°C and 100% RH.

C. ALTERNATE PATHWAY

In those tissues where respiration rises dramatically, as in the ripening of climacteric fruit, the aging of storage organ slices, the infection by parasites, or in wounding, more often than not, an increase in cyanide (CN)-resistant respiration accompanies the respiration rise. However, Laties[54] cautions against equating CN (an inhibitor of the cytochrome oxidase pathway of respiration) resistance with participation of the alternative path in the absence of CN, since CN itself elicits alternative path activity. Yoshida and Tagawa[55] found that an Arrhenius plot of succinate oxidation by mitochondria from chilling-sensitive *Cornus* callus was linear in the absence of the alternate pathway inhibitor salicylhydroxamic acid (SHAM) and broke sharply at the chilling temperature in the presence of SHAM, while mitochondria from the chilling-insensitive *Sambucus* callus exhibited a linear plot in all cases. This suggests that in the *Cornus* mitochondria, one or more components of the cytochrome path embedded in the inner mitochondrial membrane is affected negatively in conformation or function at the chilling temperature, with the result that electron flow is diverted to the alternative path. These authors further indicate that the two oxidase systems are in separate lipid domains beyond the branching point and thus affected differently by temperature, and that the diversion of electron flow in response to chilling is characteristic in those chilling-sensitive species having some alternate path. That a change in the physical structure of membrane lipid modulates these respiratory events is in agreement with the evidence recently summarized by Raison[56] on the role of thermal phase transitions in modulating the sensitivity to temperature responses in chilling-sensitive plant species (see also Chapter 9).

After 48 h at 2°C, hypocotyls from cucumber seedlings exhibited a burst of O_2 uptake, and the alternative pathway was engaged close to 45% full capacity during this postchilling respiratory burst, but only accounted for up to 50% of the increased respiratory O_2 uptake.[57] The flux through the alternative pathway returned to about the 10% level found before chilling when removed to 25°C, suggesting that there was a rapid increase in alternative pathway activity and a decrease in the cytochrome pathway activity in response to chilling temperatures. Similarly, Leopold and Musgrave[58] demonstrated that chilling involved a major reduction in the cytochrome pathway in whole axes and cotyledons from soybeans and a stimulation of the alternative pathway of respiration in cotyledon tissue. They suggested that these respiratory effects resulted from lesions induced by temperature stress during the reorganization of membranes concomitant with water uptake during imbibition.

Even though there may be problems associated with analytical techniques, such as the influence of growth temperatures, electrochemical gradient, and the pH of the assay medium, for quantification of the relative capacity of the alternate pathway to respiratory metabolism,[59] there appears to be sufficient evidence to establish that it is associated with the increased respiration observed in chilling injury and is one of the early cellular responses in the cascade of secondary events in cellular trauma.

V. CONCLUSIONS

That some anomalous change in the rate of respiration is observed in many plant tissues sensitive to chilling when exposed to low temperatures is well established. As discussed previously, when these sensitive tissues are exposed to chilling temperatures, there is first a primary response (a change in membrane lipid structure, a conformational change in some regulatory enzyme or structural protein, an alteration in the cytoskeletal structure of the cell, etc.) which is then followed by a cascade of secondary events. Changes in the normal rate or course of respiration in response to chilling are in this latter category.

Many investigators have followed the respiratory gases, i.e., O_2 uptake and CO_2 respired, in order to estimate changes in the *rate* of metabolism as a function of chilling injury in these sensitive species. The difficulty with interpreting these results — the changes in RQ or Q_{10} values, the respiratory burst, the elevated or depressed respiration — in terms of what biochemical mechanism(s) might be involved with causing injury to the tissue and symptom development is that of understanding what the "normal" metabolism is for that particular species, that particular tissue or organ, at that particular stage of development and physiology. For example, sweet potato storage roots, cucumber and tomato fruits, and leaves from sensitive species have been shown to exhibit an increased respiration when exposed to chilling treatments,[3] whereas the respiration of excised roots from wheat, pea, corn and bean seedlings,[24] and banana fruit[42] declined. Plots for respiration of meiotic rice anthers exhibited a depressed respiration at chilling temperatures, whereas those from mature anthers exhibited a normal decline with lower temperatures.[60] Studies with the ripening of climacteric fruit have shown that the respiration rate is stimulated or depressed depending on the stage of the climacteric at the time of chilling.[61]

How does one rationalize these apparently disparate observations and yet assemble some cohesive understanding of the underlying mechanisms of chilling injury? First, let us examine the time-course involved in some of these observations as an example. Data from the mitochondrial studies[45] and rice anthers[60] showing a depressed respiratory rate were established within minutes of exposure to low temperature (and this correlates with the data shown in Figure 1), whereas those from the storage roots or fruit tissue[3] were accumulated over a period of days or weeks — clearly a time period reflecting sufficient chilling to cause injury.

Second, the nature of the particular tissue or organ and its physiological state lead to inherent differences. For example, Wheaton[24] reported that the respiration of excised roots from the chilling-sensitive corn and bean was the same as the chilling-insensitive wheat and pea and exhibited a normal decline as would be expected with lower temperatures. However, his data also show that the sensitive seedlings were not injured by these treatments until they were 3 or 4 d old. Minorsky[8] suggested that the decline in respiration observed in these very young root tips and the anthers would not be as susceptible to a chilling-induced reduction in the rate of glycolysis because these highly active tissues would be chiefly respiring lipids and proteins as against those tissues dependent on sucrose for their respiratory substrate. The debate as to whether a decreased ATP supply was the cause of injury in cotton seedlings,[51] but not in cucumber leaves,[53] must center around the respective metabolic demands for energy, that is, the contrast between a rapid, demanding metabolism in the seedling vs. the maintenance demands of a mature leaf. It is also apparent that an early cellular response to chilling in some species is an immediate change in the regulation of electron distribution between the primary and the alternate pathways in respiratory metabolism. Again, the interpretation of the significance of this distribution must be interpreted in terms of the state of metabolism in the respective tissues.

In conclusion, it would appear that while measurement of respiratory gases as an approximation of changes in metabolism has been useful in the past in bringing to the fore the observation that a broad array of metabolic upsets occur as a result of chilling, more comprehensive approaches to measuring and understanding the complications that abound in intermediary metabolism must be employed to provide any additional insight as to the nature of the complex responses. A possible approach can be found in the recent advances with isothermal heat-conductance calorimetry,[62] which will allow short-term measurements of metabolic heat evolution from tissue segments and cultured cells. This method will provide for rapid evaluations of the general state of tissue metabolism and, over short time frames, test the effect of chilling temperatures of different kinds of chilling-sensitive tissues.

REFERENCES

1. **Nelson, R.,** Some storage and transportational diseases of citrus fruits apparently due to suboxidation, *J. Agric. Res.,* 46, 695, 1933.
2. **Appleman, C. O. and Smith, C. L.,** Effect of previous cold storage on the respiration of vegetables at higher temperatures, *J. Agric. Res.,* 53, 557, 1936.
3. **Lyons, J. M.,** Chilling injury in plants, *Annu. Rev. Plant Physiol.,* 24, 445, 1973.
4. **Bramlage, W. J.,** Chilling injury in crops of temperate origin, *HortScience,* 17, 165, 1982.
5. **Graham, D. and Patterson, B. D.,** Responses of plants to low, nonfreezing temperatures: proteins, metabolism and acclimation, *Annu. Rev. Plant Physiol.,* 33, 347, 1982.
6. **Wang, C. Y.,** Physiological and biochemical responses of plants to chilling stress, *HortScience,* 17, 173, 1982.
7. **Wilson, J. M. and McMurdo, A. C.,** Chilling injury in plants, in *Effects of Low Temperatures on Biological Membranes,* Morris, G. J. and Clarke, A., Eds., Academic Press, New York, 1981, 145.
8. **Minorsky, P. V.,** An heuristic hypothesis of chilling injury in plants: a role for calcium as the primary physiological transducer of injury, *Plant Cell Environ.,* 8, 75, 1985.
9. **Markhart, A. H., III,** Chilling injury: a review of possible causes, *HortScience,* 21, 1329, 1986.
10. **Lyons, J. M. and Breidenbach, R. W.,** Chilling injury, in *Postharvest Physiology of Vegetables,* Weichmann, J., Ed., Marcel Dekker, New York, 1987, chap. 15.
11. **Pantastico, E. B., Matoo, A. K., Murata, T., and Ogata, K.,** Chilling injury, in *Postharvest Physiology, Handling and Utilization of Tropical and Subtropical Fruits and Vegetables,* Pantastico, E. B., Ed., AVI Publishing, Westport, CT, 1975, 339.
12. **Ryall, A. L. and Lipton, W. J.,** *Handling, Transportation and Storage of Fruits and Vegetables,* Vol. 1, AVI Publishing, Westport, CT, 1971, 19.
13. **Levitt, J.,** *Responses of Plants to Environmental Stresses,* Vol. 1, Academic Press, New York, 1980, 23.
14. **Wills, R. B. H., Lee, T. H., Graham, D., McGlasson, W. B., and Hall, E. G.,** *Postharvest — An Introduction to the Physiology and Handling of Fruits and Vegetables,* AVI Publishing, Westport, CT, 1981, 71.
15. **Lyons, J. M., Graham, D., and Raison, J. K., Eds.,** *Low Temperature Stress in Crop Plants: The Role of the Membrane,* Academic Press, New York, 1979.
16. **Watada, A. E.,** Chilling injury of Horticultural Crops, *HortScience,* 17, 160, 1982.
17. **Raison, J. K.,** Effect of low temperature on respiration, in *The Biochemistry of Plants,* Vol. 2, Davies, D. D., Ed., Academic Press, New York, 1980, chap. 15.
18. **Raison, J. K. and Lyons, J. M.,** Chilling injury: a plea for uniform terminology, *Plant Cell Environ.,* 9, 685, 1986.
19. **Lehninger, A. L.,** *Biochemistry,* Worth Publishers, New York, 1970, chap. 16.
20. **Kader, A. A.,** Respiration and gas exchange of vegetables, in *Postharvest Physiology of Vegetables,* Weichmann, J., Ed., Marcel Dekker, New York, 1987, chap. 3.

21. **ApRees, T.,** Assessment of the contributions of metabolic pathways to plant respiration, in *The Biochemistry of Plants,* Vol. 2, Davies, D. D., Ed., Academic Press, New York, 1980, chap. 1.
22. **Dilley, D. R.,** Hormonal control of fruit ripening, *HortScience,* 4, 111, 1969.
23. **Lewis, D. A. and Morris, L. L.,** Effects of chilling storage on respiration and deterioration of several sweet potato varieties, *Proc. Am. Soc. Hortic. Sci.,* 68, 421, 1956.
24. **Wheaton, T. A.,** Physiological Comparisons of Plants Sensitive and Insensitive to Chilling Temperatures, Ph.D. thesis, University of California, Davis, 1963.
25. **Cheng, T.-S. and R. L. Shewfelt,** Effect of chilling exposure of tomatoes during subsequent ripening, *J. Food Sci.,* 53, 1101, 1988.
26. **Autio, W. R. and Bramlage, W. J.,** Chilling sensitivity of tomato fruit in relation to ripening and senescence, *J. Am. Soc. Hortic. Sci.,* 111, 201, 1986.
27. **Eaks, I. L. and Morris, L. L.,** Respiration of cucumber fruits associated with physiological injury at chilling temperatures, *Plant Physiol.,* 31, 308, 1956.
28. **Watada, A. E. and L. L., Morris,** Effect of chilling and non-chilling temperatures on snap bean fruits, *Proc. Am. Soc. Hortic. Sci.,* 89, 368, 1966.
29. **Jones, W. W.,** Respiration and chemical changes of papaya fruit in relation to temperature, *Plant Physiol.,* 17, 481, 1942.
30. **Haller, M. H., Rose, D. H., Lutz, J. M., and Harding, P. L.,** Respiration of citrus fruits after harvest, *J. Agric. Res.,* 71, 327, 1945.
31. **Platenius, H.,** Effect of temperature on the respiration rate and respiratory quotient of some vegetables, *Plant Physiol.,* 17, 179, 1962.
32. **James, W. O.,** *Plant Respiration,* Clarendon Press, Oxford, 1953.
33. **Mencarelli, F., Lipton, W. J., and Peterson, S. J.,** Responses of 'zucchini' squash to storage in low-O_2 atmospheres at chilling and nonchilling temperatures, *J. Am. Soc. Hortic. Sci.,* 108, 884, 1983.
34. **Eaks, I. L.,** Physiological studies of chilling injury in citrus fruits, *Plant Physiol.,* 35, 632, 1960.
35. **Murata, T.,** Physiological and biochemical studies of chilling injury in bananas, *Physiol. Plant.,* 22, 401, 1969.
36. **Kiener, C. M. and Bramlage, W. J.,** Temperature effects on the activity of the alternative respiratory pathway in chill-sensitive *Cucumis sativus, Plant Physiol.,* 68, 1474, 1981.
37. **Sellschop, J. P. F. and Salmon, S. C.,** The influence of chilling above the freezing point on certain crop plants, *J. Agric. Res.,* 37, 315, 1928.
38. **Christiansen, M. N.,** Influence of chilling upon seedling development of cotton, *Plant Physiol.,* 38, 520, 1963.
39. **Plank, R.,** Contribution to the theory of cold injury to fruit, *Food Res.,* 3, 175, 1938.
40. **Luyet, B. J. and Gehenio, P. M.,** The mechanisms of injury and death by low temperature, a review, *Biodynamica,* 3, 33, 1940.
41. **Fidler, J. C.,** Low temperature injury to fruits and vegetables, in Low Temperature Biology of Foodstuffs, Hawthorn, J. and Rolfe, E. J., Eds., *Recent Adv. Food Sci.,* 4, 271, 1968.
42. **Pantastico, E. B., Grierson, W., and Soule, J.,** Chilling injury in tropical fruits. I. Bananas (*Musa paradisiaca* var Sapientum cv. Lacatan), *Proc. Trop. Reg. Am. Soc. Hortic. Sci.,* 11, 83, 1967.
43. **Forward, D. C.,** Respiration: a holistic approach to metabolism, in *Plant Physiology: A Treatise,* Vol. 2, Steward, F. C. and Bidwell, R. G. S., Eds., Academic Press, New York, 1983, chap. 4.
44. **Smagula, J. M. and Bramlage, W. J.,** Acetaldehyde accumulation: is it a cause of physiological deterioration of fruits?, *HortScience,* 12, 200, 1977.
45. **Lyons, J. M. and Raison, J. K.,** Oxidative activity of mitochondria isolated from plant tissues sensitive and resistant to chilling injury, *Plant Physiol.,* 45, 386, 1970.
46. **Uritani, I. and Asahi, T.,** Respiration and related metabolic activity in wounded and infected tissues, in *The Biochemistry of Plants, A Comprehensive Treatise,* Vol. 2, Davies, D. D., Ed., Academic Press, New York, 1980, chap. 11.
47. **Shichi, H. and Uritani, I.,** Alterations of metabolism in plants at various temperatures. I. Mechanism of cold damage of sweet potato, *Bull. Agric. Chem. Soc. Jpn.,* 20, 284, 1956.
48. **Lewis, T. L. and Workman, M.,** The effect of low temperature on phosphate esterification and cell membrane permeability in tomato fruit and cabbage leaf tissue, *Aust. J. Biol. Sci.,* 17, 147, 1964.
49. **Tanczos, O. G.,** Influence of chilling on electrolyte permeability, oxygen uptake and 2,4-dinitrophenol stimulated oxygen uptake in leaf discs of the thermophilic *Cucumis sativus, Physiol. Plant.,* 41, 289, 1977.
50. **Minamikawa, T., Akazawa, T., and Uritani, I.,** Mechanisms of cold injury in sweet potatoes. II. Biochemical mechanisms of cold injury with special reference to mitochondrial studies, *Plant Cell Physiol.,* 2, 301, 1961.
51. **Stewart, J. M. and Guinn, G.,** Chilling injury and changes in adenosine triphosphate of cotton seedlings, *Plant Physiol.,* 44, 605, 1969.
52. **Stewart, J. M. and Guinn, G.,** Chilling injury and nucleotide changes in young cotton plants, *Plant Physiol.,* 48, 166, 1971.

53. **Wilson, J. M. and McMurdo, A. C.**, Chilling injury in plants, in *Effects of Low Temperatures on Biological Membranes*, Morris, G. J. and Clarke, A., Eds., Academic Press, New York, 1981, 145.

54. **Laties, G. G.**, The cyanide-resistant, alternative path in higher plant respiration, *Annu. Rev. Plant Physiol.*, 33, 519, 1982.

55. **Yoshida, S. and Tagawa, F.**, Alteration of the respiratory function in chill-sensitive callus due to low temperature stress. I. Involvement of the alternate pathway, *Plant Cell Physiol.*, 20, 1243, 1979.

56. **Raison, J. K.**, Alterations in the physical properties and thermal response of membrane lipids: correlations with acclimation to chilling and high temperature, in *Beltsville Symp. in Agricultural Research*, Vol. 9, St. John, J. B., Berlin, E., and Jackson, P. C., Eds., Rowman & Allanheld, Totowa, NJ, 1985, chap. 26.

57. **Kiener, C. M. and Bramlage, W. J.**, Temperature effects on the activity of the alternative respiratory pathway in chill-sensitive *Cucumis sativus*, *Plant Physiol.*, 68, 1474, 1981.

58. **Leopold, A. C. and Musgrave, M. E.**, Respiratory changes with chilling injury of soybeans, *Plant Physiol.*, 64, 702, 1979.

59. **Elthon, T. E., Stewart, C. R., McCoy, C. A., and Bonner, W. D., Jr.**, Alternative respiratory path capacity in plant mitochondria: effect of growth temperature, the electrochemical gradient, and assay pH, *Plant Physiol.*, 80, 378, 1986.

60. **Toriyama, K. and Hinata, K.**, Anther respiratory activity and chilling resistance in rice, *Plant Cell Physiol.*, 25, 1215, 1984.

61. **Eaks, I. L.**, Effects of chilling on respiration and ethylene production of 'Hass' avocado fruit at 20°C, *HortScience*, 18, 235, 1983.

62. **Criddle, R. S., Breidenbach, R. W., Lewis, E. A., Eatough, D. J., and Hansen, L. D.**, Effects of temperature and oxygen depletion on metabolic rates of tomato and carrot cell cultures and cuttings measured by calorimetry, *Plant Cell Environ.*, 11, 695, 1988.

Chapter 15

INFLUENCE OF CHILLING STRESS ON ETHYLENE PRODUCTION

Roger J. Field

TABLE OF CONTENTS

I. INTRODUCTION

The response of plants and detached plant organs to temperatures lower than are optimal for growth and development may be directly linked to hormonal and metabolic control processes that include ethylene production. There are three low temperature categories that are known to influence ethylene production:[1] (1) the exposure to nondamaging reduction in temperature, (2) the response to chilling in those species that are sensitive to temperatures below 10 to 12°C and above 0°C, and (3) the effect of exposure to a freezing temperature of 0°C or below.

Relatively scant attention has been paid to temperature-induced changes in ethylene production and the impact on vegetative growth and development. This is surprising in view of the observations that small variations in temperature may bring about substantial changes in ethylene production.[1,2] Bean leaf tissue (*Phaseolus vulgaris* L.) showed increases or decreases of at least 30% for basal ethylene and 50% for wound ethylene for divergencies of 5°C from the reference temperature of 20°C.[1] However, in the response of plants to chilling- and temperature-manipulated postharvest physiology, there is increasing research interest; and temperature-ethylene relationships increasingly being recognized as important.

There are two important aspects of the temperature-ethylene relationship: (1) enhanced ethylene production following exposure of chilling-sensitive species to conditions below or at the critical chilling temperature, and (2) the magnitude of ethylene production following such exposures.[3-9] Frequently, the postchilling rise in ethylene production is higher than typical basal or wound ethylene production for the species.[5,6,8] There is no universal, clear-cut interpretation of the significance of the increased ethylene. It is clearly important to establish whether ethylene promotes the physiological and biochemical changes associated with chilling damage, or if it is involved in the defense of the plant against damage, or if it is of no significance whatsoever. The latter appears unlikely in view of the positive role of ethylene in regulating other physiological processes and the use of ethylene or its metabolic precursor, 1-aminocyclopropane-1-carboxylic acid (ACC), as indicators of chilling sensitivity.[10,11]

The ability of physiologists and plant breeders to find methods of overcoming any deleterious effects of ethylene in chilling-sensitive or other species has improved since the confirmation of the principle pathway of ethylene biosynthesis in plants.[12] Detailed knowledge of the pathway, including the identification and operation of all intermediates and, to a lesser extent, the key enzymes, has provided opportunities for manipulating ethylene production.[13-16] However, at this stage, there is no universal, cost-effective treatment or method that can easily reverse the adverse effects of chilling-induced ethylene production.

II. ETHYLENE BIOSYNTHESIS

A. THE PATHWAY

Determination of the major pathway of ethylene biosynthesis in plants by Adams and Yang,[12] with confirmation by Lurssen et al.,[17] has been a critical development. Lieberman and co-workers[20] had previously identified the amino acid methionine as the starting material for ethylene biosynthesis, initially in model systems,[18,19] but later in apple fruit. Elucidation that methionine was converted to *S*-adenosylmethionine (SAM)[21] and then 1-aminocyclopropane-1-carboxylic acid (ACC)[12] came much later. The basic pathway can now be represented as

$$\text{Methionine} \rightarrow \text{SAM} \rightarrow \text{ACC} \rightarrow \text{ethylene}$$

It is now clear that the formation of ACC and, subsequently, ethylene is a small part of the methionine cycle.[22] The intermediate SAM may be converted to ACC or 5^1-methylthioadenosine *in vivo*.[22]

Conversion of methionine into SAM has been studied extensively and involves the enzyme methionine adenosyltransferase.[23,24] It has been concluded that it is unlikely that methionine adenosyltransferase becomes a rate-limiting enzyme in ethylene biosynthesis.[22] The more critical enzymatic steps are the conversion of SAM to ACC and the subsequent release of ethylene. The conversion of SAM to ACC is effected by the enzyme ACC synthase, initially extracted from tomato fruit.[25] The enzyme is soluble, believed to be a pyridoxal enzyme, and has been isolated from bean leaf tissue[26] and fruit of cucumber,[27] orange,[28] cantaloupe,[29] and numerous other plant tissues. Typically, ACC synthase levels in plant tissue are considered low, making identification and quantification difficult.[30]

It has been proposed that the enzyme coverting ACC to ethylene (ethylene-forming enzyme [EFE]) is largely constitutive and that the rate-limiting step in ethylene production is the formation of ACC.[22] The reaction of ACC to ethylene is oxygen requiring and necessitates maintenance of membrane integrity, potentially a critical point in chilling-sensitive species. Much of the early literature showing impaired ethylene production following various chemical- or environmental-induced perturbations to cell membranes are now interpreted as reduced activity of EFE.[31-33] There has been some confirmation that EFE is membrane bound,[34] although the operation *in vitro*, noncellular systems[35,36] and the existence of nonenzymatic conversion by oxidants[25,37] and free radicals[38] has not assisted in clarifying the nature of EFE or its location in the cell.

The use of isolated protoplasts and vacuoles has suggested that vacuoles, and, perhaps more specifically, the tonoplast, may be the site of EFE activity in cells induced to accumulate ACC.[39] In fruit cells where the ACC pool is small and the turnover is rapid,[12,40] ethylene formation may take place in a cytoplasmic membrane or the plasmalemma.[39,41]

B. METABOLISM OF ACC TO MALONYL-ACC

As an alternative to producing ethylene, ACC may be conjugated to 1-(malonylamino) cyclopropane-1-carboxylic acid (MACC) in plant tissues.[42-44] The enzyme responsible for the formation of MACC is constitutive and malonylation is rapid upon synthesis of ACC.[22] All of the available evidence shows that MACC formation is irreversible, is a poor producer of ethylene, and is a biologically inactive end product of ACC metabolism, not a temporary storage form.[45-49] The concentration of MACC rises in parallel with high rates of ACC synthesis and may be related to the imposition of stress.[1,47,48] While the formation of MACC represents a detoxification mechanism, reducing ethylene production potential, there is some evidence that MACC is not totally inert with respect to ethylene production. Application of MACC to vegetative tissues may release ACC and, subsequently, ethylene.[50,51] The implication of this result for interpreting the effect of chilling on ethylene production will be considered later.

There is incomplete understanding of the dynamics of the ACC to MACC conversion, its reversibility, and the implications for the formation of ethylene. There is evidence that the ethylene induction of MACC leads to the autoinhibition of ethylene synthesis.[44,52,53] Unless ethylene-producing systems are clearly characterized, the operation of such feedback systems may obscure the true relationships between external factors, such as chilling, and ethylene production.

C. CHEMICAL MANIPULATION OF ETHYLENE BIOSYNTHESIS

The understanding of the pathways of ethylene biosynthesis is limited by the lack of clarity of the enzymology and the role of endogenous chemical manipulators. The major focus has been on possible control by plant hormones and the significance of cellular carbon dioxide concentration, which may fluctuate with photosynthesis.

The interaction between auxin and ethylene was observed over 50 years ago,[54] and it was later discovered that the apparent similarity in response of some tissues to auxin and ethylene was a result of auxin-induced ethylene production.[55] More recently, auxin-induced stimulation of ethylene production has been shown to involve the conversion of SAM to ACC and the activity of ACC synthase.[55-59] The conversion of SAM to ACC is the rate-limiting step, with IAA-induced ethylene closely linked to ACC levels and ACC synthase activity.[22,57-59] The stimulation of ACC synthesis by auxin has been linked nonspecifically to RNA and protein synthesis.[22,57] The significant lag period between auxin treatment and ACC or ethylene production[6] suggests that *de novo* synthesis is occurring.

There is significant interest in nonendogenous regulators of ethylene biosynthesis, both to promote ethylene production to enhance such processes as abscission and ripening and to inhibit ethylene synthesis to control a wide range of physiological activity. Currently, there is no chemical that can be usefully employed to remove ethylene effects associated with chilling. However, the present knowledge on chemical manipulators of ethylene biosynthesis is worthy of a brief review because of their use in experimental systems.

1. Inhibitors of Ethylene Synthesis

Chemicals which act as pyridoxal enzyme inhibitors are potent inhibitors of ACC synthase. Rhizobitoxine, a β,γ-enol ether amino acid was shown to inhibit the conversion of methionine to ethylene.[60] The more recent discovery of the analogue aminoethoxyvinylglycine (AVG)[61,62] has proven to be a powerful tool in understanding ethylene biosynthesis and regulation in *in vivo* and *in vitro* systems.[12,25] The second group of pyridoxal enzyme inhibitors are hydroxylamine analogues, including aminooxyacetic acid (AOA)[16,58] and α-aminooxycarboxylic acid with an aliphatic side chain.[61]

Specific inorganic ions may have pronounced effects on ethylene biosynthesis. Cobalt (Co^{2+}) and nickel (Ni^{2+}) are effective, with Co^{2+} shown to inhibit the conversion of ACC to ethylene.[59] This effect may occur without influencing the biosynthesis of ACC.[63] The precise mode of action of Co^{2+} and Ni^{2+} is not known, although some works suggest that the inhibitory effect involves the ion complexing with sulfhydryl groups of proteins.[59,64]

General uncouplers of oxidative phosphorylation, including 2,4-dinitrophenol (DNP) and carbonylcyanide *m*-chlorophenylhydrazone (CCCP),[65,66] and nonspecific membrane-perturbing materials[33,67,68] inhibit ethylene production. The conversion of ACC to ethylene was inhibited by low concentrations of DNP, while not affecting the conversion of methionine to SAM.[69] The mechanism of DNP-induced EFE inhibition is not clear, although it may relate to ATP requirements for conversion of ACC to ethylene.[69] Some DNP effects have been attributed to membrane disruption, emphasizing a lack of specificity of action which may indicate that the use of such chemicals in the investigation of ethylene physiology is inappropriate. Similarly, lipophilic substances, such as surfactants and phospholipase D, promote general disruption of membranes,[33,67,68] but do not specifically perturb ethylene biosynthesis.

Free radical scavengers, such as sodium benzoate and *n*-propyl gallate, inhibit conversion of ACC to ethylene in vegetative and reproductive tissues.[70-72] One contention is that EFE involves a free radical reaction,[73] although others have doubted that substances such as *n*-propyl gallate act specifically as free radical scavengers in inhibiting ethylene production.[25]

Polyamines are essential growth-controlling factors in plants associated with senescence processes that include ethylene production.[74-78] Ethylene production is inhibited by polyamine-mediated inhibition of the ACC-to-ethylene step.[77,78] The role of polyamines in the endogenous control of ethylene biosynthesis requires resolution, particularly as SAM is a common precursor to ethylene and polyamines.

The confirmation of ACC as the immediate precursor of ethylene in endogenous production suggests that inhibition of ethylene could occur by competitive inhibition of ACC.

The addition of α-aminoisobutyric acid inhibits ethylene production, albeit at relatively high concentration.[68,79]

An alternative to blocking specific points of ethylene biosynthesis is to use chemicals, such as the cyclic olefin, 2,5-norbornadiene, which bind to the ethylene receptor site.[80-83] Silver ions (Ag+) behave similarly by interfering with the expression of ethylene effects rather than influencing synthesis.[84-86] Such techniques may have limited experimental value, as high concentrations of ethylene can reverse inhibitory effects.

2. Promotion of Ethylene Synthesis

Attempts to enhance ethylene production by manipulating the endogenous pathway have not been entirely effective. Addition of ACC to virtually all plant tissues stimulates ethylene release, as will the addition of the earlier precursors.[22] Indoleacetic acid and its analogues promote the conversion of SAM to ACC,[56,58,59] but generally, there is a significant lag between application and enhanced ethylene production.[1] The IAA-induced increase in ACC synthease is inhibited by nucleic acid and protein synthesis inhibitors,[57] and it is assumed that IAA exerts an effect by promoting *de novo* synthesis of ACC synthase.

The discovery of ACC prompted a search for related analogues that could be applied exogenously and release ethylene. Such a material was *N*-formyl ACC, which is not naturally occurring, but rapidly converted to ACC in plant tissues with the release of ethylene.[15] The significant synthetic plant growth regulators that release ethylene, ethephon and etacelasil, spontaneously release ethylene upon contact with water. The release is pH dependent, but not enzymic.[14,15]

Although EFE activity is oxygen dependent and carbon dioxide is a known antagonist of ethylene action,[62] increases in carbon dioxide concentration promote the conversion of ACC to ethylene.[87-90] In tobacco leaf tissue, carbon dioxide promotes both the synthesis and activity of EFE, with the latter effect being dominant.[89] The promotive effect of carbon dioxide on EFE synthesis is maintained for longer upon illumination, in part conflicting with earlier reports that light inhibited conversion of ACC to ethylene through the production of carbon dioxide.[88] The enhancement of ethylene production from ACC by bicarbonate has been demonstrated in model cell-free systems as well as in intact plants.[91] Earlier conclusions that carbon dioxide and bicarbonate effects were solely associated with induction of ethylene release from putative receptor sites[91] must be balanced against the more recent demonstration of carbon dioxide promotion of EFE synthesis.[89]

The interaction between photosynthesis and ethylene production requires clarification. It seems reasonable to assume that the inhibition of ethylene evolution by light[87-94] results from a decrease in internal carbon dioxide concentrations associated with photosynthesis.[88]

III. ETHYLENE PRODUCTION AND CHILLING TEMPERATURES

A. ETHYLENE PRODUCTION

The response of plants to temperatures lower than are optimal for growth is linked to changes in hormonal and metabolic control processes that include ethylene production. There is reasonable understanding of how temperature changes influence gross ethylene production,[1] although there is incomplete information on the control of synthesis. Principle consideration will be given to short- and long-term responses of plants to chilling temperatures and, in particular, changes associated with temperature transfers.

There is marked reduction in ethylene production when parts of chilling-sensitive species are incubated below a growing temperature of 20 to 25°C.[5,27,95] Ethylene production continues at 2.5°C in bean (*Phaseolus vulgaris* L.) leaf tissue[5] and 2°C in tomato (*Lycopersicon esculentum* Mill.) fruit plugs.[95] In the short-term, incubation of chilling-sensitive tissues at subchilling temperatures does not eliminate ethylene production. However, incubation procedures were carried out at high humidity, which is not likely to induce typical chilling responses.[96] Chilling damage to primary leaves of *P. vulgaris* induced by exposure to 5°C at 20% RH was associated with increased ethylene production and electrolyte leakage, indicating at least a partial change in membrane function.[8]

The association of chilling sensitivity with membrane perturbation and possible phase transitions below the critical chilling temperature is considered in Chapter 9 of this book. The proposal that temperatures below the critical chilling temperature affect the activity of membrane-bound enzymes, increasing their activation energy, has been applied to determinations of ethylene production.[1,5,95] Ethylene production in chilling-sensitive plants typically has two values for activation energy, with an approximate doubling of the value below the transition point.[1] Such measurements of activation energies are relatively crude and, at best, indicate differences in the rate-limiting step determining ethylene production at either side of the transition temperature point. Determination of activation energies, or Q_{10} values, below the transition point is more variable than above and attributable in part to the practical difficulties of measuring the meager ethylene production at ≤ 5°C, where it is typically less than 0.2 nl g^{-1} h^{-1}, or 5 to 10% of values at 20 to 25°C.[1,5,6] Small deviations in the slope of an Arrhenius plot can lead to significant changes in activation energy.

The discontinuities of Arrhenius plots above and below the chilling-sensitive temperature have been used to support the contention that ethylene production involves at least one membrane-bound enzyme.[15,95] The validity of this conclusion is disputed on two grounds. First, that discontinuous Arrhenius plots may not infer a membrane phase change.[97] Second, that some normally chilling-insensitive species, such as pea (*Pisum sativum* L.) and apple (*Malus domestica* Borkh.), can demonstrate temperature breaks in Arrhenius plots similar to those of similar chilling-sensitive species. It is possible that these nonacclimated, chilling-insensitive species, could either show a lack of rapid biochemical adaption with respect to ethylene synthesis, or simply have a major decline in ethylene synthesis below a certain temperature.[1] However, in apple fruit sections, the continued synthesis of ethylene at temperatures around 5°C plus the comparability with rates of reaction and activation energies for processes such as respiration, which are known to involve membrane-bound enzymes,[2] suggests that this analytical approach can support the contention that membrane-bound enzymes are involved in ethylene synthesis.

While 2°C is frequently quoted as the minimum temperature for measurable ethylene production,[1] there are reports of production at -1 to 0°C. Certain apple[98,99] and pear (*Pyrus communis* L.) cultivars[100-104] have measurable ethylene production at temperatures around 0°C, while the lowest recorded value for a flower is 1.6°C in carnation (*Dianthus caryophyllus* L.).[105] There is an increase in ACC accumulation in certain pear cultivars during low-temperature storage.[106-109] This should not be regarded as a chilling effect, but may be of significance in promoting ripening processes.

In most of the low temperature research, the greatest interest has been in the outcome for ethylene production of returning tissues to near-ambient temperatures.[1,5,6,110] Tissue response to 0°C or below, followed by exposure to ambient conditions, has usually been interpreted in terms of freezing-induced wound damage[111,112] and has little applicability to the chilling responses per se.

B. THE RATE-LIMITING STEP IN ETHYLENE BIOSYNTHESIS

There is reasonable agreement on the point of interference by chilling temperatures. The synthesis of ACC is impaired at low temperature. Wang and Adams[27] used cucumber (*Cucumis sativus* L.) and showed that the low ethylene production at 2.5°C was associated with low ACC and ACC synthase activity, suggesting that ACC synthesis was the rate-limiting step in ethylene production. Similar results were found by Field,[6] who showed enhanced ethylene production at 5°C in bean leaf tissue, following addition of exogenous ACC. Ethylene production at 2.5 to 5°C is still ACC concentration dependent.

There is no dispute that ethylene produced under chilling conditions is produced by the same pathway as that operating at ambient temperatures. Wang and Adams[27] supplied chilled and nonchilled cucumber skin tissues with ^{14}C-methionine, which was converted to ^{14}C-ACC and subsequently released ^{14}C-ethylene following treatment with a chemical degradation assay.[37]

The synthesis of ACC from SAM is considered to be the major rate-determining step in ethylene biosynthesis.[25] It is reasonable to suggest that it is this step that is influenced by low temperatures, including the response by chilling-sensitive species. The addition of exogenous ACC to apple fruit tissue incubated at temperatures from 4 to 35°C induced a constant 2.5 times increase in ethylene production over the complete temperature range.[67] One interpretation of this data is that conversion of ACC to ethylene is not markedly influenced by temperature.

C. TEMPERATURE TRANSFERS AND CHILLING-INDUCED ETHYLENE PRODUCTION

While continued exposure of chilling-sensitive species to temperatures below the critical chilling temperatures results in permanent damage and usually death, the outcome of short-duration exposure is likely to be variable (see Chapter 1). Thus, the effect of short-duration exposures to chilling temperatures upon subsequent ethylene production at nonchilling temperatures is an important area of investigation, with significant implications for the nature of chilling damage.

Using a model dwarf bean (*Phaseolus vulgaris* L.) leaf system, Field[5,6] showed that exposure to a subcritical temperature (5°C) resulted in immediate ethylene production upon transfer to 25°C. Low-temperature exposure was varied with a significant response after 0.5 h of duration, but with greater ethylene production following longer exposure. Results were obtained for a wound ethylene system immediately following the cutting of leaf discs, and for basal production using discs that had been aged for 16 h to allow wound ethylene production to decay. Both systems were characterized by the virtually lag-free production of ethylene following temperature transfer and the magnitude of ethylene production at 25°C, which exceeded normal wound production at that temperature. Similar results were obtained with cucumber fruit[27,113] tissue and with mung bean (*Vigna radiata* L.) hypocotyls,[114] although the timing of chilling events was prolonged. Cucumber fruit tissue was exposed to 2.5°C for several days before transfer to 25°C. Rapid ethylene production occurred upon transfer, provided that low-temperature exposure did not exceed 4 d, which presumably caused irrepairable damage to ethylene biosynthesis and normal cell function. Exposure to a nonchilling 13°C resulted in no significant increase in ethylene production on transfer to 25°C.

In both the dwarf bean leaf and cucumber fruit systems, transfer of tissues to the higher temperature resulted in rapid elevation of ACC levels[6,27] and ACC synthase activity.[27] In dwarf bean leaves, a 2-h exposure to 5°C resulted in an approximately tenfold increase in ACC after 4 h at 25°C, while a 4-d exposure of cucumber fruit to 2.5°C resulted in a 50-fold increase in ACC within 7 h after warming at 25°C.[6,27] In the latter system, incorporation of ^{14}C-methionine resulted in higher levels of ^{14}C-ACC and ^{14}C-ethylene from chilled than nonchilled tissue. The demonstrably greater capacity of chilled tissue to produce ACC and ethylene with relatively short (<1 h) lag times after temperature transfer has resulted in several investigations to determine the biochemical and physiological basis of the phenomenon.[6]

The use of specific inhibitors of ethylene biosynthesis has provided supporting evidence for the conclusion that formation of ACC is the step stimulated by the favorable conditions following chilling stress and is the rate-limiting step in chilling-induced ethylene production. The addition of the ACC synthesis inhibitor AVG at the end of a 2-h incubation of dwarf bean leaf tissue at 5°C does not result in increased ethylene production upon transfer to 25°C.[6] This result suggests that accumulation of ACC during chilling exposure is not responsible for the rapid "overshoot" ethylene production.[6,27]   Wang[115] has tested a large number of other ethylene synthesis inhibitors in the chilled cucumber fruit system. The free radical scavengers propyl gallate and sodium benzoate allowed ACC to accumulate, but prevented conversion to ethylene, even after exogenous ACC addition. Free radical scavengers do not inhibit ACC formation in warmed tissue that has been chilled. The behavior of the uncouplers of oxidative phosphorylation DNP and CCCP was more pronounced on the conversion of ACC to ethylene, although the inhibition of methionine to SAM resulted in lower ACC production. All of these inhibitors operated in the chilling system in a similar way to that operating under ambient conditions.[115] Their use provided no additional insights into how the magnitude of postchilling ethylene (and ACC) production is varied by the duration and intensity of the chilling experience.

D. THE BIOCHEMICAL BASIS OF "OVERSHOOT" ETHYLENE PRODUCTION

While there is little doubt that warming chilled tissues initiates ACC synthesis, there is a requirement to understand what turns on ACC synthase and, in particular, how the intensity and duration of the chilling experience is recorded by treated tissues. The latter point is important because of the relationship between chilling experience and the magnitude of subsequent ethylene production. [5,6,27] It is postulated that chilling treatments may induce the formation of a membrane-associated cytoplasmic factor, which is responsible for the enhanced ethylene production after transfer of tissue to higher temperatures.[5]

A recent investigation by Wang[115] has shown the value of using a protein synthesis inhibitor, cycloheximide, and the RNA synthesis inhibitors cordycepin and α-amanitin. Cucumbers were chilled at 2.5°C for 4 d and inhibitors vacuum infiltrated into the tissues before transfer to 30°C. In the control and RNA synthesis inhibitor-treated tissues, there was the previously reported[27] rapid increase in ACC synthase, ACC, and ethylene. Protein synthesis inhibition with cycloheximide significantly reduced the production of these components. It was concluded that the biochemical steps that could have been influenced by the two RNA synthesis inhibitors had already been stimulated by chilling. The chilling experience had induced factors or conditions required for transcription, but inhibition by cycloheximide suggested that translation of protein, presumably ACC synthase itself, had not been completed during chilling. Clearly there is still a requirement for chilling to induce transcription, and there is no clear indication of the nature of the biochemical message that is induced by chilling. Post-transcription products will include mRNA, which may accumulate and develop a potential for "overshoot" ACC synthase, ACC, and ethylene production in response to the duration and intensity of the chilling experience.

There has been no clear demonstration that rapid release of ACC and, subsequently, ethylene comes from the conjugate MACC following warming of chilled tissues. As in other stress situations,[47] the potential for MACC accumulation increases during chilling.[116] Exposure of cucumber seedlings to 5°C resulted in a two- to fourfold increase in MACC after warming at 20°C. There was no evidence of conversion of MACC to ACC, and no causal relationship between chilling experience, MACC accumulation, and ethylene production was noted. The conjugation of ACC would reduce the magnitude of potential ethylene production after tissue warming.

IV. ETHYLENE AS AN INDICATOR OF CHILLING DAMAGE

A. GENOTYPIC VARIATION

The demonstration that ethylene and ACC levels rise following chilling[5,6,27] suggests that these changes may be used as indicators of the severity of chilling damage and/or markers for variations in chilling sensitivity between species and cultivars. It can be reasonably argued that a specific physiological change such as ethylene production may be more useful and rapid than less specific assessments such as electrolyte leakage or pigmentation changes (see also Chapter 18).

Initial studies have demonstrated that genotypic differences in chilling sensitivity correlate with levels of chilling-induced ethylene. The most chill-sensitive genotypes produced the greatest amount of ethylene.[10] Storage of leaf tissue from a range of plant species at 0°C for varying periods of time, followed by warming to 20°C, resulted in significant differences in ethylene production.[10] A chilling duration of 1 h was sufficient for induction of ethylene in the highly chilling-sensitive tropical herb *Episcia reptans* Mart., while a similar response took several days with cucumber (*Cucumis sativus* L.) and various species of tropical passionfruit (*Passiflora* spp.). In the latter case, six species were tested, with ethylene production being induced by chilling for 1 to 3 d in *P. edulis flavicarpa* and, at the other extreme, greater than 15 d in *P. edulis*. The ranking of species according to ethylene production matched their known sensitivity to chilling stress.[10] Unlike some other experiments,[6,27] the most chilling-sensitive species (*Episcia reptans* and *P. edulis flavicarpa*) were shown to accumulate ACC during chilling, while this did not occur in the more chilling-resistant *P. edulis*. No account was taken of possible sequesterization of ACC as MACC.

In similar experiments, Guye et al.[11] investigated a range of *Phaseolus* spp. and cultivars. Leaves were chilled at 5°C for 24 h before measurement of ethylene at 23°C. In this case, the more chill-tolerant species and cultivars produced greater amounts of ethylene than the chill-sensitive types. The chill-sensitive genotypes also had a poorer ability to convert exogenously applied ACC to ethylene. The results have been explained in terms of greater chilling-induced water stress and dehydration of chilling-sensitive genotypes, leading to membrane disfunction and a reduced ethylene synthesis capability.

The lack of a consistent relationship between chilling sensitivity and ethylene production requires further experimentation. The use of cell-free systems[35,117,118] in combination with variations in osmotica and temperature may provide useful insight. Confirmation that increases in ACC and ethylene in cucumber seedlings were a response to the chilling experience and not a cause of the injury[119] suggests that the use of ethylene as a marker of chilling sensitivity is worthy of pursuing.

B. DEVELOPMENT OF CHILLING INJURY SYMPTOMS

Of considerable interest is whether ethylene production and chilling damage has a causal relationship or whether the two processes are independent. Given the known association between ethylene and the development of senescence symptoms[120] and the significant ethylene production following chilling,[1] it is tempting to assume that development of specific chilling symptoms is related to ethylene.

It is possible to show that elimination of ethylene production by treatment with the synthesis inhibitor AOA at the end of a chilling period does not affect the chilling injury or polyamine production on tissue warming.[119] Treatment of cucumber seedlings with ACC or ethylene before, during, or after chilling experience did not influence chilling injury symptoms.[115] While similar experiments with bean[115] and tomato[121,122] confirm the lack of a relationship, the situation is not the same with avocado (*Persea americana* Mill.), where exposure to ethylene at chilling temperatures induced chilling injury damage.[123,125]

Ethylene has been used as a partial protectant against chilling injury damage.[126,127]

Treatment of honey dew muskmelon (*Cucumis melo* L.) with ethylene at 20°C before transfer to 2.5 to 5°C significantly reduced chilling damage.[126] The ripening characteristics of fruit after subsequent retransfer to 20°C were normal. Similar responses have been noted in sweet potato (*Ipomoea batatas* L.).[128]

There is no common association between ethylene production and the promotion of, or defense against, chilling injury damage. In an attempt to make progress in this apparently confusing area, it is probably unreasonable to assume that all species will behave similarly in linking ethylene production to chilling sensitivity. The extreme range of chilling sensitivity exhibited by plants[129] suggests that it is inappropriate to strive for simple, generalized relationships.

V. CHILLING, ETHYLENE, AND PLANT DEVELOPMENT PROCESSES

All stages of plant growth are vulnerable to chilling damage, resulting in ethylene-mediated effects on development processes. The greatest interest has been focused on abscission, senescence, and ripening processes, as they are all influenced by ethylene production.[62,130] However, consideration should be given to germination, seedling development, and vegetative growth which can also be linked to effect promoted by ethylene.[1,130] Virtually all temperature-ethylene-germination relationships are associated with high temperatures and thermodormancy.[1]

A. SEEDLING DEVELOPMENT AND VEGETATIVE GROWTH

Cucumber seedlings growing at 20°C accumulated higher levels of ACC and MACC after exposure to a chilling temperature of 5°C, vs. a nonchilling 12.5°C, followed by rewarming at 20°C.[116] Exposure to 5°C for 24 or 48 h allowed seedlings to recover from chilling-induced wilting after 24 h at 20°C. Longer exposure to chilling resulted in death. The ACC levels rose during warming in all parts of the seedling, including epicotyl, cotyledon, hypocotyl, and roots. After 24 h warming at 20°C, ACC levels declined in parallel with the recovery of seedlings from chilling damage. In contrast, MACC levels were elevated during warming and remained high, confirming the nonreversibility of the conjugation process. [25,47] The observed increases in ACC and MACC are related to chilling damage, although not quantitatively. When cucumber seedlings were exposed to low or high irradiation during chilling, the ACC and MACC levels were similar on warming, although there was greater chilling damage at high irradiance.[116]

In related experiments, Wang[119] also determined changes in the polyamines, putrescine, spermidine, and spermine in chilled and warmed cucumber seedlings. The result indicated that the synthesis of polyamines and ethylene were not competitive with each other under chilling conditions.

A similar experimental approach was used with etiolated mung bean seedlings that were chilled at 0°C and subsequently warmed at 26°C.[114] Prolonged chilling (>7 d) effectively eliminated ACC-dependent ethylene production by explanted hypocotyls, while a 2-d exposure was sufficient to reduce it by 50%. Chilling for 3 d or more resulted in irreversible damage and elimination of ethylene biosynthesis. In contrast, non-chilling-sensitive pea (*Pisum sativum* L.) seedlings exposed to 0°C for 2 weeks had rapid reactivation of ACC-dependent ethylene production upon warming.[114] In unpublished studies, Clifford and Field exposed the roots of non-chilling-sensitive radish (*Raphanus sativus* L.) seedlings to 0°C for 7 h and measured significant increases in shoot ethylene production following warming. The result was interpreted as cold temperature induction of ACC in the roots, followed by its transfer to the shoots and synthesis of ethylene. Similar interpretations have been proposed for elevated shoot ethylene following waterlogging of roots.[131]

Experiments with intact seedlings provide a better system for determining relationships between chilling damage and ethylene production than organ explant systems, where chilling damage is rarely assessed and may not necessarily be induced. However, such experiments cannot be used to determine the biochemical mechanisms that may link ethylene production and chilling damage. A commonly held view is that membrane damage is the major feature linking species-specific chilling damage to ethylene formation capability.

B. SENESCENCE AND RIPENING

The senescence of vegetative tissues and the ripening of fruit involve ethylene biosynthesis.[120] While the greatest interest has focused on low-temperature storage of fruit, the possible linkage between ethylene, low temperature, and senescence phenomena is of great significance.

Enhancement of fruit ripening by low-temperature induction of ethylene was shown in attached, mature-green grapefruit fruit (*Citrus paradisa* Macf.). A diurnal 20/5°C (day/night) regime was sufficient to elevate the ethylene concentration in the air space under the peel to 100 ppb after 14 d, compared to 4 ppb in fruit held at 25/20°C.[3] After a 14-d incubation, the 20/5°C-treated fruit were ripening rapidly and had turned yellow, whereas fruit in the high-temperature regime remained green for 2 months. Similar results were found for harvested tangerine (*C. reticulata* Blanco) fruit.[3]

Low-temperature induction of ethylene production in avocado fruit may be complicated by chilling damage.[3] In a range of avocado cultivars there was an association between high ethylene production by fruit at 5°C and increased chilling sensitivity. In comparison, a chilling-tolerant cultivar showed low ethylene production at 5°C when compared to similar fruit held at 20°C. It is presumed that the differences in ethylene production were associated with parallel changes in ACC and were not related to EFE activity per se.

Chilling sensitivity in tomato fruit changes during ripening[132] and offers a useful experimental system with which to investigate associated ethylene changes.[133] Chilling sensitivity, as measured by ion leakage, initially declined (from mature green to breaker stage) as fruit ripened and then increased during the later stages (from turning to red stages). During the chilling response early in ripening, ethylene production was stimulated, but by the turning stage had reached a constant level. While high ethylene production, chilling damage, and the progress of ripening were apparently related, no causal relationships were established.[133] Earlier work had shown that ethylene treatment of mature-green and breaker-stage tomatoes did not affect chilling sensitivity.[121]

C. ABSCISSION

The premature abscission of fruits and leaves may follow a damaging exposure to low temperatures.[130] A temperature-induced overshoot in ethylene production may be responsible for prematurely initiating the abscission process.[1]

Exposure of leaves of seedling orange (*Citrus aurantium* Linn.) to freezing temperatures ($-6.1°C$) for 4 h prior to transfer to ambient conditions resulted in enhanced ethylene production and promotion of abscission.[110] There was a 12-h time lag between transfer to leaves to ambient conditions and initiation of ethylene accumulation.

Enhancement of flower senescence and the associated abscission of petals has been linked to temperature-induced ethylene production in cut flowers of carnation[105,134,135] and container-grown seed geraniums (*Pelargonium* × *hortorum* Bailey).[136] These responses are more clearly low temperature effects rather than manifestations of chilling damage.

A study using the explant abscission model of the primary leaves of dwarf bean (*Phaseolus vulgaris*)[137] has shown that in this chilling-sensitive species a 1- to 3-h exposure to 5°C at 24 or 48 h after excision reduces the time to 50% abscission by at least 24 h.[1]

It is surprising that no major studies of chilling temperature-induced abscission have been carried out. It can only be assumed that the extent of chilling damage, which often leads to plant death, is so overwhelming that abscission of leaves, fruit, and other organs is not worthy of separate study.

VI. CHILLING, ETHYLENE, AND STORAGE

Provided that excessive water loss is prevented, storage of flowers, fruit, and vegetables at chilling temperatures only leads to ethylene-related changes in tissue function on return to ambient conditions. The dominant response is for ethylene production to precede the onset of senescence and for a shortening of storage life at ambient temperatures. Few of the studies have involved chilling-sensitive species, as they are rarely stored at low temperatures.

One notable exception is a study of cucumbers (*Cucumis sativus* L.) stored at 2.5 or 13°C.[9,27] Fruit held at 2.5°C have low and constant levels of ACC, ACC synthase activity, and ethylene production[27] both before and after chilling injury development, which occurs after 3 d of storage. Chilling damage at 2.5°C was not promoted by ethylene, although the levels of ACC and ethylene production were slightly higher than at 13°C. Transfer of fruit to 25°C resulted in a rapid elevation in ACC and ethylene which may have acelerated further chilling damage.[27] Greater than 4 d of storage at 2.5°C resulted in a significant reduction in ethylene production capability on tissue warming, while ACC levels continued to rise. The result is interpreted as chilling-induced damage to EFE activity. Similar results have been obtained with stored papayas (*Carica papaya* L.).[138]

Avocado (*Persea americana* Mill.) fruit stored in air at chilling temperatures, with or without ethylene (100 ppm), showed variations in the onset of chilling damage.[124] Fruit stored with ethylene below 12°C suffered severe discoloration of tissue. Chilling sensitivity of avocado fruit increased with ethylene treatment. There was no measurement of endogenous ethylene production at chilling temperatures.

Perhaps the majority of studies linking low-temperature exposure of fruit and flowers to changes in ethylene production have involved temperate species without pronounced chilling sensitivity at 10 to 12°C. In such plants, there is variable impairment of ethylene production at low temperatures. Thus, the time taken from harvest to rapid ethylene production was shorter and more uniform at 3°C than at 18°C for pear (*Pyrus communis* L. cv. Conference) and apple (*Malus domestica* Borkh. cv. Golden Delicious) fruit.[139] Periods of alternating temperature (3 and 20°C) did not have a major influence on the mean time after harvest to onset of ethylene production by Golden Delicious apples.[139] Ethylene production in Conference pear continues to rise during storage at −1°C and may be hastened by transfer to 15°C.[140] In a similar storage study, ethylene production at −1°C was significantly greater in Beurre Bosc pear than Beurre d'Anjou pear, with the initiation of ethylene production corresponding to the appearance of specific phenolic acids in both cultivars.[141,142] Transfer of d'Anjou fruit from storage at −1.1°C to 20°C resulted in increased ethylene production which was dependent upon the duration of low-temperature storage.[143] In Eldorado pear, the rise in internal ethylene during a 4-week storage period at 0°C was paralleled by the rise in ACC.[106]

In pears, normal ripening is enhanced by postharvest storage at 0°C or below. The observed changes in fruit chemistry during cold storage and the rate of ripening on warming appear to be related to ethylene production at low temperatures.[106]

The response of cut flowers of temperate origin to low-temperature exposure involves inhibition of ethylene production and delayed senescence.[135,144-148] In carnation (*Dianthus caryophyllus* L.), there was impairment of ethylene production at 2°C, but with enhanced production upon transfer to 20°C.[135] In contrast, results obtained for cut rose (*Rosa hybrida* L.) flowers show significant ethylene production at 3°C with a further enhancement upon transfer to 22°C.[145-147]

While ethylene production can occur at low temperatures in fruit and flowers of non-chilling-sensitive species, there is no clear indication of why this is so radically different in chilling-sensitive species. The maintenance of cell membrane integrity in determining chilling sensitivity appears to be the most obvious structural/biochemical feature.

VII. CONCLUSIONS

There is incomplete information on the endogenous factors that control ethylene biosynthesis, which makes for a limited understanding of the effects of chilling temperatures on the system. One of the major limitations is a lack of a detailed understanding of the enzymology of ethylene biosynthesis from the precursor methionine and the cellular location of the intermediate steps. The linkage between induction of chilling and membrane perturbation appear to be related to changes in ethylene production.[1,115] The activity of EFE and the conversion of ACC to ethylene are particularly vulnerable to membrane damage. However, it would be unwise to assume that all membranes are equally vulnerable to chilling damage, and the identification of specific membrane-associated sites for EFE activity is crucial to a fundamental understanding of chilling-manipulated changes in ethylene production. Undoubtedly, there is value in pursuing experiments with cell-free systems and isolated membranes, provided that the results can be linked to the behavior of intact tissues.

One of the major unresolved problems is why low and/or chilling temperatures produce such a variable response in ethylene production. The ability of pear (*Pyrus communis* L.) fruit cultivars to synthesize ethylene at temperatures down to $-1°C$ is in sharp contrast to the response of highly chilling-sensitive species that cease production between 4 to $10°C$.[1,10,115] Undoubtedly, membrane function is a factor, but a clearer understanding of why the activation energy for the overall biosynthesis of ethylene is variable requires resolution.[1,41] The variations in tissue response to chilling temperatures results in the unsurprising conclusion that, overall, ethylene production is not a good indication of chilling damage.[10,11]

The association of ethylene with postharvest behavior of fruits, vegetables, and flowers is of major significance in the handling and shelf life of these commodities. Frequently, low-temperature storage is used immediately after harvest to reduce respiration and other metabolic processes that may induce rapid and undesirable senescence. There is considerable information showing that transferring plant tissues from low temperatures (2 to $5°C$) to ambient conditions initiates a rapid and large increase in ethylene production.[5,6,8,9,27] The marked overshoot in ethylene production may reduce the longevity of flowers[135] and other commodities. Resolution of the chemical basis for low-temperature-induced accumulation of a potential for high ethylene production on tissue warming is of major practical significance. It is speculated that there may be unresolved roles for ACC conjugation mechanisms, inhibition of polyamine synthesis, or regulation of post-transcriptional aspects of protein synthesis in determining the potential for high ACC and ethylene production on tissue warming. Incorporation of genetic suppression of overshoot ethylene production in some horticultural species could improve long-term storage characteristics and be preferable to the use of ethylene-inhibiting chemical treatments.

Chilling injury is not caused by ethylene, and neither is ethylene production consistently identified as a defense mechanism. Prevention of ethylene production by specific chemical inhibitors does not ameliorate the development of chilling injury. At this stage of understanding, it is impossible to establish universally applicable causal relationships between ethylene production, chilling temperature effects, and the development of symptoms of chilling injury.

REFERENCES

1. **Field, R. J.,** The effect of temperature on ethylene production in plants, in *Ethylene and Plant Development*, Roberts, J. A. and Tucker, G. A., Eds., Butterworths, London, 1985, chap. 6.
2. **Burg, S. P. and Thimann, K. V.,** The physiology of ethylene formation in apples, *Proc. Natl. Acad. Sci. U.S.A.,* 2, 335, 1959.
3. **Cooper, W. C., Rasmussen, G. K., and Waldon, E. S.,** Ethylene evolution stimulated by chilling in citrus and *Persea* sp., *Plant Physiol.,* 44, 1194, 1969.
4. **Eaks, I. L.,** Effects of chilling on respiration and volatiles of Californian lemon fruit, *J. Am. Soc. Hortic. Sci.,* 105, 865, 1980.
5. **Field, R. J.,** The effect of low temperature on ethylene production by leaf tissue of *Phaseolus vulgaris* L., *Ann. Bot.,* 47, 215, 1981.
6. **Field, R. J.,** The role of 1-aminocyclopropane-1-carboxylic acid in the control of low temperature induced ethylene production in leaf tissue of *Phaseolus vulgaris* L., *Ann. Bot.,* 54, 61, 1984.
7. **Olorunda, A. O. and Looney, N. E.,** Response of squash to ethylene and chilling, *Ann. Appl. Biol.,* 87, 465, 1977.
8. **Wright, M.,** The effect of chilling on ethylene production, membrane permeability, and water loss of leaves of *Phaseolus vulgaris, L., Planta,* 120, 63, 1974.
9. **Wang, C. Y.,** Ethylene production by chilled cucumbers (*Cucumis sativa* L.), *Plant Physiol.,* 66, 841, 1980.
10. **Chen, Y. Z. and Patterson, B. D.,** Ethylene and 1-aminocyclopropane-1-carboxylic acid as indicators of chilling sensitivity in various plant species, *Aust. J. Plant Physiol.,* 12, 377, 1985.
11. **Guye, M. G., Vigh, L., and Wilson, J. M.,** Chilling induced ethylene production in relation to chill-sensitivity in *Phaseolus* spp., *J. Exp. Bot.,* 38, 680, 1987.
12. **Adams, D. O. and Yang, S. F.,** Ethylene biosynthesis: identification of 1-aminocyclopropane-1-carboxylic acid as an intermediate in the conversion of methionine to ethylene, *Proc. Natl. Acad. Sci. U.S.A.,* 76, 170, 1979.
13. **Yang, S. F.,** Biosynthesis and action of ethylene, *HortScience,* 20, 41, 1985.
14. **Lurssen, K.,** Manipulation of crop growth by ethylene and some implications of the mode of generation., in *Chemical Manipulation of Crop Growth and Development*, McLaren, J. S., Ed., Butterworths, London, 1982, chap. 6.
15. **Lurssen, K. and Konze, J.,** Relationship between ethylene production and plant growth after application of ethylene releasing plant growth regulators., in *Ethylene and Plant Development*, Roberts, J. A. and Tucker, G. A., Eds., Butterworths, London, 1985, chap. 30.
16. **Yu, Y. B., Adams, D. O., and Yang, S. F.,** 1-aminocyclopropane-carboxylate synthase, a key enzyme in ethylene biosynthesis, *Arch. Biochem. Biophys.,* 198, 280, 1979.
17. **Lurssen, K., Naumann, K., and Schroder, R.,** 1-aminocyclopropane-1-carboxylic acid — an intermediate of the ethylene biosynthesis in higher plants, *Z. Pflanzenphysiol.,* 92, 285, 1979.
18. **Lieberman, M. and Mapson, L. W.,** Genesis and biogenesis of ethylene, *Nature (London),* 204, 343, 1979.
19. **Lieberman, M., Kunishi, A. T., Mapson, L. W., and Wardale, D. A.,** Ethylene production from methionine, *Biochem. J.,* 97, 449, 1965.
20. **Lieberman, M., Kunshi, A. T., Mapson, L. W., and Wardale, D. A.,** Stimulation of ethylene production in apple tissue slices by methionine, *Plant Physiol.,* 41, 376, 1966.
21. **Adams, D. O. and Yang, S. F.,** Methionine metabolism in apple tissue: implications of S-adenosylmethionine as an intermediate in the conversion of methionine to ethylene, *Plant Physiol.,* 35, 155, 1984.
22. **Yang, S. F. and Hoffman, N. E.,** Ethylene biosynthesis and its regulation in higher plants, *Annu. Rev. Plant Physiol.,* 35, 155, 1984.
23. **Chou, T. C., Coulter, A. W., Lombardini, J. B., Sufrin, J. R., and Talalay, P.,** The enzymatic synthesis of adenosyl-methionine, mechanism and inhibition, in *The Biochemistry of Adenosyl-methionine*, Salvatore, F., Borek, E., Zappia, V., Williams-Ashman, H. G., and Schlenk, F., Eds., Columbia University Press, New York, 1977, 18.
24. **Konze, J. R. and Kende, H.,** Interactions of methionine and selenomethionine with methionine adenosyl transferase and ethylene-generating systems, *Plant Physiol.,* 63, 507, 1979.
25. **Boller, T., Herner, R. C., and Kende, H.,** Assay for an enzymatic formation of an ethylene precursor, 1-aminocyclopropane-1-carboxylic acid, *Planta,* 145, 293, 1979.
26. **Fuhrer, J.,** Ethylene biosynthesis and cadmium toxicity in leaf tissue of beans (*Phaseolus vulgaris* L.), *Plant Physiol.,* 70, 162, 1982.
27. **Wang, C. Y. and Adams, D. O.,** Chilling-induced ethylene production in cucumbers (*Cucumis sativus* L.), *Plant Physiol.,* 69, 424, 1982.
28. **Riov, J. and Yang, S. F.,** Autoinhibition of ethylene production in citrus peel discs. Suppression of 1-aminocyclopropane-1-carboxylic acid synthesis, *Plant Physiol.,* 69, 687, 1982.

29. **Hoffman, N. E. and Yang, S. F.,** Enhancement of wound induced ethylene synthesis by ethylene in pre-climacteric cantaloupe, *Plant Physiol.,* 69, 317, 1982.

30. **Acaster, M. A. and Kende, H.,** Properties and partial purification of ACC synthase, *Plant Physiol.,* 72, 139, 1983.

31. **Field, R. J.,** A relationship between membrane permeability and ethylene production at high temperature in leaf tissue of *Phaseolus vulgaris* L., *Ann. Bot.,* 48, 33, 1981.

32. **Imaseki, H. and Watanabe, A.,** Inhibition of ethylene production by osmotic shock. Further evidence for membrane control of ethylene production, *Plant Cell Physiol.,* 19, 345, 1978.

33. **Odawara, S., Watanabe, A., and Imaseki, H.,** Involvement of cellular membrane in regulation of ethylene production, *Plant Cell Physiol.,* 18, 569, 1977.

34. **John, P.,** The coupling of ethylene biosynthesis to a transmembrane electrogenic proton flux, *FEBS Lett.,* 152, 141, 1983.

35. **Konze, J. R. and Kende, E.,** Ethylene formation from 1-aminocyclopropane-1-carboxylic acid and in homogenates of etiolated pea seedlings, *Planta,* 146, 293, 1979.

36. **Mayak, S., Legge, R. L., and Thompson, J. E.,** Ethylene formation from 1-aminocyclopropane-1-carboxylic acid by microsomal membranes from senescing carnation flowers, *Planta,* 153, 49, 1981.

37. **Lizada, M. C. C., and Yang, S. F.,** A simple and sensitive assay for 1-aminocyclopropane-1-carboxylic acid, *Anal. Bichem.,* 100, 140, 1979.

38. **Legge, R. L., Thompson, J. E., and Baker, J. E.,** Free radical-mediated formation of ethylene from 1-aminocyclopropane-1-carboxylic acid: a spin trap study, *Plant Cell Physiol.,* 23, 171, 1982.

39. **Guy, M. and Kende, H.,** Conversion of 1-aminocyclopropane-1-carboxylic acid to ethylene by isolated vacuoles of *Pisum sativum* L., *Planta,* 160, 281, 1984.

40. **Kende, H. and Boller, T.,** Wound ethylene and 1-aminocyclopropane-1-carboxylate synthase in ripening tomato fruit, *Planta,* 151, 476, 1981.

41. **Mattoo, A. K., Chalutz, E., and Lieberman, M.,** Effects of lipophilic and water-soluble membrane probes on ethylene synthesis in apple and *Penicillium digitatum, Plant Cell Physiol.,* 20, 1097, 1979.

42. **Amrhein, N., Schneebeck, D., Skorupka, H., Tophof, S., and Stockigt, J.,** Identification of a major metabolite of the ethylene precursor 1-aminocyclopropane-1-carboxylic acid in higher plants, *Naturwissenschaften,* 68, 619, 1981.

43. **Hoffman, N. E. and Yang, S. F.,** Identification of 1-(malonylamino) cyclopropane-1-carboxylic acid, an ethylene precursor in higher plants, *Biochem. Biophys. Res. Commun.,* 104, 765, 1982.

44. **Yang, S. F., Liu, Y., Su, L., Peiser, G. D., Hoffman, N. E., and Mckeon, T.,** Metabolism of 1-aminocyclopropane-1-carboxylic acid, in *Ethylene and Plant Development,* Roberts, J. A. and Tucker, G. A., Eds., Butterworths, London, 1985, chap. 2.

45. **Amrheim, N., Breuing, F., Eberle, J., Skorupka, H., and Tophof, S.,** The identification and metabolism of 1-(malonylamino) cyclopropane-1-carboxylic acid, in *Plant Growth Substances 1982,* Wareing, P. F., Ed., Academic Press, London, 1982, 249.

46. **Hoffman, N. E., Fu, J. R., and Yang, S. F.,** Identification and metabolism of 1-(malonylamino) cyclopropane-1-carboxylic acid in germinating peanut seeds, *Plant Physiol.,* 71, 197, 1983.

47. **Hoffman, N. E., Liu, Y., and Yang, S. F.,** Changes in 1-(malonylamino) cyclopropane-1-carboxylic acid content in wilted wheat leaves in relation to their ethylene production rates and 1-aminocyclopropane-1-carboxylic acid content, *Planta,* 157, 518, 1983.

48. **Van Loon, L. C. and Fontaine, J. J. H.,** Accumulation of 1-(malonylamino) cyclopropane-1-carboxylic acid in ethylene-synthesizing tobacco leaves, *Plant Growth Regul.,* 2, 227, 1984.

49. **Satoh, S. and Etashi, Y.,** *In vivo* formation of 1-malonylaminocyclopropane-1-carboxylic acid and its relationship to ethylene production in cocklebur seeds segments: a tracer study with 1-amino-2-ethylcyclopropane-1-carboxylic acid, *Phytochemistry,* 23, 1561, 1984.

50. **Matern, U., Feser, C., and Heller, W.,** N-malonyltransferases from peanut, *Arch. Biochem. Biophys.,* 235, 218, 1984.

51. **Jiao, X.-Z., Philosoph-Hadas, S., Su, L.-Y., and Yang, S. F.,** The conversion of 1-(malonylamino) cyclopropane-1-carboxylic acid to 1-aminocyclopropane-1-carboxylic acid in plant tissues, *Plant Physiol.,* 81, 637, 1986.

52. **Liu, Y., Hoffman, N. E., and Yang, S. F.,** Ethylene-promoted malonylation of 1-aminocyclopropane-1-carboxylic acid participates in auto inhibition of ethylene synthesis in grapefruit flavedo discs, *Planta,* 164, 565, 1985.

53. **Su, L., McKeon, T., Grierson, D., Cantwell, M., and Yang, S. F.,** Development of 1-aminocyclopropane-1-carboxylic acid synthase and polygalacturonase activities during the maturation and ripening of tomato fruits, *HortScience,* 19, 576, 1984.

54. **Zimmerman, P. W. and Wilcoxon, F.,** Several chemical growth substances which cause initiation of roots and other responses in plants, *Contrib. Boyce Thompson Inst.,* 7, 209, 1935.

55. **Jones, J. F. and Kende, H.,** Auxin-induced ethylene biosynthesis in subapical stem sections of etiolated seedlings of *Pisum sativum* L., *Planta,* 146, 649, 1979.

56. **Yoshii, H. and Imaseki, H.,** Biosynthesis of auxin-induced ethylene, effects of indole-3-acetic acid, benzyladenine and abscisic acid on endogenous levels of 1-aminocyclopropane-1-carboxylic acid (ACC) and ACC synthase, *Plant Cell Physiol.,* 22, 369, 1981.

57. **Yoshii, H. and Imaseki, H.,** Regulation of auxin-induced ethylene biosynthesis. Repression of inductive formation of 1-aminocyclopropane-1-carboxylate synthase by ethylene, *Plant Cell Physiol.,* 23, 639, 1982.

58. **Yu, Y. B., Adams, D. O., and Yang, S. F.,** Regulation of auxin-induced ethylene production in mung bean hypocotyls. Role of 1-aminocyclopropane-1-carboxylic acid, *Plant Physiol.,* 63, 589, 1979.

59. **Yu, Y. B. and Yang, S. F.,** Auxin-induced ethylene production and its inhibition by aminoethoxyvinyl-glycine and cobalt ion, *Plant Physiol.,* 64, 1074, 1979.

60. **Giovanelli, J., Owens, L. D., and Mudd, S. H.,** Mechanism of inhibition of spinach b-cystathionase by rhizobitoxine, *Biochem. Biophys. Acta,* 227, 671, 1971.

61. **Amrhein, N. and Wenker, D.,** Novel inhibitors of ethylene production in higher plants, *Plant Cell Physiol.,* 20, 1635, 1979.

62. **Lieberman, M.,** Biosynthesis and action of ethylene, *Annu. Rev. Plant Physiol.,* 30, 533, 1979.

63. **Bradford, K. J., Hsiao, T. C., and Yang, S. F.,** Inhibition of ethylene synthesis in tomato plants subjected to anaerobic root stress, *Plant Physiol.,* 70, 1503, 1982.

64. **Thimman, K. V.,** Studies on the growth and inhibition of isolated plant parts. V. The effects of cobalt and other metals, *Am. J. Bot.,* 43, 241, 1956.

65. **Lau, O. L., Murr, D. G., and Yang, S. F.,** Effect of 2,4-dinitrophenol on auxin-induced ethylene production and auxin conjugation in mung bean tissue, *Plant Physiol.,* 54, 182, 1974.

66. **Murr, D. P. and Yang, S. F.,** Inhibition of *in vivo* conversion of methionine to ethylene by L-canaline and 2,4-dinitrophenol, *Plant Physiol.,* 55, 79, 1975.

67. **Apelbaum, A., Burgoom, A. C., Anderson, J. D., Solomos, T., and Lieberman, M.,** Some characteristics of the system converting 1-aminocyclopropane-1-carboxylic acid to ethylene, *Plant Physiol.,* 67, 80, 1981.

68. **Kende, H., Acaster, M. A., and Guy, M.,** Studies on the enzymes of ethylene biosynthesis, in *Ethylene and Plant Development,* Roberts, J. A. and Tucker, G. A., Eds., Butterworths, London, 1985, chap. 3.

69. **Yu, Y. B., Adams, D. O., and Yang, S. F.,** Inhibition of ethylene production by 2,4-dinitrophenol and high temperature, *Plant Physiol.,* 66, 286, 1980.

70. **Apelbaum, A., Wang, S. Y., Burgoon, A. C., Baker, J. E., and Lieberman, M.,** Inhibition of the conversion of 1-aminocyclopropane-1-carboxylic acid to ethylene by structural analogs, inhibitors of electron transfer, uncouplers of oxidative phosphorylation, and free radical scavengers, *Plant Physiol.,* 67, 74, 1981.

71. **Baker, J. E., Lieberman, M., and Anderson, J. D.,** Inhibition of ethylene production in fruit slices by a rhizobitoxine analog and free radical scavengers, *Plant Physiol.,* 61, 886, 1978.

72. **Baker, J. E., Wang, C. Y., Lieberman, M., and Hardenburg, R.,** Delay of senescence in carnations by rhizobitoxine analog and sodium benzoate, *HortScience,* 12, 38, 1977.

73. **Konze, J. R., Jones, J. F., Boller, T., and Kende, H.,** Effect of 1-aminocyclopropane-1-carboxylic acid on the production of ethylene in senescing plants of *Ipomoea tricolor* Cav., *Plant Physiol.,* 66, 566, 1980.

74. **Shih, L. M., Kaur-Sawhney, R., Fuhrer, J., Samanta, S., and Galston, A. W.,** Effects of exogenous 1,3-diaminopropane and spermidine on senescence of oat leaves. I. Inhibition of protease activity, ethylene production and chlorophyll loss as related to polyamine content, *Plant Physiol.,* 70, 1592, 1982.

75. **Smith, T. A.,** Recent advances in the biochemistry of plant amines, *Phytochemistry,* 14, 865, 1975.

76. **Suttle, J. C.,** Effects of polyamines on ethylene production, *Phytochemistry,* 20, 1477, 1981.

77. **Apelbaum, A., Burgoon, A. C., Anderson, J. D., Lieberman, M., Ben-Arie, R., and Mattoo, A. K.,** Polyamines inhibit biosynthesis of ethylene in higher plant tissue and fruit protoplasts, *Plant Physiol.,* 68, 453, 1981.

78. **Ben-Arie, R., Lurie, S., and Mattoo, A. K.,** Temperature-dependant inhibitory effects of calcium and spermine on ethylene biosynthesis in apple discs correlate with changes in microsomal membrane viscosity, *Plant Sci. Lett.,* 24, 239, 1982.

79. **Satoh, S. and Esashi, Y.,** α-Aminoisobutyric acid: a probable competitive inhibitor of conversion of 1-aminocyclopropane-1-carboxylic acid to ethylene, *Plant Cell Physiol.,* 21, 939, 1980.

80. **Bleeker, A. B., Rose-John, S., and Kende, H.,** An evaluation of 2,5-norborn adiene as a reversible inhibitor of ethylene action in deep-water rice, *Plant Physiol.,* 84, 395, 1987.

81. **Sisler, E. C., Goren, R., and Huberman, M.,** Effect of 2,5-norbornadiene on abscission and ethylene production in citrus leaf explants, *Plant Physiol.,* 63, 114, 1985.

82. **Sisler, E. C. and Pian, A.,** Effect of ethylene and cyclic olefins on tobacco leaves, *Tob. Sci.,* 17, 68, 1973.

83. **Sisler, E. C. and Yang, S. F.,** Anti-ethylene effects of cis-2-butene and cyclic olefins, *Phytochemistry,* 23, 2765, 1984.

84. **Beyer, E. M.,** A potent inhibitor of ethylene action in plants, *Plant Physiol.,* 58, 268, 1976.

85. **Veen, H.,** Effects of silver on ethylene synthesis and action in cut carnations, *Planta,* 145, 467, 1979.

86. **Reid, M. S., Paul, J. L., Farhoomand, M. B., Kofranek, A. M., and Stalby, G. L.,** Pulse treatments with the silver thiosulfate complex extend the vase life of cut carnations, *J. Am. Soc. Hortic. Sci.,* 105, 25, 1980.

87. **Horton, R. F. and Saville, B. J.,** Carbon dioxide enrichment, transpiration and 1-aminocyclopropane-1-carboxylic acid-dependant ethylene release from oat leaves, *Plant Sci. Lett.,* 36, 131, 1984.

88. **Kao, C. H. and Yang, S. F.,** Light inhibition of the conversion of 1-aminocyclopropane-1-carboxylic acid to ethylene in leaves is mediated through carbon dioxide, *Planta,* 155, 261, 1982.

89. **Philosoph-Hadas, S., Aharoni, N., and Yang, S. F.,** Carbon dioxide enhances the development of ethylene forming enzyme in tobacco leaf discs, *Plant Physiol.,* 82, 925, 1986.

90. **Preger, R. and Gepstein, S.,** Carbon dioxide-independant and dependant components of light inhibition of the conversion of 1-aminocyclopropane-1-carboxylic acid to ethylene in oat leaves, *Physiol. Plant.,* 60, 187, 1984.

91. **McRae, D. G., Coker, J. A., Legge, R. L., and Thompson, J.,** Bicarbonate/CO_2-facilitated conversion of 1-aminocyclopropane-1-carboxylic acid to ethylene in model systems and intact tissues, *Plant Physiol.,* 73, 784, 1983.

92. **De Laat, A. M. M., Bradenburg, D. C. C., and Van Loon, L. C.,** The modulation of the conversion of 1-aminocyclopropane-1-carboxylic acid to ethylene by light, *Planta,* 153, 172, 1981.

93. **Gepstein, S. and Thimann, K. V.,** The effect of light on the production of ethylene from 1-aminocyclopropane-1-carboxylic acid by leaves, *Planta,* 149, 196, 1980.

94. **Wright, S. T. C.,** The effect of light and dark periods on the production of ethylene from water-stressed wheat leaves, *Planta,* 153, 172, 1981.

95. **Mattoo, A. K. and Lieberman, M.,** Localization of the ethylene-synthesizing system in apple tissue, *Plant Physiol.,* 60, 794, 1977.

96. **Simon, E. W.,** Phospholipids and plant membrane permeability, *New Phytol.,* 73, 377, 420.

97. **Bagnall, D. J. and Wolfe, J. A.,** Chilling sensitivity in plants: do the activation energies of growth processes show an abrupt change at a critical temperature, *J. Exp. Bot.,* 29, 1231, 1978.

98. **Hansen, E.,** Quantitative study of ethylene production in apple varieties, *Plant Physiol.,* 20, 631, 1945.

99. **Knee, M.,** Evaluating the practical significance of ethylene in fruit storage, in *Ethylene and Plant Development,* Roberts, J. A. and Tucker, G. A., Eds., Butterworths, London, 1985, chap. 24.

100. **Ulrich, R.,** Temperature and maturation: pears require preliminary cold treatment, *Recent Adv. Bot.,* 1172, 1961.

101. **Hansen, E.,** Postharvest physiology of fruits, *Annu. Rev. Plant Physiol.,* 17, 459, 1966.

102. **Wang, C. Y., Mellenthin, W. M., and Hansen, E.,** Effect of temperature on development of premature ripening in Bartlett pears, *J. Am. Soc. Hortic. Sci.,* 96, 122, 1971.

103. **Looney, N. E.,** Interaction of harvest maturity, cold storage and two growth regulators on ripening in Bartlett pears, *J. Am. Soc. Hortic. Sci.,* 97, 81, 1972.

104. **Sfakiotakis, E. M. and Dilley, D. R.,** Induction of ethylene production in 'Bosc' pears by postharvest cold stress, *HortScience,* 9, 336, 1974.

105. **Nichols, R.,** Ethylene, production during senescence of flowering, *J. Hortic. Sci.,* 41, 279, 1966.

106. **Wang, C. Y., Sams, C. E., and Gross, K. C.,** Ethylene, ACC, soluble polyuronide, and cell wall noncellulosic neutral sugar content in 'Eldorado' pears during cold storage and ripening, *J. Am. Soc. Hortic. Sci.,* 110, 687, 1985.

107. **Blankenship, S. M. and Richardson, D. G.,** Development of ethylene biosynthesis and ethylene induced ripening in 'd'Anjou' pears in air and controlled atmosphere storage, *HortScience,* 21, 1020, 1986.

108. **Blankenship, S. M. and Richardson, D. G.,** ACC and ethylene levels in 'd'Anjou' pears in air and controlled-atmosphere storage, *HortScience,* 21, 1020, 1986.

109. **Wang, C. Y.,** Pear fruit maturity, harvesting, storage, and ripening, in *The Pear,* van der Zwet, T. and Childers, N. F., Eds., Horticulture Publishing, Gainesville, FL, 1982, 431.

110. **Young, R. E. and Meredith, F.,** Effect of exposure to subfreezing temperatures on ethylene evolution and leaf abscission in citrus, *Plant Physiol.,* 48, 724, 1971.

111. **Elstner, E. F. and Konze, J. R.,** Effect of point freezing on ethylene and ethane production by sugar beet leaf disks, *Nature (London),* 263, 351, 1976.

112. **Kimmerer, T. W. and Kozlowski, T. T.,** Ethylene, ethane, acetaldehyde, and ethanol production by plants under stress, *Plant Physiol.,* 69, 840, 1982.

113. **Anderson, C. R. and Kent, M. W.,** Storage of cucumber fruit at chilling temperature, *HortScience,* 17, 527, 1982.

114. **Etani, S. and Yoshida, S.,** Reversible and irreversible reduction of ACC dependant ethylene formation in mung bean (*Vigna radiata* L. Wilczek) hypocotyls caused by chilling, *Plant Cell Physiol.,* 28, 83, 1987.

115. **Wang, C. Y.,** Relation of chilling stress to ethylene production, in *Low Temperature Stress Physiology in Crops,* Li, P., Ed., CRC Press, Boca Raton, FL, 1989, 177.

116. **Wang, C. Y.,** Effect of temperature and light on ACC and MACC in chilled cucumber seedlings, *Sci. Hortic.,* 30, 47, 1986.

117. **Vinkler, C. and Apelbaum, A.,** Ethylene formation in plant mitochondria. Dependance on the transport of 1-aminocyclopropane-1-carboxylic acid across the membrane, *Physiol. Plant.,* 63, 387, 1985.

118. **Mayne, R. G. and Kende, H.,** Ethylene biosynthesis in isolated vacuoles of *Vicia faba* L. — requirements for membrane integrity, *Planta,* 167, 159, 1986.

119. **Wang, C. Y.,** Changes of polyamines and ethylene in cucumber seedlings in response to chilling stress, *Physiol. Plant.,* 69, 253, 1987.

120. **Roberts, J. A.,** Ethylene and foliar senescence, in *Ethylene and Plant Development,* Roberts, J. A. and Tucker, G. A., Eds., Butterworths, London, 1985, chap. 21.

121. **Kader, A. A. and Morris, L. L.,** Amelioration of chilling injury symptoms on tomato fruits, *HortScience,* 10, 324, 1975.

122. **Ogura, N., Hayashi, R., Ogishima, T., Abe, Y., Nakagawa, H., and Takehana, H.,** Ethylene production by tomato fruits at various temperatures and effect of ethylene on the fruits, *Nippon Nogeikagaku Kaishi,* 50, 519, 1976.

123. **Young, R. E. and Lee, S. K.,** The effect of ethylene on chilling and high temperature injury of avocado fruit, *HortScience,* 14, 421, 1979.

124. **Lee, S. K. and Young, R. E.,** Temperature sensitivity of avocado fruit in relation to C_2H_4 treatment, *J. Am. Soc. Hortic. Sci.,* 109, 689, 1984.

125. **Chaplin, G. R., Wills, R. B. H., and Graham, D.,** Induction of chilling injury in stored avocados with exogenous ethylene, *HortScience,* 18, 952, 1983.

126. **Lipton, W. J., and Aharoni, Y.,** Chilling injury and ripening of 'honey dew' muskmelons stored at 2.5° or 5°C after ethylene treatment at 20°C, *J. Am. Soc. Hortic. Sci.,* 104, 327, 1979.

127. **Lipton, W. J., and Aharoni, Y.,** Rates of CO_2 and ethylene production and of ripening of 'honey dew' muskmelons at a chilling temperature after pretreatment with ethylene, *J. Am. Soc. Hortic. Sci.,* 104, 846, 1979.

128. **Buescher, R. W.,** Hard core in sweet potato roots are influenced by cultivar, curing, and ethylene, *HortScience,* 12, 326, 1977.

129. **Graham, D. and Patterson, B. D.,** Responses of plants to low, nonfreezing temperature: proteins, metabolism and acclimation, *Annu. Rev. Plant Physiol.,* 33, 347, 1982.

130. **Abeles, F. B.,** *Ethylene in Plant Biology,* Academic Press, New York, 1973.

131. **Jackson, M. B.,** Ethylene and the responses of plants to excess water in their environment — a review, in *Ethylene and Plant Development,* Roberts, J. A. and Tucker, G. A., Eds., Butterworths, London, chap. 20.

132. **Cook, C. C., Parsons, C. S., and McColloch, L. P.,** Methods to extend storage of fresh vegetables aboard ships of the U.S. Navy, *Food Technol. (Chicago),* 12, 548, 1958.

133. **Autio, W. R. and Bramlage, W. J.,** Chilling sensitivity of tomato fruit in relation to ripening and senescence, *J. Am. Soc. Hortic. Sci.,* 111, 201, 1986.

134. **Maxie, E. C., Farnham, D. S., Mitchell, F. G., Sommer, N. F., Parsons, R. A., Snyder, R. G., and Rae, H. L.,** Temperature and ethylene effects on cut flowers of carnation (*Dianthus caryophyllus* L.), *J. Am. Soc. Hortic. Sci.,* 98, 568, 1973.

135. **Barrowclough, P.,** The Effect of Coolstorage and Silver Ions on the Subsequent Ethylene Production and Vase Life of Cut Carnations, B. Hort. Sc. Honours Diss., Lincoln College, University of Canterbury, England, 1986.

136. **Armitage, A. M., Heins, R., Deans, S., and Carlson, W.,** Factors influencing flower petal abscission in the seed-propagated Geranium, *J. Am. Soc. Hortic. Sci.,* 105, 562, 1980.

137. **Jackson, M. B. and Osborne, D. J.,** Ethylene, the natural regulator of leaf abscission, *Nature (London),* 225, 1019, 1970.

138. **Chan, H. T., Sanxter, S., and Couey, H. M.,** Electrolyte leakage and ethylene production induced by chilling injury of papayas, *HortScience,* 20, 1070, 1985.

139. **Knee, M., Looney, N. E., Hatfield, S. G. S., and Smith, S. M.,** Indication of rapid ethylene synthesis by apple and pear fruits in relation to storage temperature, *J. Exp. Bot.,* 34, 1207, 1983.

140. **Knee, M.,** Development of ethylene biosynthesis in pear fruits at -1°C, *J. Exp. Bot.,* 38, 1724, 1987.

141. **Blankenship, S. M. and Richardson, D. G.,** Changes in phenolic acids and internal ethylene during long-term cold storage of pears, *J. Am. Soc. Hortic. Sci.,* 110, 336, 1985.

142. **Blankenship, S. M. and Richardson, D. G.,** Development of ethylene biosynthesis and ethylene-induced ripening in 'd'Anjou' pears during the cold requirement for ripening, *J. Am. Soc. Hortic. Sci.,* 110, 520, 1985.

143. **Chen, P. M., Mellenthin, W. M., and Borgic, D. M.,** Changes in ripening behaviour of 'd'Anjou' pears (*Pyrus communis*) after cold storage, *Sci. Hortic.,* 21, 137, 1983.

144. **Borochov, A., Itzhaki, H., and Spiegelstein, H.,** Effect of temperature on ethylene biosynthesis in carnation petals, *Plant Growth Regul.,* 3, 159, 1985.

145. **Faragher, J. D. and Mayak, S.,** Physiological responses of cut rose flowers to exposure to low temperature: changes in membrane permeability and ethylene production, *J. Exp. Bot.,* 35, 965, 1984.

146. **Faragher, J. D., Mayak, S., and Tirosh, T.,** Physiological response of cut rose flowers to cold storage, *Physiol. Plant.,* 67, 205, 1986.
147. **Mayak, S. and Faragher, J. D.,** Storage environment related stresses and flower senescence, *Acta Hortic.,* 181, 33, 1986.
148. **Paulin, A., Kerhardy, F., and Maestri, B.,** Effect of drought and prolonged refrigeration on senescence in cut carnation *Dianthus caryophyllus), Physiol. Plant.,* 64, 535, 1985.

Part IV. Prevention and Reduction of
Chilling Injury

Chapter 16

INFLUENCE OF CONTROLLED ATMOSPHERES AND PACKAGING ON CHILLING SENSITIVITY

Charles F. Forney and Werner J. Lipton

TABLE OF CONTENTS

I. INTRODUCTION

Ever since injury has been attributed to holding fresh fruits and vegetables at chilling temperatures, efforts have been made to find ways to alleviate chilling injury (CI) so that the beneficial effects of cold storage could be fully attained. Modification of the storage atmosphere, which increases the storage life of many commodities, has been studied extensively to determine its influence on the development of CI.

We will discuss the influence of atmosphere modification with respect to oxygen (O_2), carbon dioxide (CO_2), ethylene, and water vapor concentration on the response of fruits and vegetables to chilling exposures. Included will be results based on tests in which the atmosphere within an entire container of produce was modified or in which modification was achieved by the use of reduced atmospheric pressure (hypobaric storage), various coating materials, such as wax, or packaging techniques, such as wrapping in plastic films. Modification of the various components of the storage atmosphere has evoked mixed results in attempts to mitigate chilling injury.

II. BULK MODIFICATION OF ATMOSPHERE

Bulk modification encompasses large volumes, such as stationary storages, holds of ships, and truck trailers; smaller volumes involve overwrapped pallet stacks or bins and individual shipping boxes. Large or small, the resulting controlled atmosphere (CA) may alter the response of a commodity to chilling temperatures.

A. OXYGEN AND CARBON DIOXIDE

Kidd et al.[1] appear to have been the first to test the influence of CA storage on CI. Internal breakdown (IB), a symptom of CI of some cultivars of apples, was a problem at 0°C, whereas ripening was too rapid at higher temperatures. They found that low O_2 levels in storage inhibited the formation of IB, while high CO_2 levels tended to increase the incidence and severity of IB.

Atmospheric modification successfully inhibits the development of CI in avocado (*Persea americana*). Oxygen levels lowered to 2% decreased CI in several cultivars held 3 or more weeks at 4.5 or 7°C[2,3] and were more effective with high CO_2. Carbon dioxide levels of 10% decreased or prevented CI. Thus, Fuchs and Waldin avocados held 3 to 4 weeks at 7°C in 2% O_2 and 10% CO_2 showed only traces of CI, while those held in air were severely injured.[2] Similarly, CI was reduced in Booth 8 and Lula avocados held up to 60 d under these same conditions.[4]

Stone fruits held at temperatures below 7.5°C for 2 or more weeks may develop IB. It is manifested in browning of the flesh around the seed, dry, mealy texture (wooliness) of the flesh, and development of off-flavors. The development of IB can be delayed by the use of CA storage in some cultivars. Six peach and two nectarine cultivars held for 9 weeks at 0°C in 1% O_2 and 5% CO_2 and then ripened in air at 18°C developed little or no IB.[5-7] Injury was severe in fruit held in air. Similar results were reported with the storage of Stark's Red Gold nectarines stored 8 weeks at −0.5°C.[8] Fruit stored in 2.5 or 21% O_2 with 2.5 or 5% CO_2 retained good quality, while that stored in air or 2.5% O_2 was dry and mealy. Okubo peaches stored 3 weeks at 1°C in 3% O_2 and 3% CO_2 and then ripened at 20°C also exhibited less IB than fruit held in air.[9] Oxygen levels below 1%[6] and CO_2 levels of 10%, however, were reported to induce off-flavors.[10] J. H. Hale peaches (*Prunus persica* (L.) Batsch), which tolerate high CO_2 levels very well, ripened normally at 20°C after being held at 1°C for 42 d in 20% CO_2 added to air, while those in air failed to ripen normally after only 21 d.[11] In addition, 5% CO_2 was reported to suppress CI expressed as surface pitting and browning in Japanese apricot (*Prunus mume*, Sieb. et Zucc.) held at 0 or 5°C.[12]

In all studies that reported a reduction in IB by storage in CA, elevated levels of CO_2 were more effective than reduced levels of O_2. In addition, CA has been reported to be more effective in preventing IB when used in conjunction with delayed storage[10] or intermittent warming.[7]

In several cultivars of stone fruit, however, CA storage was reported to have no beneficial effect on preventing IB. In Independence and Flamekist nectarines and Santa Rosa and Songold plums, atmospheres of 2 to 21% O_2 with or without 2 to 10% CO_2 did not reduce IB relative to controls after 21 or 42 d at $-0.5°C$.[13] In Fay Elberta peaches, storage atmospheres containing 5% CO_2 had increased IB after 2 weeks at 5°C when compared with fruit held in air.[14]

Low O_2 helped to reduce the expression of CI in several fruit. Storing pineapples at 8°C in air, and then exposing them at 22°C to 3% O_2 for 1 week reduced the incidence of internal browning 65% relative to samples held at 22°C in air.[15] However, 2% O_2 during chilling and 5% CO_2 during and/or after chilling had no effect on expression of CI. The severity of skin scald and internal injury of papaya caused by chilling was slightly reduced by holding the fruit in 1.5% O_2 and 2% CO_2.[16] The severity of CI in Zucchini squash was reduced by 1 to 4% O_2 during 2 weeks of storage at 2.5°C; however, severity increased to a level similar to that in the control when the fruit was subsequently held several days at 10°C in air.[17] Elevated levels of CO_2, however, were more successful than low O_2 in preventing CI in Zucchini.[18] Squash held in 5 or 10% CO_2 in air for 19 d at 5°C followed by 4 d of aeration at 13°C showed no CI, while 30 to 60% of those held in air were injured.

The influence of CA on the development of CI in citrus appears to be variable. Holding fruit in low-O_2 atmospheres has shown some promise, although results are variable. Storage in 5 to 15% O_2 reduced the severity of CI of grapefruit, Valencia and Temple oranges, and Persian limes relative to storage in air.[19-21] However, 5 to 15% O_2 increased the amount of severe pitting on grapefruit stored under conditions that decreased pitting in another study.[22] The authors provided no explanation for this difference. Miller,[23] in a review of earlier work, found that most attempts to store citrus in high CO_2 resulted in rind injury, off-flavors, or increases in decay. In a more recent work, 3 weeks in 10% CO_2 followed by 4 weeks in air at about 4.5°C reduced the incidence of CI in grapefruit about 50% compared with fruit held continuously in air.[3] However, any benefit of reduced CI was negated during longer storage by an increase in the proportion of decayed grapefruit and oranges.[19,21,22]

Controlled atmosphere does not appear to increase chilling tolerance in many crops and, in some cases, may lead to additional injury. Mature-green and breaker tomatoes showed no reduction in CI when held 10 d at 0°C in O_2 concentrations that ranged from 3 to 50%.[24] This chilling exposure, however, was so severe that any amelioration would have been very surprising. In addition, 5 to 20% CO_2 injured tomatoes, and the degree of injury was additive to that resulting from chilling alone.[24,25] Cucumbers respond in a similar manner, with 3 to 75% CO_2 leading to increased pitting at chilling temperatures. However, the level of O_2 had no effect, except at extreme levels (less than 1 or 100%), where it enhanced pitting.[26] Carbon dioxide at 5 to 20% increased the incidence of sheet pitting and decay in bell peppers when compared with pods stored in air.[27] Asparagus spears stored 7 d at 3°C in 20% CO_2 or more plus 2 d at 15°C in air developed more CI than spears held in air.[28] Controlled atmosphere storage did not affect the development of CI in okra.[29]

B. CARBON DIOXIDE PRETREATMENTS

Brooks and McColloch[30] reported that short treatments (20 to 48 h) with high concentrations of CO_2 (40 to 45%) prior to placing grapefruit in low-temperature storage resulted in a decrease in the development of CI. Much work has been done since that time to try to develop CO_2 treatments that could be used commercially. Nevertheless, the effect has failed due to the variable response of diverse samples of grapefruit. A 14-d treatment with 20%

CO_2 applied to midseason (harvested in February) Florida grapefruit at 4.5°C reduced CI during 8 or 12 weeks of subsequent storage at 4.5°C.[31] In another study, 10, 20, and 30% CO_2 applied for 1 or 7 d reduced CI in early-season (harvested in November) grapefruit during 4 weeks of storage at 4°C, but not during longer periods.[32] In both studies, CO_2 treatments often either were not beneficial or actually increased pitting of the peel. Early- and late-season Florida grapefruit were found to be injured more readily than midseason fruit by treatments with high levels of CO_2.[31,32] CO_2 treatments before to storage appeared to be much more effective in reducing CI when applied at nonchilling than chilling temperatures.[33-35] Most, if not all, of this increased tolerance seems to have been attributable to holding the fruit at the warm temperature.

Brief treatments with high CO_2 (20% for 2 d at 4°C) decreased CI in avocados,[36] had no substantial effect on suppressing CI in Japanese apricots held at 5°C,[12] and increased CI in lemons (40% CO_2 for 3 d at 21°C)[37] and limes (30 or 40% CO_2 for 1 d at 21°C).[38]

C. ETHYLENE

The influence of ethylene on ripening and senescence of fruit and other plant tissues and the interaction of these processes with CI has been addressed in several studies. Gassing horticulturally mature honeydew melons for 24 h with 1000 ppm ethylene induced ripening and virtually prevented CI in the melons that subsequently were stored 2 weeks at 2.5°C.[39,40] The use of 1 to 10 ppm ethylene to degreen grapefruit has had variable effects on the chilling tolerance of the fruit. Grierson[41] found that early-season Florida grapefruit exposed at 29°C to 2.5 ppm ethylene for 48 h were more resistant to chilling-induced pitting than untreated fruit, while midseason fruit treated with ethylene became more sensitive. He concluded that ethylene increased tolerance in green fruit, while it reduced tolerance in yellow fruit. In another study, exposure of early grapefruit at 29°C to 5 ppm ethylene for up to 3 d resulted in an increase in CI on fruit subsequently held 21 d at 1°C.[42] Likewise, degreening of Bearss lemons with ethylene increased susceptibility to CI.[37]

Avocados exposed to ethylene (1 to 100 ppm) during storage at 1.5°C developed more CI during 4 or 6 weeks than fruit stored without ethylene.[43,44] These results may explain how CA reduces CI in avocados stored at low temperatures. Reduced O_2 and elevated CO_2 levels are known to inhibit ethylene synthesis and its action, respectively. In this manner, CA may inhibit the synthesis and/or action of endogenous ethylene and thus prevent development of ethylene-enhanced CI.

D. RELATIVE HUMIDITY

Water loss has been associated with the development of CI in many commodities. One way to decrease the rate of water loss is to store the fruit or vegetable in an atmosphere high in relative humidity (RH). Morris and Platenius[45] demonstrated that pitting of cucumbers and bell peppers that was associated with CI could be reduced by storage in atmosphere with RH near 100%. Conversely, pitting was increased at lower RH. Cucumbers held in 54% RH for 10 d at 1°C were severely pitted, while those held in 99% RH were slightly or not at all pitted. McColloch[46] also observed more pitting on bell peppers held in 88 to 90% RH rather than in 96 to 98% RH for 12, 15, or 18 d at 0°C. This amelioration of pitting was maintained even during a subsequent 4 d at 18°C.

High humidity slowed the ripening and largely prevented CI in bananas stored 10 d at 5°C.[47] Pitting affected only about 4% of bananas held in 100% RH, but affected 36 and 84% of fruit stored in 75 or 50% RH, respectively.

It has long been realized that low RH in cold storage increases the rate of development of CI in citrus.[23] Brooks and McColloch[30] observed that the area of pitting on grapefruit stored at 4.5°C was nearly 4 times greater at 65 to 75% RH than at 85 to 90% RH after 4 to 10 weeks. In a more recent study, 10% of the surface of grapefruit held 5 weeks at 4.5°C was pitted when held in 100% RH, while twice this area was affected in 75 and 50% RH.[48] Limes demonstrated a greater response to RH than grapefruit; 6% of the fruit were injured in 100% RH, but nearly 98% were injured in 50% RH after 4 weeks at 4.5°C.

In contrast, a rapid water loss (0.5%) from eggplant prior to storage for 7 d at 5°C followed by holding at 20°C reduced the development of CI.[49] This chilling tolerance was retained after recovery to initial weight by water absorption. However, CI was not reduced when weight loss was more than 1%. The authors suggest that this reduction in chilling sensitivity is associated not only with water content, but with movement of water within fruit tissue. Controlled water loss from fruit may be involved in the increased chilling tolerance obtained during curing and should be investigated.

Greater water loss (2.5 vs. 0.5%) due to storage in low vs. high RH also was correlated with a decrease in IB of apples after 12 or 13 weeks at −1°C.[50] The author suggests that increasing water loss in apples helps to remove volatile acetate esters that may be involved in the development of IB. Simon[51] suggested that water soaking of the intercellular spaces is an early stage of IB in apples and that it is followed by hydrostatic rupture of the protoplasts. Evaporation of intercellular water would raise the tonicity of the remaining solution and, therefore, reduce this postulated rupture of the protoplast, and, therewith, development of IB.

III. HYPOBARIC STORAGE

The influence of low atmospheric pressure on the development of CI in fresh fruits and vegetables varies with the commodity. Commodities that respond favorably to low O_2 and/or low ethylene also appear to respond similarly to low atmospheric pressure. Chilling injury in avocado, which is ameliorated by low O_2 and low ethylene, is also ameliorated by hypobaric storage.[52,53] Avocados held 5 weeks at 4.5°C at 220 mmHg were pitted on 11% of the fruit surface vs. 31% when held at 760 mmHg.[53] Chilling injury in grapefruit, which low O_2 and exposure to ethylene may influence, also responds to low pressures. Thus, only 4% of the surface was pitted in fruit held at 220 mmHg for 7 weeks at 4.5°C, but 23% was pitted in atmospheric pressure.[48] In another study, 380 mmHg had no effect on the development of CI of grapefruit.[41] Limes[48] also showed a lower degree of CI when held at 220 mmHg than when held at 760 mmHg. Low atmospheric pressure (220 mmHg) decreased the incidence of peel pitting (27 vs. 92% at 760 mmHg) and inhibited ripening of bananas held 4 weeks at 5°C,[53] but low atmospheric pressures have been reported to have no effect on the development of CI in peaches[9,11] or tomatoes.[24]

IV. WAX AND OTHER COATINGS AND PACKAGING

The benefits that can be achieved by altering the bulk storage atmosphere also can be obtained by the use of wax or other coatings and various packaging films. These effects became known in the 1930s when researchers observed that waxing fruit decreased CI.[30,54] At the same time, it was also observed that wrapping grapefruit in oiled paper,[30,54] waxed crystalline paper,[55] or cellophane sheets[30] all decreased pitting of the peel when fruit were held at chilling temperatures. Many new waxes and polymer materials have been developed since the 1930s, and many of them have been shown to influence the development of CI.

A. WAX AND OTHER COATINGS

Waxes and other coatings have been applied to the surface of fresh fruits and vegetables for many years to improve their appearance and reduce water loss. In the 1930s, Brooks and McColloch[30] found that waxing grapefruit reduced the incidence of chilling-induced pitting by 82%. Since that report, waxing repeatedly has been reported to decrease the expression of CI in grapefruit.[22,41,56,57] In addition, covering grapefruit with a polyethylene emulsion reduced CI.[58] Chilling injury was also reduced in Valencia[59,60] and Temple[19] oranges by waxing. However, waxing has been shown to increase CI in limes.[48,57] Limes treated with one, two, or three applications of Flavorseal 93 wax developed 8, 32, and 41% more pitting of the peel, respectively, than unwaxed fruit after being held 4 weeks at 4.5°C.[48] High concentrations of CO_2 increased the development of CI in limes.[38] High levels of CO_2 that accumulate in waxed fruit may be responsible for this increased CI. Waxing has also been found to reduce the incidence and severity of CI in cucumber[45] and pineapple.[15,61]

Waxing may reduce CI by modification of internal O_2 and CO_2 concentration[60] or by reducing water loss.[45] It is not clear which of these two mechanisms may be influencing expression of CI in various commodities, but it appears that either or both may be acting, depending on the commodity.

B. PLASTIC FILMS

In recent years, a large variety of polymer materials have become available for use in the packaging of fresh produce. These films are made from a wide variety of materials including polyethylene (PE), polyvinylchloride (PVC), and various copolymers. Sealing fresh fruits or vegetables in these materials has improved their keeping quality, including the suppression of CI. These polymeric films influence the development of CI in fresh produce by creating a modified atmosphere around the commodity and by slowing water loss. The degree to which a film influences each factor depends on the formulation of the film, its thickness, and how it is applied.

Methods of application include plastic bags, enclosing some type of container, such as pressboard trays, and individual shrink-wrapping of the item. In the remainder of this chapter we will review how these packaging techniques influence the development of CI in a variety of commodities.

In general, fruits and vegetables in which CI is reduced by CA and/or high RH benefit from packaging in various types of films. Avocados held between 4 and 6°C in PE bags expressed little or no CI after being stored up to 58 d, while control fruit held in ambient air were severely injured.[62,63] Thus, fruit in bags lost minimal weight after 14 d, while controls lost 16 to 26%.[62] In the studies of Scott and Chaplin,[63] the atmospheres inside the bags were reported to contain 2 to 6% O_2 and 3 to 7% CO_2.

Bananas responded favorably to being placed in PE bags for the prevention of CI in storage.[64] As indicated earlier, CI is inhibited in bananas by modified atmospheres and high humidity. Scott and Gandanegara[64] showed that individual fruit held in 0.04-mm-thick PE bags at 10°C could be held for 70 d with development of only slight CI, while fruit held in air were damaged severely after only 22 d. The atmospheres in these bags contained 8 to 16% O_2 and 2 to 5% CO_2 throughout the storage period. The influence of these PE bags on water loss, however, was not reported.

Citrus has a more variable response to film packaging than other fruit. The response of grapefruit depended on the physiology of the fruit when they were placed into containers covered with films that significantly altered the O_2 and CO_2 levels of the atmosphere within. According to Wardowski et al.,[57] most lots of fruit which were held for about 7.5 weeks at 4.5°C in film-lined cartons in which 4 to 9% CO_2 accumulated developed less CI than similar fruit which were held in perforated film-lined cartons. Grapefruit stored 7.5 weeks at 4.5°C in film-covered drums which accumulated up to 20% CO_2 also developed less CI than paper-covered drums, except with fruit that was harvested after the onset of bloom.[65] Grierson[66] has suggested that susceptibility to CI is related to plant hormone activity at harvest, which could explain the altered response of later-harvested fruit. Grierson[41] lined boxes, covered plastic tubs, and wrapped trays with three types of film: PVC, polypropylene, or rubber hydrochloride. None of these films greatly altered the O_2 or CO_2 concentrations around the fruit, but they all reduced CI. Grierson suggested that high humidity in the film-wrapped packages may explain this reduction in CI.

Grapefruit individually shrink-wrapped in PE film developed less CI than unwrapped fruit.[67-69] Since PE films caused a minimal modification of the internal atmosphere of grapefruit,[68] the major benefit from these films appears to be prevention of water loss from the fruit, as Ben-Yehoshua et al.[70] and Purvis[69,71] concluded. The former also suggested that the high level of RH in the microatmosphere of the fruit might maintain a juvenile hormonal balance inside the fruit which could impart tolerance to chilling temperatures. Reduction in CI due to film wrapping of fruit has also been reported in oranges,[60] lemons,[37,67] and limes.[57]

Film packaging of other fruits has been reported to reduce CI. Japanese apricots (*Prunus mume* Sieb. et Zucc.) held in PE bags at 0 or 5°C were injured less than controls held in perforated PE bags.[12] The response depended on cultivar and was attributed to accumulation of CO_2, reduction of O_2, and high RH in the nonperforated bags. Individually wrapping Tendral melons in PE prevented the development of CI in fruit held up to 75 d at 7 to 8°C, while 20% of unwrapped melons developed CI.[72] Likewise honeydew melons[73] wrapped with a PVC shrink film and stored for 21 d at 2.5°C had a 30% lower incidence of CI than nonwrapped melons. After 21 d, the atmosphere in the cavity of wrapped melons contained 17% O_2 and 6% CO_2, but that of the controls contained 20% O_2 and 1% CO_2. In addition, wrapped fruit lost only 20% as much weight as the unwrapped fruit. Eggplant wrapped in high-density PE and stored up to 17 d at 8 to 9°C were less severely injured than fruit held in paper bags, low-density PE, or a heat-shrinkable copolymer.[74] While all plastic films were equally effective in reducing weight loss, the authors suggest that high-density PE was most effective in creating an atmosphere that inhibited CI. On the other hand, CI was not significantly reduced in papayas stored 14 d at 2°C in PE wraps[16] or PE bags.[75]

Water loss in cucumbers and peppers due to storage in low RH increases their suscep-tibility to CI.[45] Wrapping these fruit in films is a practical method to decrease water loss and CI during storage. Severity of CI in cucumbers was reduced when they were wrapped with PE[76,77] or PVC[76] shrink films. Perforated PE bags and PVC-wrapped trays were reported to decrease CI in 9 of 19 cultivars of sweet pepper stored 18 d at 4 to 6°C.[78] Miller and Risse[79] reported that wrapping unspecified cultivars of bell pepper in a heat-shrinkable copolymer film had no effect on the expression of CI when pods were held for 21 d at 1, 4, or 7°C.

Hobson[80,81] reports that CI can be reduced in orange-green (early breaker) tomatoes by holding them in PE bags. Chilling injury in tomatoes held 6 d in PE pags at 2°C[80] or in PE bags that were first flushed with 5% O_2 at 5°C[81] and then ripened at 20°C was less than fruit held in air. However, CI was not reduced when orange-green tomatoes were wrapped with PVC cling wrap and held 6 days at 8°C.[81]

Storing okra in perforated or nonperforated PE bags provided no protection against CI during 10 d at 1 or 6°C.[82] In fact, storage of pods in nonperforated bags increased injury, and the authors suggest that elevated levels of CO_2 in the bag may have been responsible for this increased injury.

V. CONCLUSIONS

The variable responses that diverse commodities have to modification of the storage atmosphere in relation to amelioration of CI make it difficult to offer any general assessment of the mechanism involved or to propose generally applicable recommendations. Beneficial effects from low-O_2 atmosphere appear to be primarily from the suppression of symptom development and, in some cases, are negated after the commodity is placed in air. Elevated levels of CO_2 are more effective than low O_2 in many commodities in reducing CI. However, the response of fruits and vegetables to CO_2 is variable. A given concentration of CO_2 may result in amelioration or enhancement of CI, depending on the physiological status of the commodity. This response to CO_2 appears to be complex, depending on a host of physiological and environmental factors. Research that would explain what determines the nature of the response of a commodity to chilling temperatures during or after exposure to CA is needed before we can predict the interaction of CA and CI.

Ethylene seems to help prevent CI in many climacteric fruit, such as melons, by stimulating ripening. In other fruits and vegetables, however, the mechanism by which ethylene influences the chilling response is yet to be explained.

Storage in high RH generally inhibits the development of CI. Reduction of water loss by high RH may reduce the development of CI symptoms, such as surface pitting, but in many cases it appears that these atmospheres maintain the commodity in a physiological state that is more resistant to CI.

The development of a wide variety of film formulations and thicknesses available make it possible to select a packaging material suitable for almost any commodity. Decreasing water loss, which usually decreases the expression of CI, is easily obtained by the use of most films, even when they are perforated. Maintaining a favorable O_2 and CO_2 concentration can also be achieved by choosing films having properties that account for the respiration rate of the commodity to be wrapped.[83] Optimal composition of the atmosphere, however, needs to be determined for each fruit or vegetable until we better understand the factor determining the chilling sensitivity of a commodity.

REFERENCES

1. **Kidd, F., West, C., and Kidd, M. N.,** *Gas Storage of Fruit,* Special Rep. No. 30, Department of Science and Industrial Research, London, 1927, 87.
2. **Spalding, D. H. and Reeder, W. F.,** Low-oxygen high-carbon dioxide controlled atmosphere storage for control of anthracnose and chilling injury of avocados, *Phytopathology,* 65, 458, 1975.
3. **Vakis, N., Grierson, W., and Soule, J.** Chilling injury in tropical and subtropical fruits. III. The role of CO_2 in suppressing chilling injury of grapefruit and avocados, *Proc. Trop. Reg. Am. Soc. Hortic. Sci.,* 14, 89, 1970.
4. **Spalding, D. H. and Reeder, W. R.,** Quality of 'Booth 8' and 'Lula' avocados stored in a controlled atmosphere, *Proc. Fla. State Hortic. Soc.,* 85, 337, 1972.
5. **Anderson, R. E. and Penney, R. W.,** Intermittent warming of peaches and nectarines stored in a controlled atmosphere or air, *J. Am. Soc. Hortic. Sci.,* 100, 151, 1975.
6. **Anderson, R. E., Parsons, C. S., and Smith, W. L., Jr.,** Controlled atmosphere storage of eastern-grown peaches and nectarines *U.S. Dep. Agric. Mark. Res. Rep.,* 836, 1969, 19.
7. **Anderson, R. E., Penney, R. W., and Smith, W. L., Jr.,** Combining CA storage with intermittent warming extends the storage life of peaches and nectarines, in *Controlled Atmospheres for Storage and Transportation of Perishable Agricultural Commodities,* Horticultural Rep. No. 28, Dewey, D. H., Ed., Michigan State University, East Lansing, 1977, 149.
8. **Olsen, K. L. and Schomer, H. A.,** Influence of controlled-atmospheres on the quality and condition of stored nectarines, *HortScience,* 10, 582, 1975.
9. **Kajiura, I.,** CA storage and hypobaric storage of white peach 'Okuba', *Sci. Hortic.,* 3, 179, 1975.
10. **O'Reilly, H. J.,** Peach storage in modified atmospheres, *Proc. Am. Soc. Hortic. Sci.,* 49, 99, 1947.
11. **Wade, N. L.,** Effects of storage atmosphere, temperature and calcium on low-temperature injury of peach fruit, *Sci. Hortic.,* 15, 145, 1981.
12. **Iwata, T. and Yoshida, T.,** Chilling injury in Japanese apricot (Mume, *Prunus mume* Sieb. et Zucc.) fruits and preventive measures, *Stud. Inst. Hortic. Kyoto Univ.,* 9, 135, 1979.
13. **Eksteen, G. J., Visagie, T. R., and Laszlo, J. C.,** Controlled-atmosphere storage of South African grown nectarines and plums, *Decid. Fruit Grower,* 36, 128, 1986.

14. **Kader, A. A., El-Goorani, M. A., and Sommer, N. F.,** Postharvest decay, respiration, ethylene production, and quality of peaches held in controlled atmospheres with added carbon monoxide, *J. Am. Soc. Hortic. Sci.,* 107, 856, 1982.

15. **Paull, R. E. and Rohrback, K. G.,** Symptom development of chilling injury in pineapple fruit, *J. Am. Soc. Hortic. Sci.,* 110, 100, 1985.

16. **Chen, N. and Paull, R. E.,** Development and prevention of chilling injury in papaya fruit, *J. Am. Soc. Hortic. Sci.,* 111, 639, 1986.

17. **Mencarelli, F., Lipton, W. J., and Peterson, S. J.,** Responses of 'Zucchini' squash to storage in low-O_2 atmospheres at chilling and nonchilling temperatures, *J. Am. Soc. Hortic. Sci.,* 108, 884, 1983.

18. **Mencarelli, F.,** Effect of high CO_2 atmospheres on stored zucchini squash, *J. Am. Soc. Hortic. Sci.,* 112, 985, 1987.

19. **Chace, W. G., Jr.,** Controlled-atmosphere storage of Florida citrus fruits, *Proc. 1st Int. Citrus Symp.,* 3, 1365, 1969.

20. **Pantastico, E. B., Mattoo, A. K., Murata, T., and Ogata, K.,** Chilling injury, in *Postharvest Physiology, Handling and Utilization of Tropical and Subtropical Fruits and Vegetables,* Pantastico, E. B., Ed., AVI Publishing, Westport, CT, 1975, 339.

21. **Scholz, E. W., Johnson, H. B., and Buford, W. R.,** Storage of Texas red grapefruit in modified atmospheres, A progress report, *U.S. Dep. Agric. Res. Serv. Mark. Res. Rep.,* 414, 1960, 11.

22. **Davis, P. L., Roe, B., and Bruemmer, J. H.,** Biochemical changes in citrus fruits during controlled-atmosphere storage, *J. Food Sci.,* 38, 225, 1973.

23. **Miller, E. V.,** Physiology of citrus fruit in storage, *Bot. Rev.,* 12, 393, 1946.

24. **Morris, L. L. and Kader, A. A.,** Postharvest physiology of tomato fruits, *Univ. Calif. Davis Veg. Crops Ser.,* 171, 36, 1975.

25. **Morris, L. L. and Kader, A. A.,** Postharvest physiology of tomato fruits, *Annu. Rep. Calif. Fresh Market Research Board 1976—77,* California Fresh Market Research Board, 1977, 46.

26. **Eaks, I. L.,** Effect of modified atmospheres on cucumbers at chilling and non-chilling temperatures, *Proc. Am. Soc. Hortic. Sci.,* 67, 473, 1956.

27. **Cappellini, M. C., Lachance, P. A., and Hudson, D. E.,** Effect of temperature and carbon dioxide atmospheres on the market quality of green bell peppers, *J. Food Qual.,* 7, 17, 1984.

28. **Lipton, W. J.,** Post-harvest responses of asparagus spears to high carbon dioxide and low oxygen atmospheres, *Proc. Am. Soc. Hortic. Sci.,* 86, 347, 1965.

29. **Ogata, K., Yamauchi, N., and Minamide, T.,** Physiological and chemical studies on ascorbic acid of fruits and vegetables. I. Changes of ascorbic acid content during maturation and storage period in okras, *J. Jpn. Soc. Hortic. Sci.,* 44, 192, 1975.

30. **Brooks, C. and McColloch, L. P.,** Some storage diseases of grapefruit, *J. Agric. Res.,* 52, 319, 1936.

31. **Hatton, T. T., Jr., Smoot, J. J., and Cubbedge, R. H.,** Influence of carbon dioxide exposure on stored mid- and late-season 'Marsh' grapefruit, *Proc. Trop. Reg. Am. Soc. Hortic. Sci.,* 16, 49, 1972.

32. **Hatton, T. T. and Cubbedge, R. H.,** Influence of initial carbon dioxide exposure on stored early-, mid- and late-season 'Marsh' grapefruit, *Proc. Fla. State Hortic. Soc.,* 87, 227, 1974.

33. **Hatton, T. T. and Cubbedge, R. H.,** Effects of prestorage carbon dioxide treatments and delayed storage on chilling injury of 'Marsh' grapefruit, *Proc. Fla. State Hortic. Soc.,* 88, 335, 1975.

34. **Hatton, T. T. and Cubbedge, R. H.,** Conditioning Florida grapefruit to reduce chilling injury during low-temperature storage, *J. Am. Soc. Hortic. Sci.,* 107, 57, 1982.

35. **Anon.,** Influence of high levels of CO_2 in reducing chilling injury of grapefruit stored at 277.5 K (45°C), Publ. Cold Storage Res. Lab., Isr. Fruit Growers Assoc., 1979, 16.

36. **Marcellin, P. and Chaves, A.,** Effects of intermittent high CO_2 treatment on storage life of avocado fruits in relation to respiration and ethylene production, *Acta Hortic.,* 138, 155, 1983.

37. **McDonald, R. E.,** Effects of vegetable oils, CO_2, and film wrapping on chilling injury and decay of lemons, *HortScience,* 21, 476, 1986.

38. **Spalding, D. H. and Reeder, W. F.,** Conditioning 'Tahiti' limes to reduce chilling injury, *Proc. Fla. State Hortic. Soc.,* 96, 231, 1983.

39. **Lipton, W. J. and Mackey, B. E.,** Ethylene and low-temperature treatments of honeydew melons to facilitate long-distance shipment, *U.S. Dep. Agric., Agric. Res. Serv. Rep.,* 10, 1984, 28.

40. **Lipton, W. J., Aharoni, Y., and Elliston, E.,** Rates of CO_2 and ethylene production and of ripening of 'Honey Dew' muskmelons at a chilling temperature after pretreatment with ethylene, *J. Am. Soc. Hortic. Sci.,* 104, 846, 1979.

41. **Grierson, W.,** Chilling injury in tropical and subtropical fruits. IV. The role of packaging and waxing in minimizing chilling injury of grapefruit, *Proc. Trop. Reg. Am. Soc. Hortic. Sci.,* 15, 76, 1971.

42. **Hatton, T. T. and Cubbedge, R. H.,** Effects of ethylene and chilling injury and subsequent decay of conditioned early 'Marsh' grapefruit during low-temperature storage, *HortScience,* 16, 783, 1981.

43. **Chaplin, G. R., Wills, R. B. H., and Graham, D.,** Induction of chilling injury in stored avocados with exogenous ethylene, *HortScience,* 18, 952, 1983.

44. **Chaplin, G. R.,** Ethylene and chilling injury in stored avocados, Res. Rep. 1983/1984, CSIRO Division of Food Research, Australia, 1984, 46.

45. **Morris, L. L. and Platenius, H.,** Low temperature injury to certain vegetables after harvest, *Proc. Am. Soc. Hortic. Sci.,* 36, 609, 1938.

46. **McColloch, L. P.,** Chilling injury and Alternaria rot of bell peppers, *U.S. Dep. Agric. Mark. Res. Rep.,* 536, 1962, 16.

47. **Pantastico, E. B., Grierson, W., and Soule, J.,** Chilling injury in tropical fruits. I. Bananas (*Musa paradisiaca* var. *Sapientum* cv. Lacatan), *Proc. Trop. Reg. Am. Soc. Hortic. Sci.,* 11, 82, 1967.

48. **Pantastico, E. B., Soule, J., and Grierson, W.,** Chilling injury in tropical and subtropical fruits. II. Limes and grapefruit, *Proc. Trop. Reg. Am. Soc. Hortic. Sci.,* 12, 171, 1968.

49. **Nakamura, R., Fujii, S., Inaba, A., and Ito, T.,** Effect of weight loss prior to cold storage on chilling sensitivity in eggplant fruit, *Sci. Rep. Fac. Agric. Okayama Univ.,* 68, 19, 1986.

50. **Wills, R. B. H.,** Influence of water loss on the loss of volatiles by apples, *J. Sci. Food Agric.,* 19, 354, 1968.

51. **Simon, E. W.,** The symptoms of calcium deficiency in plants, *New Phytol.,* 80, 1, 1978.

52. **Lyons, J. M.,** Chilling injury in plants, *Annu. Rev. Plant Physiol.,* 24, 445, 1973.

53. **Pantastico, E. B.,** Postharvest physiology of fruits. I. Chilling injury, *Philipp. Agric.,* 51, 697, 1968.

54. **Nelson, R.,** Some storage and transportational diseases of citrus fruits apparently due to suboxidation, *J. Agric. Res.,* 46, 695, 1933.

55. **van der Plank, J. E. and Davies, R.,** Temperature-cold injury curves of fruit, *J. Pomol. Hortic. Sci.,* 15, 226, 1937.

56. **Grierson, W.,** Physiological disorders of citrus fruits, *Proc. Int. Soc. Citric.,* 2, 764, 1981.

57. **Wardowski, W. F., Grierson, W., and Edwards, G. J.,** Chilling injury of stored limes and grapefruit as affected by differentially permeable packaging films, *HortScience,* 8, 173, 1973.

58. **Davis, P. L. and Harding, P. L.,** The reduction of rind breakdown of Marsh grapefruit by polyethylene emulsion treatments, *Proc. Am. Soc. Hortic. Sci.,* 75, 271, 1960.

59. **Chace, W. G., Jr., Davis, P. L., and Smoot, J. J.,** Response of citrus fruit to controlled atmosphere storage, *XII Int. Congr. Refrig. (Madrid),* 3, 383, 1967.

60. **Martinez-Javega, J. M., Jimenez-Cuesta, M., and Cuquerella, J.,** Utilization of polyvinyl chloride (PVC) film for individual seal packaging of citrus fruit, *Proc. Int. Soc. Citric.,* 2, 722, 1981.

61. **Rohrbach, K. G. and Paull, R. E.,** Incidence and severity of chilling induced internal browning of waxed 'Smooth Cayenne' pineapple, *J. Am. Soc. Hortic. Sci.,* 107, 453, 1982.

62. **Adikaram, N. K. B. and Theivendirarajah, K.,** Studies on the storage of avocado fruits and their spoilage organisms, *Ceylon J. Sci. Biol. Sci.,* 14, 83, 1981.

63. **Scott, K. J. and Chaplin, G. R.,** Reduction of chilling injury in avocados stored in sealed polyethylene bags, *Trop. Agric. (Trinidad),* 55, 87, 1978.

64. **Scott, K. J. and Gandanegara, S.,** Effect of temperature on the storage life of bananas held in polyethylene bags with ethylene absorbent, *Trop. Agric. (Trinidad),* 51, 23, 1974.

65. **Wardowski, W., Albrigo, L., Grierson, W., Barmore, C., and Wheaton, T.,** Chilling injury and decay of grapefruit as affected by thiabendazole, benomyl, and CO_2, *HortScience,* 10 (Abstr.), 381, 1975.

66. **Grierson, W.,** Chilling injury in tropical and subtropical fruits: effect of harvest date, degreening, delayed storage and peel color on chilling injury of grapefruit, *Proc. Trop. Reg. Am. Soc. Hortic. Sci.,* 18, 66, 1974.

67. **Ben-Yehoshua, S., Kobiler, I., and Shapiro, B.,** Effects of cooling versus seal-packaging with high-density polyethylene on keeping qualities of various citrus cultivars, *J. Am. Soc. Hortic. Sci.,* 105, 536, 1981.

68. **Kawada, K.,** Use of polymeric films to extend postharvest life and improve marketability of fruits and vegetables — UNIPACK: individually wrapped storage of tomatoes, oriental persimmons and grapefruit, in *Controlled Atmospheres for Storage and Transport of Perishable Agricultural Commodities,* Oregon State Univ. Symp. Ser. No. 1, Richardson, D. G. and Meheriuk, M., Eds., Oregon State University, Corvallis, 1982, 87.

69. **Purvis, A. C.,** Relationship between chilling injury of grapefruit and moisture loss during storage: amelioration by polyethylene shrink film, *J. Am. Soc. Hortic. Sci.,* 110, 385, 1985.

70. **Ben-Yehoshua, S., Shapiro, B, Chen, Z. E., and Lurie, S.,** Mode of action of plastic film in extending life of lemon and bell pepper fruits by alleviation of water stress, *Plant Physiol.,* 73, 87, 1983.

71. **Purvis, A. C.,** Moisture loss and juice quality from waxed and individually seal-packaged citrus fruits, *Proc. Fla. State Hortic. Soc.,* 96, 327, 1983.

72. **Martinez-Javega, J. M., Jimenez-Cuesta, M., and Cuquerella, J.,** Conservacion frigorifica del melon 'Tendral' (Cold storage of 'Tendral' melons), *An. Inst. Nac. Invest. Agrar. Agric.,* 23, 111, 1983.

73. **Rij, R. E. and Ross, S. R.,** Effects of shrink wrap on internal gas concentrations, chilling injury and ripening of 'Honey Dew' melons, *J. Food Qual.,* 11, 175, 1988.

74. **Mohammed, M. and Sealy, L.,** Extending the shelf-life of melongene (*Solanum melongena* L.) using polymeric films, *Trop. Agric. (Trinidad),* 63, 36, 1986.

75. **Thompson, A. K. and Lee, G. R.,** Factors affecting the storage behavior of papaya fruit, *J. Hortic. Sci.,* 46, 511, 1971.

76. **Adamicki, F.,** Effects of storage temperature and wrapping on the keeping quality of cucumber fruits, *Acta Hortic.,* 156, 269, 1984.

77. **Wiebe, H. J.,** Haltbarkeit von Salatgurken nach Verpackung in Schrumpffolie, *Gemuese,* 24, 54, 1969.

78. **Gorini, F. L., Eccher Zerbini, P., and Uncini, L.,** Storage suitability of some sweet pepper cultivars *Capsicum annuum* L. as affected by temperature and packaging, *Acta Hortic.,* 62, 131, 1977.

79. **Miller, W. R. and Risse, L. A.,** Film wrapping to alleviate chilling injury of bell peppers during cold storage, *HortScience* 21, 467, 1986.

80. **Hobson, G. E.,** The short term storage of tomato fruit, *J. Hortic. Sci.,* 56, 363, 1981.

81. **Hobson, G. E.,** Low-temperature injury and the storage of ripening tomatoes, *J. Hortic. Sci.,* 62, 55, 1987.

82. **Tamura, J. and Minamide, T.,** Harvesting maturity, handling, storage of okra pods [sic], *Bull. Univ. Osaka Prefect. Ser. B.,* 36, 97, 1984.

83. **Rij, R. E. and Mackey, B.,** Transmission rates of CO_2 at 0°, 10°, or 22°C through flexible films used for marketing fruits and vegetables, *HortScience,* 21, 284, 1986.

Chapter 17

REDUCTION OF CHILLING INJURY WITH TEMPERATURE MANIPULATION

T. T. Hatton

TABLE OF CONTENTS

I. INTRODUCTION

Chilling injury (CI) is one of the great limitations of extended storage for many horticultural crops. CI differs from freeze injury in that it occurs above the freezing point for the commodity. CI may be described and manifested in numerous forms, depending on commodity and severity of injury. A number of horticultural commodities, especially those from tropical and subtropical origins, may display symptoms of CI when exposed to temperatures as high as 13°C.[1,2] Solving the problem of CI would revolutionize extended storage, shipping, and marketing for many horticultural commodities.

The manipulation of temperature for the purpose of storing fruits and vegetables is probably the major approach presently used for reducing CI. Proper temperatures provide conditions that avoid or reduce CI and deterioration until the commodity reaches its destination. Presently, tremendous losses occur throughout the food distribution system because of improper storage temperatures, even though mechanical refrigeration is readily available. Often CI losses are due to inadequate knowledge of proper storage temperatures specific for the commodity.

Low storage temperatures above freezing result in lowering respiration rates and, thus, extend storage life. The limitation to low temperature is the development of CI and the decay that usually follows CI. Conversely, high temperatures result in high respiration rates and, thus, a shortening of storage life due to rapid deterioration. Deterioration of perishable commodities may be due to one or more factors such as shrivel, senescence, spoilage, sprouting, and other undesirable changes.

Three types of temperature manipulations for fruits and vegetables are available to reduce CI: (1) optimum storage at temperatures at which a commodity does not develop CI, which is by far the most important commercial practice; (2) prestorage temperature conditioning; and (3) intermittent warming during storage.

II. OPTIMUM STORAGE TEMPERATURES

Optimum storage temperaures are usually established for safe, long-term storage and are documented for numerous horticultural commodities, fruits and vegetables, and some specific cultivars.[3-5] Similar information is also available for horticultural crops and other commodities exposed to long-term transport environments.[6] Regardless of recommended storage or transit conditions, maximum storage life is enhanced by storage of high-quality commodities as soon after harvest as possible.

A. PRECOOLING

Precooling, which is the rapid removal of field heat from freshly harvested commodities before storage, shipment, or processing, is necessary for many perishable horticultural crops. Precooling is the initial action in proper temperature management. Delays experienced between harvest and the time of precooling will increase deterioration. Rapid cooling retards shrivel and slows decay and senescence for most horticultural commodities.

Storage facilities are usually constructed to maintain reasonably low temperatures, but generally do not have the refrigeration capacity nor the air movement necessary for quick cooling. For this reason, precooling is mostly an adjunct operation with special equipment. The primary methods of precooling are by forced air, icing, hydrocooling, and vacuum cooling. Sometimes combinations of these are used. Regardless of the method of precooling, much of the benefit will be lost if commodities are not cooled promptly after harvesting.[7,8]

B. TEMPERATURE MAINTENANCE

The temperature of the commodity must be maintained within the recommended range. Tropical fruits such as avocados, mangos, and papayas will soften or ripen if the temperature

TABLE 1
Fruits and Vegetables Susceptible to Chilling Injury when Stored at Moderately Low but Nonfreezing Temperatures

Commodity	Approximate lowest safe temperature (°C)	Character of injury when stored between 0°C and safe temperatures[a]
Apples, certain cultivars[b]	2—3	Internal browning, brown core, soggy breakdown, soft scald
Asparagus	0—2	Dull, gray-green, and limp tips
Avocados, certain cultivars[b]	4.5—13	Grayish-brown discoloration of flesh
Bananas, green or ripe[b]	11.5—13	Dull color when ripened
Beans, snap[b]	7	Pitting and russeting
Cranberries	2	Rubbery texture, red flesh
Cucumbers	7	Pitting, water-soaked spots
Eggplants	7	Surface scald, alternaria rot, blackening of seeds
Limes	7—9	Pitting, turning tan with time
Mangos, certain cultivars[b]	10—13	Grayish, scald-like discoloration of skin, uneven ripening
Okra	7	Discoloration, water soaked areas, pitting, decay
Papayas	7	Pitting, failure to ripen, off-flavor, decay
Pepper, sweet	7	Sheet pitting, alternaria rot on pods and calyxes, darkening of seed
Potatoes	3	Mahogany browning (Chippewa and Sebago), sweetening
Sweet potatoes	13	Decay, pitting, internal discoloration, hard core when cooked
Tomatoes		
Ripe[b]	7—10	Water-soaking and softening, decay
Mature-green	13	Poor color when ripe, alternaria rot
Watermelons	4.5	Pitting, objectionable flavor

[a] Symptoms often apparent only after removal to warm temperatures, as in marketing.

[b] Depends on the cultivar.

From Hardenburg, R. E., Watada, A. E., and Wang, C. Y., *U.S. Dep. Agric. Agric. Handb.*, 66, 1986.

is too high and develop CI if it is too low. As long as the temperature is compatible with the commodity, the low end of any recommended temperature range should be considered, especially for long storage or transit periods.

The characteristics of CI are described along with a listing of respective lowest safe temperature for representative commodities (Table 1). Some commodities susceptible to CI can sometimes be stored at temperatures lower than optimum for a few days without developing CI, but a number of exceptions exist, such as bananas, cucumbers, eggplant, okra, and mature-green tomatoes. Recommended optimum temperatures must be strictly followed for these commodities and those similarly susceptible to CI.

Storage life may be altered or commodities may respond differently to their environment if they are overmature or immature; consequently, it is important that they be harvested at the optimum stage of maturity for storage purposes. A classic example relating to CI is the chilling sensitivity of avocados as a function of the stage of the climacteric.[9] The least sensitive stage is postclimacteric, where fruit can be kept at 2°C for 6 to 7 weeks. Avocados on the climacteric rise and at the climacteric peak were most sensitive to CI and showed injury after 19 d at 2°C. Postclimacteric fruit could be transferred to 2°C at 36 to 48 h after the climacteric peak.

The storage life of fruits and vegetables usually varies not only because of differences in cultivars and stages of maturity, but because of the exposure to climatic and soil conditions, cultural practices, and handling procedures.[10]

C. RELATIVE HUMIDITY

Relative humidity (RH) is the amount of moisture in the air at a specific temperature in relation to saturation at the specific temperature. Temperature manipulation is not independent from RH, and the two must be considered together. The RH in storage facilities, like temperature, directly affects shelf life and quality of the stored commodities[11].

Maintaining high RH is usually a problem in most commercial cold storages because the water-holding capacity of air decreases as the temperature falls. For example, air of 95% RH at 0°C contains much less water by weight than air of the same RH at 20°C.

Relative humidity of 90 to 95% is recommended for most horticultural commodities to reduce moisture loss in the stored commodity.[3-5] When the RH is too low, shrivel or wilt will likely occur in most horticultural commodities; if the RH is too high, the risk of decay and carton collapse increases. Raising the RH to 100% was reported to reduce CI in cucumbers and peppers.[12] Similarly, peppers, stored for 12 d at 0°C and 88 to 90% RH had 67% CI, compared to only 33% CI at 96 to 98% RH for the same exposure period and temperature.[13] CI of 'Tahiti' limes was reduced by raising the RH to 100%.[14]

D. QUARANTINE USE

The manipulation of temperatures near freezing for specified periods of time has long been used as a quarantine procedure to disinfest insect pests from horticultural commodities. "Cold treatment" is usable for a number of fruits that are not susceptible to CI: apple, apricot, cherry, nectarine, peach, pear, plum, quince, grape, kiwi, and persimmon.[14] Most citrus fruits, including the citron and some orange and mandarin cultivars, are compatible with cold treatments, whereas grapefruit, lemon and lime show little tolerance to cold treatment. Most tropical fruits are intolerant of the cold treatment for the time required for disinfestation of the insect.

Temperatures for the cold treatment can range from slightly lower than freezing to just above 2°C and are graduated for specific periods of time according to the insect in question[15] (Table 2). Temperature fluctuations during the actual cold treatment are undesirable because certification of specified exposure time of the commodity to the required temperature is strictly enforced. The warmest temperature during cold treatment is used to determine exposure time required.

E. MIXED LOADS

Sometimes, precise temperature maintenance must be compromised. Shippers and receivers of fresh fruits and vegetables often handle shipments that consist of more than one commodity. In such mixed loads, it is imperative to combine only commodities that are compatible. Compatible commodities are those which can be shipped together without adverse effects on any one commodity during the usual maximum transit period for the most perishable commodity in the load. Circumstances dictate when such commodities would be shipped together. The recommended conditions for shipment of each group of commodities merely represent a satisfactory compromise[16]

III. PRESTORAGE TEMPERATURE CONDITIONING

The manipulation of temperature for the purpose of conditioning or curing prior to cold storage is a commercial practice for potatoes and sweet potatoes. The main purpose of curing for potatoes and sweet potatoes is to reduce decay rather than to reduce CI, which is the

TABLE 2
The Cold Treatment as a Quarantine Measure for Several Species of Fruit Flies

Insect	Exposure time (d)	Temperature[a] °C	°F
Mediterranean fruit fly (*Ceratitus capitata*)	10	0	32
	11	0.55	33
	12	1.11	34
	14	1.66	35
	16	2.22	36
Mexican fruit fly (*Anastrepha ludens*)	18	0.53	33
	20	1.11	34
	22	1.66	35
Other species of *Anastrepha*	11	0	32
	13	0.55	33
	15	1.11	34
	17	1.66	35
Queensland fruit fly (*Dacus tryoni*)	13	0	32
	14	0.55	33
	18	1.11	34
	20	1.66	35
	22	2.22	36
False codling moth (*Cryptophlebia leucotreta*)	22	−0.55	31

[a] Pulp of the fruit must be at or below the indicated temperature at time of beginning treatment.

From U.S. Department of Agriculture Plant Protection and Quarantine Treatment Manual, Sect. T 107, U.S. Department of Agriculture, Washington, D.C., 1987.

main adverse factor for some commodities. The term "curing" is often used in commercial trade, while "conditioning" is mostly used elsewhere.

A. SOME COMMERCIAL USES OF CURING
Whenever cold storage will be used for late-crop potatoes, they should be cured by holding at 10 to 15.5°C and high RH (95%) for 10 to 14 d to allow suberization and wound periderm formation, which is the healing of bruises and cuts.[17] Wound periderm formation is most rapid at 21°C; however, lower temperatures are used because decay is less likely to occur. Curing reduces weight loss during subsequent storage and also the hazard of rot by preventing the entry of organisms. After potatoes are cured, they should be placed at 3.5 to 4.5°C. Storage at 4°C is considered optimum for a number of cultivars for maximum storage life, because sprout growth is negligible, shrinkage is low, and other losses are minimized. Potato tubers tend to develop an undesirable sweet flavor, a result of CI, when exposed to temperatures below 3°C

Sweet potatoes should be cured at 29°C and 90 to 95% RH for 4 to 7 d.[18] Curing helps prevent the entrance of decay organisms through injuries received during harvesting and handling operations. If the curing temperature and RH are lower than 29°C, healing is slower and less effective in preventing subsequent decay in storage. After curing, the temperature should be lowered to 13 to 16°C. This is usually by ventilation with outside air.[19,20] Temperatures above 16°C stimulate development of sprouts. CI symptoms were recently reported to be greater for noncured than cured roots of sweet potatoes held for 1, 2, 3, and 4 weeks at 7.2°C, followed by long-term storage at 15.6°C.[21]

Fresh lemons need curing to develop color, juice content, and flavor. Those from relatively dry, Mediterranean-type climates, such as California, are cured in refrigerated facilities at 13 to 15.5°C and 85 to 90% RH. Such lemons in good ventilated storage can keep from 1 to 4, or sometimes even 6 months. Long-term storage at temperatures below

13°C may cause development of CI, which is manifested as rind pitting and staining of the membranes (''membranosis'') separating the segments. Temperatures higher than 15.5°C favor the growth of decay organisms and shorten the shelf life.[22]

Lemons of good storage potential can be held for extended periods at terminal markets at 7 to 13°C for less than 4 weeks or for longer periods at 11 to 13°C to avoid CI.[23]

B. QUARANTINE USE

As a result of the ban on ethylene dibromide (EDB) fumigation by the Environmental Protection Agency (EPA), alternative quarantine treatments, preferably nonchemical, are needed for numerous commodities. For many years, fumigation with EDB was satisfactory and accepted throughout the world for numerous kinds of insects infesting various commodities; consequently, little research was undertaken to develop alternatives. Insects indigenous to the tropics and subtropics tend to be more readily eradicated by the cold treatment than insects found in the temperate climates.

A recent and important commercial use of temperature manipulation to avoid or reduce CI is the prestorage conditioning (curing) of Florida grapefruit destined for Japan and other locations requiring this quarantine measure.

Recommended storage and transit temperatures for Florida grapefruit are 10°C for mid- and late-season fruit and 16°C for early-season fruit; grapefruit often sustain CI when exposed to temperatures below these. CI in grapefruit is of two types, sometimes observed together: rind pitting, which prevails near 4°C, and brown staining of the rind, common near 0°C. Figure 1 shows CI of grapefruit in the form of rind pitting. Both forms of CI often result in decay when the fruit is removed from cold storage and placed at higher ambient temperatures.

Constant storage of grapefruit at 1°C for 28 d resulted in excessive CI; however, conditioning fruit for 7 d at 10, 16, or 21°C before cold treatment reduced CI during 21 d of storage at 1°C under high-humidity conditions.[24,25] Tests also showed that conditioning for 2 and 4 d was not as effective in reducing CI as conditioning for 7 d. Conditioning at 16°C resulted in less CI than conditioning at 21°C, or 27°C in minimizing CI of fruit later stored at 1°C[26] (Table 3).

Subsequently, in large-scale stationary tests, as well as in commercial test shipments to Japan, grapefruit were successfully conditioned by holding for 7 d at 16°C before storing or shipping at approximately 1°C up to 21 d.[27,28] Now several million boxes of Florida grapefruit are shipped annually to Japan using the cold treatment. The actual cold treatment is accomplished aboard ship en route to destination. Most of the conditioning is done in refrigerated facilities on land prior to loading; however, in some instances, partial or total conditioning is done aboard ship. Considerable time is needed for conditioning, pulling the fruit temperature down to the cold temperature required, and in the cold treatment itself. In fact, the total time required could, for all these processes, exceed transit time; in which case conditioning of the fruit should be done on land before loading, followed by precooling the commodity so that desired low temperatures are reached quickly.

Preharvest conditions in the grove, as well as postharvest handling procedures, can also directly affect the extent of CI in stored grapefruit.[10] That portion of the grapefruit crop harvested from the outer canopy of the tree is more susceptible to CI than fruit harvested from elsewhere on the tree.[29] Pre- and postharvest fungicidal applications of benomyl and postharvest applications of thiabendazole reduced CI.[30,31] By combining thiabendazole with cooling, susceptibility of grapefruit to CI can be reduced, and the cold treatment can be practiced with a low CI risk.[32] Waxing grapefruit and packaging in film minimizes CI[33] Raising the RH to 100% during storage greatly reduces CI.[14]

Lemons stored for 21 d at 1°C and held 14 d at 21°C sustained 15% CI compared to 1% after a 10°C storage and a 21°C holding period.[34] Holding lemons for 3 d at 21°C before

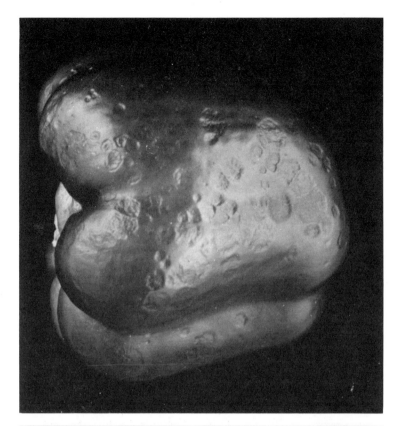

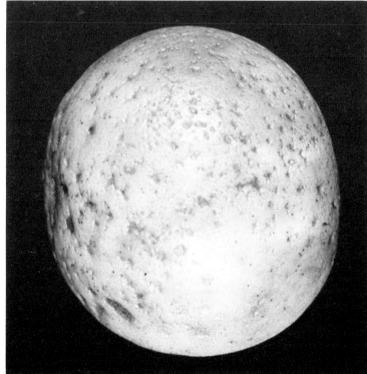

FIGURE 1.   Chilling injury: pitting of grapefruit and pepper.

TABLE 3
Chilling Injury of Conditioned and Stored Marsh
Grapefruit[a]

Conditioning (°C)	Storage[c] (°C)	Chilling injury[b]	
		Immediately after storage (%)	After holding at 21°C for 7 d (%)
None	28 d, (1)	23 a[d]	27 a
None	28 d, (10)	0 c	0 c
7 d, (16)	21 d, (1)	0 c	0 c
7 d, (21)	21 d, (1)	11 b	13 b
7 d, (27)	21 d, (1)	11 b	11 b

[a] Each value represents 360 fruit (120 from each of 3 sources) collected April 15, 1981. Regardless of surface area affected, values were derived from number of affected fruit.
[b] Mostly brown staining of the rind, with some pitting.
[c] Relative humidity ranged from 88 to 92% for fruit at 1 to 10°C and from 80 to 92% for fruit at 16°C and above.
[d] Means in each column not followed by the same letter indicate significant difference at 5% level by Duncan's multiple range test.

From Hatton, T. T. and Cubbedge, R. H., *HortScience,* 18 (5), 721, 1983.

storing at 1°C for 21 d reduced CI.[35] Lemons conditioned at 21 or 27°C for 7 d following degreening at 16°C developed CI when stored at 1°C for 21 d, regardless of conditioning treatment. Similar fruit held at 10°C did not develop CI.[36] Limes conditioned by holding at various temperatures for 1 week, prior to storage for 2 weeks at 2°C, developed less CI than nonconditioned limes.[37]

Bananas transferred directly from 21°C to 5°C show more CI than bananas which have their temperature step lowered by steps of 3°C at 12-h intervals.[38]

There was less CI in papayas conditioned for 4 d at 12.5°C before storage for 12 or 14 d at 2.0°C than in nonconditioned fruit. The decreases in CI with conditioning was associated with partial fruit ripening.[39]

Effective conditioning of two curcurbits, cucumber and watermelon, has been reported. Cucumber fruit were conditioned at 18°C for 1, 2, 3, 6, and 9 d before storage at 5°C. CI of the fruit was reduced by prestorage conditioning for more than 6 d.[40] In addition, prestorage warming for 24 h at 36 to 40°C, but not lower than 30°C, reduced CI of cucumber fruit during subsequent storage at 5°C. Postharvest conditioning of cucumber fruit at 15°C for 1 d prior to storage at 6.5°C was effective in lowering CI, but conditioning in double steps at 15 and at 10°C seemed to be more effective.[41] Symptoms of external CI in watermelons, primarily brown staining of the rind, were reduced by conditioning fruit at 26°C for 4 d before storage at 0 or 7°C.[42] The amount of brown staining intensified during 4- and 8-d holding periods at 21°C after removal from storage at 0°C for 8 d or longer, but was always less in conditioned fruit. Less CI occurred in fruit stored at 7°C than at 0°C.

Reduction of CI has been reported in three solanaceous crops, eggplant, pepper, and tomato. Prestorage conditioning at 10 or 20°C for 5, 10, or 15 d extended shelf life of eggplant fruit stored at 1°C by retarding CI.[43] The occurrence of pitting, the first symptom of CI, was delayed by 2 to 3 d. Postharvest conditioning at 15°C for 1 d prior to storage at 6.5°C was effective in lowering CI for eggplant fruit, but conditioning at 15°C and then at 10°C seemed to be more effective.[41] Peppers are readily conditioned against CI, which is in the form of sheet pitting (Figure 1), by 5 to 10 d of conditioning at 10°C.[11] Red ripe peppers appear to be resistant to CI. Peppers stored continuously at 1.7°C for 2 d or longer developed CI, whereas the injury was delayed if they were held at 5 or 10°C prior to storage.[44] The rate of development of CI was reduced by holding for 4 d or more at 5 or 10°C; 4 or 8 d at 5°C or 8 d at 10°C reduced the amount and severity of CI during subsequent storage at 1.7°C. Peppers which were nonconditioned and conditioned at 10°C for 5 d were stored at the recommended 7°C or at 1°C for 20 d; CI appeared only in the nonconditioned peppers stored at 1°C and was seen after 3 d of storage.[36] Conditioning dramatically reduced CI, even after 20 d of 1°C storage. Tomatoes harvested hot (e.g., 32°C) and chilled for 7 d at 7°C, or harvested cool (e.g., 19°C) and held at 37°C for 7 h before chilling at 2.5°C for 4 d, ripened more slowly than those that were either harvested cool or held at 12.5°C for 7 h before chilling.[45]

IV. INTERMITTENT WARMING

Intermittent warming is the same as fluctuating or manipulating the temperature from low to high and then back to low one or more times for various periods of time. CI can be avoided in many tissues, such as fruits, seeds, and seedlings, if they are returned to a warmer temperature before degenerative changes occur[2] (see also Chapter 18).

Commercial prestorage conditioning or curing is already used on grapefruit to be treated with cold treatments for quarantine purposes, and on potato tubers as well as sweet potato roots destined for long-term storage. These commodities also respond to intermittent warming. However, the use of intermittent warming for these commodities is not warranted unless some advantage can be shown over the more simple prestorage conditioning.

Apples susceptible to a disorder known as "superficial scald" can have the injury controlled when the fruit are subjected to brief intermittent warming treatments before storage at 3°C.[46]

Cranberries stored constantly at about 0°C had 26% breakdown. Berries removed to 21°C for 1 d of each 4-week storage period had only 11% breakdown. Storage was up to 20 weeks.[47]

Discoloration, in the form of browning of stems and pedicels of grapes, was accelerated by using fluctuating temperatures (5, 3, and 1°C for 6 and 12 h), but there was little difference in discoloration rate between clusters stored at fluctuating temperatures and those stored constantly at 6°C.[48]

Chilling injury of grapefruit stored at 0 and 2°C and removed to 21°C for 1 d after 1 week, and again at the end of 2 weeks of storage, was lessened.[49] Grapefruit warmed to 21°C for 8 h/week were not injured by CI during 4- and 8-week storage periods at 4°C.[50-52]

Intermittent warming also shows potential in the storage of peaches and nectarines. In peaches, control of CI in the form of internal breakdown was obtained by removing the fruit to ambient temperatures (23 to 25°C) for 48 h after 2 and 4 weeks of storage at 0°C; a 6-week storage life was thus obtained.[53] Peaches and nectarines shifted to 0°C after storage for 1 or 2 weeks at 5°C developed less breakdown than those shifted from 0 to 5°C.[54] Fruit that were warmed for 2 d at about 18.3°C during low-temperature storage developed less breakdown than comparable fruit that were not warmed (Table 4).

Shelf life of plums can be extended as much as 50 to 75% if fruit stored at 0°C or slightly less are warmed to 18°C for 2 d between the 15th and 20th days in storage. The storage is thereby extended for another 15 or 20 d, and internal browning, a form of CI, is prevented.[55]

Moving potato tubers to 16°C for 1 week after every 3rd week during storage at 0°C eliminated all forms of CI, except skin browning.[56]

In sweet potatoes, the chilling effect brought about by 2 weeks of storage at 7.5°C was

TABLE 4
Internal Appearance Rating of Peaches and Nectarines After 4 Weeks of Storage Plus Ripening[a]

Starting storage temperature (°C)	Shifting-warming schedule[b]	Not shifted		Shifted	
		Not warmed	Warmed	Not warmed	Warmed
0	After 1 week	77 cd[c]	95 ab	37 e	82 bcd
	After 2 weeks		98 a	46 e	94 ab
5	After 1 week	48 e	80 bcd	72 cd	94 ab
	After 2 weeks		86 abc	66 d	94 ab

[a]   Rating of 100 indicates excellent appearance (no breakdown); fruit rated 70 or less probably unmarketable because of breakdown.
[b]   Shifting = transfer of fruit from 0 to 5°C or from 5 to 0°C. Warming treatment = 2 d at 18.3°C.
[c]   Mean separation by Duncan's multiple range test, 5% level.

From Anderson, R. E., *J. Am. Soc. Hortic. Sci.,* 104, 459, 1979.

reversed by transfer to 15°C. However, after 4 and 6 weeks at the chilling temperature, removal to 15°C did not reverse the CI.[57]

Cucumbers and peppers stored at 2.5°C and intermittently warmed to 20°C for 24 h at 3-d intervals did not develop CI.[58] In other tests with eggplant, snap bean, cucumber, and pepper, the fluctuating temperatures (5, 3, 2, or 1°C with the standard at 6°C and one cycle every 6, 12, or 24 h) promoted the amelioration of CI of snap bean and eggplant, but not cucumber and pepper.[59] Similar tests show that a steady low temperature retains the quality of asparagus, broad bean, shallot, and some mushrooms, whereas fluctuating temperatures reduced quality; edible podded pea was less affected by temperature fluctuations.[59]

CI was alleviated in okra by exposing the pods to 0 or 5°C for 4 to 8 d with intermittent warming before transfer to a steady 20°C.[60]

V. OUTLOOK

Temperature manipulation to alleviate CI of fruits and vegetables presents several opportunities. Prestorage conditioning, as well as conditioning by intermittent warming, also shows promise for extending the storage life of commodities. Such conditioning has the possibility of expanding the list of fruits and vegetables compatible with one another in a mixed-load situation. One important potential is the use of prestorage conditioning as a tool to permit commodities to tolerate the cold treatment as a quarantine procedure.

Warning should be given that recommended optimum storage temperatures and conditions for fresh fruits and vegetables may change as new cultivars are developed and as more information is gained on storage requirements of these commodities.

REFERENCES

1. **Watada, A. E., Morris, L. L., Couey, H. M., Bramlage, W. J., Wolk, W. D., Herner, R. C., and Wang, G. Y.,** Chilling injury of horticultural crops, *HortScience,* 17 (2),160, 1982.
2. **Lyons, J. M.,** Chilling injury in plants, *Annu. Rev. Plant Physiol.,* 24, 445, 1973.
3. **Hardenburg, R. E.,** Storage recommendations, shelf life, and respiration rates for horticultural crops in *Handbook of Transportation and Marketing in Agriculture,* Vol. 1, Finney, E. E., Jr., Ed., CRC Press, Boca Raton, FL, 1981, 261.

4. **Hardenburg, R. E., Watada, A. E., and Wang, C. Y.,** The commercial storage of fruits, vegetables, and florist and nursery stocks, *U.S. Dep. Agric. Agric. Handb.,* 66, 1986.

5. **The Refrigeration Research Foundation,** *Commodity Storage Manual, Refrigeration Research Foundation,* Bethesda, MD, 1983.

6. **Nicholas, C. J.,** Export handbook for U.S. agricultural products, *U.S. Dep. Agric. Agric. Handb.,* 593, 1985.

7. **Grierson, W., Bennett, A. H., and Bowman, E. K.,** Forced-air precooling of citrus fruit on a moving conveyer, *U.S. Dep. Agric. Agric. Res. Serv. Rep.,* 52-40, 1970.

8. **Redit, W. H.,** Protection of rail shipments of fruits and vegetables, *U.S. Dep. Agric. Agric. Handb.,* 195, 1969.

9. **Kosiyachinda, S. and Young, R. E.,** Chilling sensitivity of avocado fruit at different stages of the respiratory climacteric, *J. Am. Soc. Hortic. Sci.* 101 (6), 665, 1976.

10. **Grierson, W. and Hatton, T. T.,** Factors involved in storage of citrus fruit: a new evaluation, *Proc. Int. Soc. Citric.* 1, 227, 1977.

11. **Lipton, W. J., Gaffney, J. J., Band, L. A. M., Webster, B. D., Cook, R. J., Papendick, R. I., van den Berg, L., Lentz, C. P., Grierson, W., and Wardowski, W. F.,** Relative humidity — physical realities and horticultural implications, *HortScience,* 13 (5), 550, 1978.

12. **Morris, L. L. and Platenius, H.,** Low temperature injury to certain vegetables, *Proc. Am. Soc. Hortic. Sci.,* 36, 609, 1938.

13. **McColloch, L. P.,** Chilling injury and alternaria rot of bell peppers, *U.S. Dep. Agric. Mark. Res. Rep.,* 536, 1962.

14. **Pantastico, E. B., Soule, J., and Grierson, W.,** Chilling injury in tropical and subtropical fruits. II. Limes and grapefruit, *Proc. Trop. Reg. Am. Soc. Hortic. Sci.,* 12, 171, 1968.

15. U.S. Department of Agriculture Plant Protection and Quarantine Treatment Manual, Sect. T107, U.S. Department of Agriculture, Washington, DC, 1987.

16. **Lipton, W. J. and Harvey, J. M.,** Compatibility of fruits and vegetables during transport in mixed loads, *U.S. Dep. Agric. Mark. Res. Rep.,* 1070, 1977.

17. **Artschwager, E.,** Wound periderm formation in the potato as affected by temperature and humidity, *J. Agric. Res.,* 35, 995, 1927.

18. **Kushman, L. J.,** Effect of injury and relative humidity during curing on weight and volume loss of sweet potatoes during curing and storage, *HortScience,* 10 (3), 275, 1975.

19. **Cooley, J. S., Kushman, T. J., and Smart, H. J.,** Effect of temperature and duration of storage on quality of stored sweet potatoes, *Econ. Bot.,* 8, 21, 1954.

20. **Kushman, L. J. and Wright, F. S.,** Sweet potato storage, *U.S. Dep. Agric. Agric. Handb.,* 358, 1969.

21. **Picha, D. H.,** Chilling injury and low temperature sugar changes in sweet potato roots, *HortScience,* 19 (3), 592, Abstr., 1984.

22. **Smoot, J. J., Houck, L. G., and Johnson, H. B.,** Market diseases of citrus and other subtropical fruits, *U.S. Dep. Agric. Agric. Handb.,* 398, 1983.

23. **Eaks, I. L.,** Effect of temperature and holding periods on physical and chemical characteristics of lemon fruit, *J. Food Sci.,* 26, 593, 1961.

24. **Hatton, T. T. and Cubbedge, R. H.,** Conditioning Florida grapefruit to reduce chilling injury during low-temperature storage, *J. Am. Soc. Hortic. Sci.,* 107 (1), 57, 1982.

25. **Hatton, T. T. and Cubbedge, R. H.,** Reducing chilling injury in grapefruit by prestorage conditioning, *U.S. Dep. Agric. Agric. Res. Serv. Adv. Agric. Technol.,* ATT-S-25, 1982.

26. **Hatton, T. T. and Cubbedge, R. H.,** Preferred temperature for prestorage conditioning of 'Marsh' grapefruit to prevent chilling injury at low temperatures, *HortScience,* 18 (5), 721, 1983.

27. **Hatton, T. T. and Cubbedge, R. H.,** Quarantine research to disinfest Florida citrus fruit of the Caribbean fruit fly, *Citrus Ind. Mag.* 67, 2, 1986.

28. **Ismail, M. A., Hatton, T. T., Dezman, D. J., and Miller, W. R.,** In transit cold treatment of Florida grapefruit shipped to Japan in refrigerated van containers: problems and recommendations, *Proc. Fla. State Hortic. Soc.,* 99, 117, 1986.

29. **Purvis, A. C.,** Influence of canopy depth on susceptibility of 'Marsh' grapefruit to chilling injury, *HortScience,* 15 (6), 731, 1980.

30. **Wardowski, W. F., Albrigo, L. G., Grierson, W., Barmore, C. R., and Wheaton, T. A.,** Chilling injury and decay of grapefruit as affected by thiabendazole, benomyl, and CO_2, *HortScience,* 10 (4), 381, 1975.

31. **Schiffman-Nadel, M., Chalutz, E., Waks, J., and Dagan, M.,** Reduction of chilling injury in grapefruit by thiabendazole and benomyl during long-term storage, *J. Am. Soc. Hortic Sci.,* 100 (3), 270, 1975.

32. **Chalutz, E., Waks, J., and Schiffman-Nadel, M.,** Reducing susceptibility of grapefruit to chilling injury during cold treatment, *HortScience,* 20 (2), 226, 1985.

33. **Purvis, A. C.,** Relationship between chilling injury of grapefruit and moisture loss during storage: amelioration of polyethylene shrink film, *J. Am. Soc. Hortic. Sci.,* 110 (3), 385, 1985.

34. **McDonald, R. E., Hatton, T. T., and Cubbedge, R. H.,** Chilling injury and decay of lemons as affected by ethylene, low temperature, and optimal storage, *HortScience,* 20 (1), 92, 1985.

35. **McDonald, R. E.,** Effects of vegetable oils, CO_2, and film wrapping on chilling injury and decay of lemons, *HortScience,* 21 (3), 476, 1986.

36. **King, A. I.,** Effect of conditioning on chilling injury in lemon and bell pepper fruits, *HortScience,* 21 (3; Abstr.) 862, 1986.

37. **Spalding, D. H. and Reeder, W. F.,** Conditioning 'Tahiti' limes to reduce chilling injury, *Proc. Fla. State Hortic. Soc.,* 96, 231, 1983.

38. **Pantastico, E. B., Grierson, W., and Soule, J.,** Chilling injury in tropical fruits. I. Bananas (*Musa paradisiaca* var. Sapientum cv. Lacatan), *Proc. Trop. Reg. Am. Soc. Hortic. Sci.,* 11, 82, 1967.

39. **Chen, N. and Paull, R. E.,** Development and prevention of chilling injury in papaya fruit, *J. Am. Soc. Hortic. Sci.,* 111 (4), 639, 1986.

40. **Hirose, T.,** Effects of pre- and interposed warming on chilling injury, respiratory rate and membrane permeability of cucumber fruits during cold storage, *J. Jpn. Soc. Hortic. Sci.,* 53, 439, 1985.

41. **Nakamura, R., Inaba, A., and Ito, T.,** Effect of cultivating conditions and postharvest stepwise cooling on the chilling sensitivity of eggplant and cucumber fruits, *Sci. Rep. Fac. Agric. Okayama Univ.,* 66, 19, 1985.

42. **Picha, D. H.,** Postharvest fruit conditioning reduces chilling injury in watermelons, *HortScience,* 21 (6), 1407, 1986.

43. **Abe, K. and Chachin, K.,** Influence of conditioning on the occurrence of chilling injury and the changes of surface structure of eggplant fruits, *J. Jpn. Soc. Hortic Sci.,* 54, 247, 1985.

44. **Thompson, B. D.,** Chilling injury and physical damage of peppers, *Proc. Fla. State Hortic. Soc.,* 91, 144, 1978.

45. **Saltveit, M. E., Jr. and Cabresa, M.,** Tomato fruit temperature before chilling influences ripening after chilling, *HortScience,* 22 (3), 452, 1987.

46. **Kidd, F. and West, C.,** The cause and control of superficial scald of apples, *G. B. Dep. Sci. Ins. Res. Food Invest. Board Spec. Rep., 1934,* 111, 1935.

47. **Hruscha, H. W.,** Physiological breakdown in cranberries — inhibition by intermittent warming during cold storage, *Plant Dis. Rep.,* 54, 219, 1970.

48. **Ito, T. and Nakamura, R.,** Fluctuating temperature tolerance of fresh fruits and vegetables, *J. Jpn. Soc. Hortic. Sci.,* 54 (2; Abstr.) 257, 1985.

49. **Brooks, C. and McColloch, L. P.,** Some storage diseases of grapefruit exported from Florida, *J. Agric. Res.,* 52, 319, 1936.

50. **Davis, P. L.,** Intermittent warming of grapefruit to avoid rind injury during storage, *Proc. Fla. State Hortic. Soc.,* 86, 280, 1973.

51. **Davis, P. L. and Hofmann, R. C.,** Reduction of chilling injury of citrus fruits in cold storage by intermittent warming, *J. Food Sci.,* 38, 871, 1973.

52. **Hatton, T. T., Davis, P. L., Cubbedge, R. H., and Munroe, K. A.,** Temperature management and carbon dioxide treatments that reduce chilling injury in grapefruit stored at low temperatures, *Proc. Int. Soc. Citric.,* 2, 728, 1981.

53. **Ben-Arie, R., Lavee, S., and Guelfat-Reich, S.,** Control of woolly breakdown in 'Elberta' peaches in cold storage by intermittent exposure to room temperature, *J. Am. Soc. Hortic. Sci.,* 95 (6), 801, 1970.

54. **Anderson, R. E.,** The influence of storage temperature and warming during storage on peach and nectarine fruit quality, *J. Am. Soc. Hortic. Sci.,* 104 (4), 459, 1979.

55. **Smith W. H.,** Control of low-temperature injury in the Victoria plum, *Nature (London),* 159, 541, 1947.

56. **Hruscha, H. W., Smith, W. L., and Baker, J. E.,** Reducing chilling of potatoes by intermittent warming, *Am. Potato J.,* 46, 38, 1969.

57. **Lieberman, M., Craft, C. C., and Wilcox, M. S.** Effect of chilling on the chlorogenic acid and ascorbic acid content of Puerto Rico sweet potatoes, *Proc. Am. Soc. Hortic Sci.,* 74, 642, 1959.

58. **Wang, C. Y. and Baker, J. E.,** Effects of two free radical scavengers and intermittent warming on chilling injury and polar lipid composition of cucumber and sweet pepper fruits, *Plant Cell Physiol.,* 20 (1), 243, 1979.

59. **Ito, T. and Nakamura, R.,** The effect of fluctuating temperature on chilling injury of several kinds of vegetables, *J. Jpn. Soc. Hortic. Sci.,* 53 (2), 202, 1948.

60. **Ilker, Y. and Morris, L. L.,** Alleviation of chilling injury of okra, *HortScience,* 10 (3; Abstr.), 324, 1975.

Chapter 18

ALLEVIATION OF CHILLING INJURY OF HORTICULTURAL CROPS

Chien Yi Wang

TABLE OF CONTENTS

I. INTRODUCTION

Many germinating seeds and seedlings of tropical and subtropical origin are sensitive to chilling temperatures. Injury induced by chilling temperatures shortens the growing season and restricts the geographic distribution of these crops. Alleviation of chilling injury will not only lengthen the growing season and allow a wider distribution, but will also reduce erratic germination and prevent yield reduction during cool seasons.

Refrigerated storage is considered to be the most effective method for preserving the quality of harvested horticultural products because low temperature retards aging, respiration, ripening, decay, and other undesirable metabolic changes.[1] Other postharvest techniques, such as controlled atmosphere storage, waxing, film packaging, chemical treatments, and irradiation, cannot substitute for proper refrigeration in retarding spoilage and are thus considered to be supplements to refrigeration. Unfortunately, for crops susceptible to chilling injury, low-temperature storage is often more detrimental than beneficial. Chilling injury results in a loss of quality manifested as pitting, discoloration, water soaking, internal breakdown, uneven or incomplete ripening, off-flavor, and, perhaps of greatest importance, a weakening of the tissues which renders the commodity very susceptible to pathogenic decay. Thus, the advantage of refrigeration cannot be used for the preservation of chilling-sensitive crops. On the other hand, these crops deteriorate rapidly and have a short storage life when not refrigerated. As a consequence, tremendous losses occur after harvest. Therefore, it is important to recognize the magnitude and economic impact of chilling injury and the need to search for ways to alleviate this problem.

II. GENETIC MODIFICATION

One of the most important factors affecting the sensitivity of plants to chilling injury is the genotype. The genetic makeup of the species or cultivar governs the responses of the plant to chilling stress. Great genetic variation in plants provides opportunities for modifying chilling sensitivity by breeding.[2] Species from high altitudes or latitudes are usually less sensitive to chilling injury than those from low altitudes or latitudes. Identifying and isolating the genes responsible for chilling resistance and the subsequent transfer of these genes into susceptible species can provide us with a promising method for modifying chilling injury. Many researchers have expressed optimism in genetic control and have concurred that breeding for chilling-resistant cultivars may be the best approach in solving the chilling injury problem.[3-8]

Cultivar differences in chilling resistance have been observed in tomato,[9] *Passiflora* spp.,[10] cotton,[11] corn,[12] rice,[13] sorghum,[14] cucumber,[15] summer squash,[16] snap bean,[17,18] soybean,[19] avocado,[20] banana,[21] and *Aglaonema*.[22] The establishment of chilling-resistant hybrids is possible by selecting resistant plants as parents. Some of the genes regulating the fruit characteristics relating to water loss may also be important in determining their susceptibility to chilling injury.[16] The possibility of breeding for chilling tolerance has been shown in tomato,[23,26-30] cotton,[24] corn,[25,33] cucumber,[31] and sorghum.[32]

Several plant breeders have attempted to improve the germination of seeds at low temperatures in order to plant the crops earlier in the season and to prevent delay of emergence following seeding in soils cooler than 15°C.[34-41] There has been some success in obtaining progeny with a higher germinating ability at low temperatures. However, the ability of seeds to germinate at suboptimal temperatures and seedling tolerance to chilling temperatures are not consistently correlated.[35] It appears that a polygenic system with important additive effects is involved in the inheritance of low-temperature sprouting ability, and the maternal parent plays an important role in determining the sprouting behavior of F_1 hybrids in tomato.[36-39]

It is difficult to breed directly for chilling resistance because chilling sensitivity appears to be controlled by a number of genes which may be scattered throughout the genome, making it difficult to transfer them all at once (see also Chapter 6). Another problem with breeding is that many domestic crops do not have a compatible line which can serve as a source of genes for chilling resistance. Wild types, which tend to have a large degree of genetic variability, need to be used to provide the necessary genes for chilling resistance.[27-30] Many crosses may be required in order to acquire the desirable chilling-resistant trait and still retain high organoleptic quality for commercial use.

Another possible way of reducing chilling injury is the transfer of genes for resistance into sensitive species by genetic engineering methods. The first step in these techniques is the identification of the genes which are responsible for chilling tolerance. Structural information from cDNA clones of mRNA coding for the chilling-resistant genes, or from classical genetic mapping techniques, is required to narrow down the genome and the region on the chromosome which carries the resistant trait. Immunological screening or antisense RNA technology may be applied for characterization of the genes. The success of this method would also depend upon proper transfer of the genes and the successful expression of the transferred genes. Theoretically, this goal is attainable. However, each step of these complicated procedures requires extensive research, especially if a number of scattered genes are involved in regulating the chilling sensitivity. All of these genes would have to be located, identified, and isolated before further steps could be taken. Nevertheless, although many obstacles must be overcome, this approach seems to be very promising, and the results would be useful in the market place.

III. CHEMICAL TREATMENT

A number of chemicals have been shown to be effective in reducing chilling injury. The effect of calcium on chilling injury has been investigated extensively. Chaplin and Scott[42] reported that the capacity of avocado fruit to withstand chilling injury was positively correlated to the endogenous calcium levels. They found that the proximal end of the fruit mesocarp had the highest level of calcium and the lowest sensitivity to chilling injury, whereas the opposite was true for the distal end. They also showed that vacuum infiltration with 1 to 7.5% calcium into avocados before low-temperature storage reduced the chilling-induced darkening of the vascular bundles in the fruit. In some temperate zone crops, such as certain varieties of apples, fruits with a higher concentration of calcium are less susceptible to low-temperature breakdown during storage than those with a lower concentration of calcium.[43] Postharvest application of calcium chloride by dipping or by vacuum infiltration reduced the incidence of low-temperature breakdown in apples[44,45] and peaches.[46] Treatment with calcium and potassium salts also alleviated chilling injury in okra.[47]

In some fruits, calcium content in the fruit is not closely related to chilling sensitivity. Purvis[48] reported that the variation in the susceptibility of grapefruit to chilling injury did not appear to be associated with the endogenous calcium content of the flavedo tissue. He found that although fruit harvested from the exterior canopy positions pitted more severely during storage at 5°C than did fruit harvested from the interior canopy of the same tree, the flavedo tissue of all the fruit had similar calcium contents. Dipping or vacuum infiltration with 1% $CaCl_2$ did not ameliorate chilling injury symptoms in tomatoes.[49] However, at higher concentrations, calcium uptake affected ripening rate and reduced chilling injury of tomatoes.[50] There are other examples of calcium not being beneficial in reducing chilling injury. Chen and Paull[51] have shown that calcium treatment led to increased chilling injury of papaya fruit. Thus, calcium may reduce, have no effect on, or increase chilling injury, depending upon the type of fruit and the concentration used.

Treatment with chemicals which act as free radical scavengers or oxygen-quenching agents has been shown to reduce chilling injury.[52-55] The application of ethoxyquin or sodium benzoate to cucumbers and sweet peppers resulted in the maintenance of a relatively high degree of unsaturation of fatty acids in polar lipids and effectively reduced the severity of chilling injury in these crops.[52] Ethoxyquin and sodium benzoate are known to be antioxidants and free radical scavengers.[56] Ethoxyquin has also been shown to be an effective inhibitor of superficial scald in apples and pears.[1,57,58] Other antioxidants, such as butylated hydroxytoluene and butylated hydroxyanisole have also been used as free radical scavengers to prevent oxidation of unsaturated fatty acids in membrane lipids and to reduce low temperature-induced disorders.[59-61]

Treatment with the triazole uniconazole prevented a chilling-induced loss of phospholipid, retarded the accumulation of free fatty acids, and alleviated symptoms of chilling injury such as loss of turgor, chlorosis, and increased solute leakage in tomato seedlings.[53] All of these protections against chilling injury were attributed to the free radical scavenging property of the chemical. Biological membranes containing abundant unsaturated lipids are often bathed in an oxygen-rich, metal containing fluid and are susceptible to peroxidative attacks involving free radicals.[56] Many free radicals are highly reactive chemically and can induce oxidative breakdown of the double bonds in the fatty acids of membrane lipids. Such oxidation of membranes by free radicals may lead to fatty acid peroxidation[62] and phospholipid deesterification.[63]. This process can be inhibited by scavengers of free radicals, which can protect the membranes from oxidative damage and contribute to chilling tolerance.

Chemical modification of lipids in chilling-sensitive species has led to changes in plant responses to chilling. Treatment of tomato seedlings with ethanolamine induced an alteration of phospholipid composition by increasing the relative amount of C18 acyl chain in phosphatidylethanolamine and phosphatidylcholine at the expense of C16:0.[64] The treatment also reduced the symptoms of chilling injury. A microscopic examination of chilled tomato seedlings after ethanolamine treatment revealed that the cotyledonary tissues maintained better cell turgor and cell shape and showed less chilling damage in the cytoplasm, mitochondria, and cell walls.[65] In contrast, treatment of cotton seedlings with 4-chloro-5-(dimethylamino)-2-phenyl-3 (2H)-pyridazinone, a pyridazinone analogue found to inhibit the production of linolenic acids,[66] increased the ratio of linoleic acid to linolenic acid in polar lipids in root tips and resulted in an increase in chilling sensitivity.[67,68]

Treatment with choline chloride or ethanolamine also inhibited the chilling-induced photooxidative damage to leaves of cucumber seedlings.[69] The effectiveness of these amino alcohols in reducing chilling injury was attributed to their enhancement of total phospholipid content, particularly phosphatidylcholine and phosphatidylethanolamine. It was suggested that choline chloride and ethanolamine may act as precursors for phosphatidylcholine and phosphatidylethanolamine, respectively. An increase in total phospholipid content and an accumulation of phosphatidylcholine and phosphatidylethanolamine have been observed after low-temperature hardening.[70-79] These changes in the membrane phospholipid content may be part of a defense mechanism against chilling injury.

A group of membrane stabilizers and glycerol homologues were reported to alleviate chilling injury of sorghum seedlings by Tajima and Shimizu.[80] They proposed that these stabilizers reduced chilling injury by protecting biological membranes from damage induced by environmental stresses, by inhibiting the increase in membrane permeability, and by controlling the fluidity of cellular membranes, and that the glycerol homologues prevented chilling injury by inhibiting low-temperature-induced denaturation of proteins.

Jones et al.[81] tested several chemical agents for their ability to prevent chilling injury and found that postharvest treatments with dimethylpolysiloxane, safflower oil, and mineral oil were effective in preventing chilling-induced underpeel discoloration of bananas at 9°C. Coating grapefruit with vegetable oil or vegetable oil-water emulsions before storage at 3°C has also been shown to delay and reduce symptoms of chilling injury.[82] How these natural oils reduce chilling injury is not clear, but it is probably due to their effect of reducing water loss from these commodities. Applications of antitranspirants, which also reduce water loss, prevented chilling-induced wilting of cotton seedlings.[83]

Application of a thin coat of lanolin on chilling-sensitive seeds, such as soybean and cotton, provided protection against imbibitional chilling injury.[84] The germination of soybean seeds at chilling temperatures caused membrane-related lesions which led to increased leakage of solutes[85] and decreased cytochrome oxidase-mediated respiration.[86] This injury was believed to result from rapid imbibition of cold water.[87-89] The use of lanolin greatly reduced the rate of imbibition and thereby significantly ameliorated chilling injury.

The incidence of chilling injury in grapefruit has been reported to be reduced by treatments with thiabendazole or benomyl.[90-92] Peaches and nectarines treated repeatedly with benomyl have also been shown to retain higher acidity and sustain less internal breakdown than untreated fruit after long term storage at low temperature.[93] Thiabendazole and benomyl are primarily used as fungicides to reduce postharvest diseases of fruits. Their protection against chilling injury is considered to be an additional beneficial effect and is commercially valuable. The cold treatment (0°C for 10 d or 2.2°C for 16 d) required to eliminate the Mediterranean fruit fly frequently caused chilling injury to citrus fruit, particularly in certain seasons when the fruit are more susceptible because of the growing condition. The incorporation of the fungicide thiabendazole in the wax coating applied to the fruit greatly reduced the incidence of chilling injury.[94] The combination of a thiabendazole treatment with delayed cooling was demonstrated to be even more effective in reducing the susceptibility of grapefruit to chilling injury during cold treatment.[94] These treatments are very useful and can allow cold treatment of grapefruit to be practiced without adversely affecting fruit quality.

A synthetic plant growth regulator, paclobutrazol [(2RS,3RS)-1-(4-chlorophenyl)-4,4-dimethyl-2-(1,2,4-triazol-1-yl)-pentan-3-ol], has been shown to increase the tolerance of plants to chilling stress in seedlings of cucumber and zucchini squash[95,96] and snap beans.[97] Paclobutrazol is a broad-spectrum plant growth retardant.[98-105] It acts as an inhibitor of gibberellic acid (GA) synthesis through its inhibition of the enzyme kaurene oxidase.[106] Treatment of plants with paclobutrazol has been shown to have profound morphological, physiological, and biochemical effects. Morphological effects include a reduction in internode elongation and leaf expansion, with a concomitant thickening of the leaves.[95,100-105] Physiological effects include shifting assimilate partitioning from leaves to roots, enhancement of carbohydrate levels in all parts of apple seedlings, augmentation of chlorophyll, soluble protein, and mineral element concentration in leaf tissue, and increase in root respiration in apple.[107,108] Biochemical effects of paclobutrazol treatment include a delay in the onset of dark-induced senescence which correlated with altered levels or activities of several key metabolic enzymes in soybean leaves.[109,110]

The increase in chilling tolerance by paclobutrazol treatment could be related to its morphological and metabolic effects. Seedlings of cucumber and zucchini squash treated with paclobutrazol exhibited shorter leaf and stem length, thicker leaves, and higher chlorophyll content.[95] They also suffered less wilting and desiccation following chilling exposure. Although the paclobutrazol treatment did not significantly change the polar lipid fatty acid composition or free sterol content, it reduced chilling-induced degradation of leaf membrane lipids.[96] Maintenance of membrane integrity is important in preventing chilling injury.[111,112] The prevention of membrane lipid loss by paclobutrazol is similar to the effect of uniconazole treatment.[53] An application of uniconazole was shown to increase total lipid soluble antioxidants, α-tocopherol, and ascorbic acid levels in leaves. These triazole compounds may protect the membrane components from oxidative damage and lipid peroxidation during chilling by increasing the defense mechanism of the tissue against free radical attack. Since paclobutrazol inhibits GA biosynthesis,[98,99,106] it is possible that it may also enhance chilling resistance through modification of the hormonal balance. Changes in the hormonal balance have been shown to alter the capacity of plants to withstand chilling stress (see Section IV).

Triadimefon, another triazole compound, has also been reported to protect barley and bean plants from chilling injury.[113,114] Chilling treatment induced an increase in electrolyte leakage and a decrease in chlorophyll level in leaves of bean seedlings. All of these symptoms of chilling injury were either delayed or prevented by root application of triadimeforn. Fletcher and Hofstra[113] proposed that triadimefon may stimulate the production of abscisic acid (ABA) and thereby confer protection against various kinds of stresses. Asare-Boamah et al.[115] later showed that triadimefon induced a transient rise of ABA levels in bean plants. Triadimefon has also been reported to be an inhibitor of GA biosynthesis in higher plants.[116] It is likely that triadimefon shifts the balance of important plant hormones, particularly the levels of ABA and GA, toward a ratio which favors the increase in the capacity of tissues to withstand stresses.

IV. HORMONAL REGULATION

Studies have shown that resistance to low-temperature stress, including chilling and freezing injury, is related to the balance of growth inhibitors and growth promotors.[117-121] The interaction of all biologically active molecules is important in determining the sensitivity of tissues to low-temperature stress.[122] Growth regulators often affect many metabolic processes which, in turn, may also influence the susceptibility of crops to chilling injury. Observations of the seasonal variation of the sensitivity of grapefruit to chilling injury have led Grierson and co-workers[123-125] to postulate that the susceptibility of grapefruit to chilling injury is related to growth regulator levels at the time of harvest. A change of hormone balance, induced by environmental stresses, may thus affect the metabolic balance which, in turn, modifies postharvest resistance to chilling injury.

Applications of benzyladenine, GA and 2,4,-dichlorophenoxyacetic acid significantly altered the susceptibility of grapefruit to chilling injury.[124] Treatment with auxin greatly reduced chilling-induced leaf wilting of sorghum seedlings.[80] It was postulated that auxin may increase the 3',5'-cyclic AMP content and maintain membrane stability, thereby diminishing membrane impairment by chilling temperatures.[80,126] An increase in resistance to low temperatures has been associated with an increase in the ABA/GA ratio[127] or a decrease in GA content.[128-130] However, some studies have also shown that GA applications increased the capacity of plants to tolerate low-temperature stress.[131-133]

Ethylene treatment reduces chilling injury in some crops, but increases chilling injury in others. Ethylene treatment may also have no effect on the development of chilling injury in certain crops. For example, the gassing of mature-green tomatoes with ethylene before or after storage at a chilling temperature did not affect the development of chilling injury symptoms.[49,134] Likewise, treatment of cucumber seedlings with ethylene before, during, or after chilling exposure also did not affect chilling injury.[135] The exposure of sweet potatoes to ethylene during the postchilling induction period did not reduce the incidence of hard core, but significantly reduced its severity.[136] The exposure of honeydew muskmelons to 1000 ppm of ethylene at 20°C for 24 h before storage at 2.5°C for 2.5 weeks significantly reduced the incidence of chilling injury.[137-139] However, in avocados, very slight chilling injury was observed in fruit kept in air at 6°C, while very severe chilling injury was found in fruit treated with ethylene at both 6 and 9°C.[140,141] The exposure of avocados to ethylene during storage at 1.5°C also increased severity of chilling injury.[142] Similarly, the use of ethylene to degreen lemons at 1°C markedly increased chilling injury and decay compared to degreening without ethylene.[143] In early-season grapefruit, degreening with ethylene before cold storage reduced chilling injury, whereas ethylene tended to increase chilling injury in midseason grapefruit.[123] Ethylene may alter the maturity and physiology of fruits and indirectly change the sensitivity of these crops to chilling. Whether the effect is beneficial or detrimental depends upon the stage of development and the kind of fruit involved.

The relationship between polyamines and chilling injury has been receiving increasing attention. The apparent involvement of membrane damage in chilling injury[121,144] and the ability of polyamines to stabilize membranes[145,146] have generated the hypothesis that polyamines may play a role in reducing chilling injury.[147,148] Correlations between increased resistance to chilling injury and increased polyamine levels have been reported in several plant species.[149-152]

Polyamines are a group of low-molecular-weight organic cations. Although it is not clear whether polyamines can be classified as plant hormones, there is considerable evidence that polyamines actively regulate plant growth, development, and senescence.[146,153-157] It has also been suggested that polyamines can act as second messengers and mediate the effects of endogenous plant hormones.[156] The polyamine titer fluctuates in response to light, temperature, and stress. Polyamine levels in plant tissues change in response to various kinds of stresses,[158] including water stress,[159,160] acid stress, [161,162] osmotic stress,[163] mineral deficiency,[164,165] and chilling stress.[149-152,166] Most of these stresses result in an increase in the putrescine level. It is not known whether this increase serves as a protective mechanism. However, polyamines have been shown to affect DNA and RNA synthesis and degradation; to regulate the rate of transcription; to inhibit activities of protease, ribonuclease, and peroxidase; to stabilize ribosomal structures; and to maintain membrane integrity.[167-171] Application of polyamines prevented lysis of isolated protoplasts and reduced the breakdown of macromolecules.[172] Exogenous polyamines have been shown to delay senescence in the excised leaves or leaf segments of a number of monocots and dicots.[168,173,174] Application of polyamines also alleviated symptoms of stress caused by ozone.[175]

Treatments which elevated polyamine levels were shown to reduce chilling injury. The use of temperature preconditioning or low-oxygen storage significantly increased spermidine and spermine levels and retarded the development of chilling injury symptoms in zucchini squash during storage at 2.5°C.[147,148] Polyamine levels also increased markedly upon chilling in bean plants[150] and cucumber seedlings.[176] Spermidine and spermine in citrus leaves were found to increase one- to fourfold after 1 week of exposure to cold-hardening temperatures (15.6°C day and 4.4°C night).[151] These increases were positively correlated with the ability of the citrus plants to withstand low-temperature stress. It has been suggested that an increase in polyamines during stress may help to maintain cell membrane thermostability by minimizing changes in fluidity and solute leakage.[151,177] Direct treatment of zucchini squash with spermine after harvest, but before storage at chilling temperature, increased the endogenous polyamine levels and reduced chilling injury.[149] These results indicate that polyamines may play a role in reducing chilling injury.

Since polyamines are strong cations, they may interact with the anionic components of the membrane, such as phospholipids.[178-180] This interaction could serve to stabilize the lipid bilayer surface and thus retard membrane deterioration. Polyamines have also been shown to have free radical scavenging properties[181] and to protect microsomal membranes from peroxidation.[182]

The exposure of zucchini squash to chilling temperature resulted in an increase in chloroform-soluble fluorescent products,[177] which are by-products of lipid peroxidation.[183] In the preconditioning treatment, a concomitant reduction of fluorescent products was associated with high polyamine levels in the treated samples. Since the accumulation of chloroform soluble-fluorescent products has been shown to be correlated with oxidative damage,[183] these results may indicate that polyamines can act to prevent chilling injury in squash by a mechanism which involves protection of membrane lipids from peroxidation. Thus, polyamines may inhibit the development of chilling injury symptoms by interactions with membranes, as well as through their antioxidant acitivity.

Polyamines and ethylene have opposite effects in plants. Ethylene tends to hasten senescence, whereas polyamines show antisenescent activity. It is interesting to note that both polyamine bisosynthesis and ethylene biosynthesis use *S*-adenosylmethionine as a common precursor[170,184] and produce 5'-methylthioadenosine as a common by-product.[146,185,186] However, evidence has suggested that the two biosynthetic pathways do not actively compete with each other for the same substrate either during fruit ripening[187] or under chilling stress conditions.[176] Therefore, the reduction of chilling injury by high levels of polyamines is not necessarily associated with low levels of ethylene. The role of polyamines in reducing chilling injury is more likely related to their antioxidant activity and their effect on membrane stability.

The beneficial effect of ABA treatment in reducing chilling injury has been shown in grapefruit,[125] tomato seedlings,[188] cotton seedlings,[189,190] cucumber cotyledons and plants,[191-193] and a red-pigmented cultivar of coleus.[191] Increasing evidence has shown that high endogenous levels of ABA are related to increased chilling tolerance. The protection of rice seedlings and corn leaves against chilling injury by mefluidide was found to be mediated through its effect on ABA levels.[194-197] Mefluidide was capable of triggering an increase in endogenous ABA content in maize leaves when the plants were grown in a nonchilling environment with sufficient water supply.[195] It was suggested that this increase in endogenous ABA before chilling could be an essential step in activating a protecting mechanism against chilling injury during low-temperature exposure.[194]

The mechanism through which ABA reduces chilling injury is not fully understood. It is known that high ABA levels induce stomatal closure[198-201] and reduce water loss.[202,203] Since wilting is one of the major symptoms of chilling in seedlings,[204-206] the reduction of water loss should reduce chilling injury. Exogenous ABA treatment has been employed to induce stomatal closure before exposure to low temperatures and, thus, avoid dehydration and chilling injury in seedlings of cotton, cucumber, and bean.[206-208] However, these studies do not explain the reduction of chilling injury by ABA in tissues, such as tobacco callus,[209] which do not possess stomata.

Markhart and co-workers[210,211] suggested that ABA might directly affect the membrane that limits water flow through the root. The protective effect of ABA during the chilling of soybean seedlings was independent of its effect on stomata, because stomatal resistances were lower and leaf water potentials were higher in the ABA-treated plants than in the nontreated plants when roots were at 10°C and shoots at 25°C.[202] It was suggested that the mechanism of action of ABA in the response of plants to chilling stress involves membrane alterations.[202,212] Rikin et al.[190,213] reported that depolymerization of the microtubular network was involved in the development of chilling injury in cotton seedlings and that ABA decreased chilling injury by stabilizing the microtubular network. ABA has also been reported to suppress chilling-induced ion leakage[214] and to prevent the loss of reduced glutathione and membrane phospholipids.[189]

V. TEMPERATURE MANAGEMENT

A. TEMPERATURE PRECONDITIONING

Plants may be conditioned or hardened by exposure to temperatures slightly above the critical chilling range.[215-217] For example, tomato seedlings conditioned at 12.5°C for 3 h or more were more resistant to subsequent chilling at 1°C than seedlings that were not conditioned.[216] Corn seedlings grown on a 16°C-day and 6°C-night temperature regime were more tolerant to chilling at 0°C than plants grown on a 20°C-day and 15°C-night regime.[218] In a separate study, seedlings of chilling-susceptible maize genotypes conditioned at 10°C for 4 d recovered better after chilling at 5°C for 2 d than those not conditioned.[219] Gradual chilling of germinated Stokes aster achenes also resulted in a significant reduction in damage compared to abrupt chilling.[220] Exposure of sweet peppers to 10°C for 5 or 10 d reduced chilling injury at 0°C.[215] A 7 d exposure of grapefruit to 10 or 15°C prevented or significantly reduced chilling injury during storage at 0° or 1°C.[217,221] Zucchini squash preconditioned at

10 or 15°C for 2 d were found to have less chilling injury during subsequent storage at 2.5 or 5°C.[147,311] Preconditioning papaya fruit for 4 d at 12.5°C before storage for 14 d at 2°C reduced chilling injury.[51] The decrease in chilling sensitivity in papayas with preconditioning treatment was thought to be associated with partial fruit ripening. Allowing papayas to ripen at 24°C has also been shown to decrease chilling injury during subsequent storage at 5°C.[222] The reduction of chilling injury by prestorage temperature conditioning has also been demonstrated in lemons,[143,223] cucumbers,[224] eggplants,[225] tomatoes,[308,309] and watermelons[226] (see also Chapter 17).

The beneficial effects of temperature preconditioning have been attributed to various physiological changes. Increases in the degree of unsaturation of fatty acids in response to hardening or chilling temperatures have been demonstrated in a number of plants, including cotton seedlings,[67] narcissus bulbs,[227] peach bark,[228] sweet peppers and cucumbers,[52] tomatoes, [229] wheat,[230,231] alfalfa roots and leaves,[232,233] and seeds of rape, flax, and sunflower.[234,235] These increases in fatty acid unsaturation during low-temperature conditioning have been proposed to be a result of altered fatty acid desaturase activity and not of preferential bisosynthesis of individual phospholipids.[227,236] Conditioning at 12°C and 85% RH also increased the degree of unsaturation of fatty acids in phospholipids and prevented chilling injury in the leaves of *Phaseolus vulgaris, Cucumis sativus,* and *Gossypium hirsutum.*[237]

Increases in sugars and starch and decreases in RNA, protein, and lipid-soluble phosphate were found in cotton plants during exposure to hardening temperatures of 15°C (day) and 10°C (night).[238] This conditioning also reduced leakage of metabolites and prevented subsequent chilling injury in cotton seedlings at 5°C[239] Similar reduction of electrolyte leakage and chilling injury by low-temperature hardening was also found in *Maranta leuconeura* plants.[240]

B. INTERMITTENT WARMING

Intermittent warming involves the interruption of chilling exposure with brief warm periods. Raising temperatures above the critical point for chilling injury during low-temperature storage can either allow the tissue to recover from a stressed condition or accelerate the degradative processes, depending upon the stage of chilling injury. When it is still at the reversible stage, raising the temperature usually induces higher metabolic activities and allows the tissue to metabolize excess intermediates accumulated during chilling or to replenish any substances which were depleted or were not able to be synthesized during chilling. Warming of the chilled tissues may also repair chilling-induced damage to membranes, organelles, or metabolic pathways.[241] Chilling-induced inhibition of the translocation process in sensitive species is rapidly reversible upon transfer of the plants from chilling to nonchilling temperatures.[242] The effect of chilling on translocation is attributed to a perturbation of membrane function, rather than a blocking of sieve plate pores. Shifting of temperatures from cold to warm, and then from warm to cold, probably induces a rapid readjustment of metabolism that may also include increased synthesis of unsaturated fatty acids,[52,121] thus making tissues more resistant to chilling injury. Intermittent warming has been shown to be effective in reducing chilling injury in the following crops: apples,[243,244] citrus fruit,[245-247] cranberries,[248] cucumbers,[52] nectarines,[93,249] okra,[47] peaches,[93,249-251] plums,[252] potatoes,[253] sweet peppers,[52] sweet potatoes,[254] tomatoes, [255] and zucchini squash[311] (see also Chapter 17). A successful commercial use of intermittent warming was reported for lemon[256]

C. CHILLING INJURY ASSAYS FOR INTERMITTENT WARMING TREATMENT

When chilling injury has progressed to an irreversible stage, exposure of the chilled commodities to warmer temperatures enhances the degradative processes and accelerates the development of the injury symptoms. Therefore, it is important to initiate the intermittent warming treatment before irreversible damage takes place. However, if the warming treatment is applied too early, too frequently, or too long, tissues may become excessively soft and vulnerable to bruise or decay. Thus, the timing and duration of the warming treatment are critical. An early and rapid assay for detecting and quantitatively measuring the chilling responses in plants is needed to help determine the correct time for application of intermittent warming treatments. The assay should accurately reflect the degree of chilling exposure in tissues before the occurrence of irreversible changes.

A number of physiological measurements have been suggested for use as indices of the severity of chilling injury. These measurements include the rate of electrolyte leakage,[10,257,258] the rate of respiration,[259,260] the degrees of unsaturation of membrane lipids,[261] the change in levels of certain metabolites,[262] and colorimetric measurement of internal browning.[263] However, most of these methods of detection are either insensitive to the early effects of chilling or are time consuming, destructive, and inconvenient to use.

One of the earliest quantitative responses to chilling exposure is the alteration of photosynthetic activity.[264,265] A change in photosynthetic metabolism is likely to change the yield of fluorescence emitted from the chlorophyll of photosystem II. Measurement of chlorophyll fluorescence has been shown to be a rapid and nondestructive assay for providing information about the chilling stress response of plants.[218,266-270] More detailed information on chlorophyll fluorescence can be found in Chapter 8 and, therefore, will not be repeated here.

Another method that has been shown to reflect changes caused by chilling stress involves the measurement of refreshed delayed light emission.[271,272] Refreshed delayed light emission is different from chlorophyll fluorescence in that the latter results from the relaxation of a chlorophyll molecule directly excited by light and occurs within a few nanoseconds after excitation, whereas the former results primarily from relaxation of a chlorophyll molecule which has been excited by the back reaction of photosystem II[273] and may be detected for seconds or even hours after light exposure.[271] Since changes in chloroplast membranes should affect refreshed delayed light emission, chilling-induced alterations of those membranes should be reflected in emission patterns over time, or in levels at a specific time. Abbott and Massie[271] showed that the maximum levels of refreshed delayed light emission of chilled cucumber and bell pepper fruit were lower than those of nonchilled fruit. They suggested that the reduction in the amount of refreshed delayed light emission by chilling might be due to an impairment of chloroplast activity rather than a simple loss of chlorophyll. This method provides a nondestructive and rapid indication of chilling injury. However, both measurements of chlorophyll fluorescence and refreshed delayed light emission involve chlorophyll molecules and require the presence of an active photosynthetic system. These methods are suitable for measuring the chilling-induced changes in photosystem II in green tissues, such as seedlings, but are not applicable to use for commodities which do not contain chlorophyll.

The changes of 1-aminocyclopropane-1-carboxylic acid (ACC) levels or ethylene production in tissues have been suggested as an indicator of chilling injury[274-276] (see also Chapter 15). Increases in ACC levels and ethylene production have been shown to be a response of chilling-sensitive tissues to low-temperature stress.[277,278] The accumulation of ACC can be detected as early as 1 h after chilling exposure in leaves of the chilling-sensitive tropical herb *Episcia reptans*.[276] In cucumber skin tissue, an increase in ACC levels was detected at 20°C after 24 h at 2.5°C or 36 h at 5 or 7.5°C.[135] ACC levels increased linearly with increasing length of the chilling exposure. In papayas, fruit stored at a chilling temperature produced more ethylene upon transfer to warmer temperatures than did fruit stored at a nonchilling temperature.[279] Differences in ethylene production between chilled and nonchilled fruit increased with length of storage. In sweet peppers, the length of chilling treatment required to induce increases of ACC and ethylene production was closely correlated with the chilling sensitivity of pods harvested in different seasons.[280] In *Passiflora,* the duration of chilling required for the burst of ethylene production was also closely correlated with the chilling susceptibility of different species.[276] In honeydew melons, ACC levels in the rind from the ground spot were consistently higher than those from the top of melons after chilling, and the susceptibility to chilling injury was also higher at the bottom than at the top of the melons.[281,282] Since tissues from chilling-resistant species and nonchilled tissues from chilling-sensitive species do not contain any significant amount of ACC and ethylene, the endogenous ACC levels and ethylene production stimulated by chilling in the sensitive tissues can be used as an indicator of chilling exposure.

Recently, a technique utilizing pulse nuclear magnetic resonance (NMR) spectroscopy was reported to be useful as a nondestructive and sensitive diagnostic tool for monitoring the response of cells to chilling.[283] Kaku and Iwaya-Inoue[284] showed that the profiles of thermal hysteresis of the T_1 relaxation time of water protons in gloxinia leaves varied with the chilling sensitivity and degree of injury. They further demonstrated that the sum of the T_1 ratio can be considered as a quantitative index for evaluating chilling sensitivity and injury.[283]

All of the above assays measure the extent of chilling exposure. These assays show promise for use in evaluating the degree of chilling injury and will help to determine the correct timing and duration of intermittent warming.

VI. OTHER METHODS

Other techniques which have also been reported to be effective in reducing chilling injury include controlled-atmosphere storage, high relative humidity, film packaging, waxing, and hypobaric storage. Refer to Chapter 16 for more detailed information on these subjects.

The efficiency of controlled atmospheres in ameliorating chilling injury symptoms is dependent upon the commodity, the concentrations of O_2, and CO_2, the timing and duration of the treatment, and the storage temperature. Beneficial effects of controlled atmospheres in reducing chilling injury have been demonstrated in avocados,[285,286] grapefruit,[287,288] Japanese apricots,[289] okra,[47] papayas,[51] peaches and nectarines,[46,93,290] pineapples,[291] and zucchini squash.[292-294]

Chilling injury can be prevented in many crops by maintaining a high relative humidity in the atmosphere around the commodities. One of the first symptoms of chilling injury to the leaves of many important tropical and subtropical species is a rapid leaf wilting within a few hours of the start of chilling.[295] In many crop plants, such as *Phaseolus vulgaris,* leaf wilting at 5°C can be prevented by enclosing the plant inside a plastic bag and maintaining the relative humidity at 100% surrounding the leaves.[208] The development of chilling injury in fruits and vegetables can be delayed by decreasing the vapor pressure deficit and reducing the moisture loss from the commodities.[296-297] Maintaining high relative humidity has been demonstrated to reduce chilling injury of citrus fruit,[297,298] cucumbers and bell peppers,[296,299] and bananas.[21]

Prevention of chilling injury by film packaging is believed to be related to the maintenance of a high relative humidity around the commodity during storage and reduction of water loss. In addition, the modified atmosphere within the package may also have some effect on the expression of chilling injury symptoms in some crops. This method has been reported to be effective in preventing chilling injury in papayas,[51] avocados,[300] bananas,[301] grapefruit,[302-305] lemons,[143,304] limes,[303] Japanese apricots,[289] cucumbers,[306] honeydew melons,[307] and tomatoes.[308,309]

Waxing and film packaging have similar effects on fresh produce. Waxing reduces the transpiration rate and modifies the internal O_2 and CO_2 concentrations of the commodities.

The reduction of chilling injury by waxing is attributed to these factors. Waxing has been shown to be effective in preventing chilling injury in cucumbers,[296] pineapples,[291] papayas,[51] grapefruit,[298,302] and oranges.[310]

VII. CONCLUSIONS

The ultimate objective of studying chilling injury is to develop methods to reduce the losses which result from this phenomenon. Various approaches have been taken in an attempt to achieve this objective. These techniques include genetic modification, chemical treatment, hormonal regulation, temperature preconditioning, intermittent warming, controlled atmospheres, hypobaric storage, film packaging, waxing, and maintaining a high relative humidity in the storage environment. The effectiveness of these methods in reducing chilling injury often varies with the commodity and is affected by the maturity and physiological state of the commodity as well as a number of other factors. Most of these methods are either effective in alleviating the development of certain specific chilling symptoms or effective only on some specific crops, but none of the available treatments can stop or prevent chilling injury in all crops in all circumstances. Probably these treatments retard only the expression of some specific chilling injury symptoms or affect secondary rather than primary events of the chilling response.

Other than genetic modification, any treatment expected to be universally effective and to confer prolonged protection against chilling injury must be able to directly block the primary events that occur at the very beginning of chilling exposure so that chilling-induced alterations are stopped from the outset, before the destructive chain of events is set in motion. A better understanding of the basic mechanism underlying chilling injury will help us devise such a treatment.

REFERENCES

1. **Hardenburg, R. E., Watada, A. E., and Wang, C. Y.,** The commercial storage of fruits, vegetables, and florist and nursery stocks, *U.S. Dep. Agric. Agric. Handb.,* 66, 1, 1986.
2. **Vallejos, C. E.,** Genetic diversity of plants for response to low temperatures and its potential use in crop plants, in *Low Temperature Stress in Crop Plants: The Role of the Membrane,* Lyons, J. M., Graham, D., and Raison, J. K., Eds., Academic Press, New York, 1979, 473.
3. **McWilliam, J. R.,** Physiological basis for chilling stress and the consequences for crop protection, in *Crop Reactions to Water and Temperature Stresses in Humid, Temperate Climates,* Raper, C. D., Jr. and Kramer, P. J., Eds., Westview Press, Boulder, CO, 1983, chap. 8.
4. **Graham, D. and Patterson, B. D.,** Responses of plants to low, nonfreezing temperatures: proteins, metabolism, and acclimation, *Annu. Rev. Plant Physiol.,* 33, 347, 1982.
5. **Couey, H. M.,** Chilling injury of crops of tropical and subtropical origin, *HortScience,* 17, 162, 1982.
6. **Bramlage, W. J.,** Chilling injury of crops of temperate origin, *HortScience,* 17, 165, 1982.
7. **Wolk, W. D. and Herner, R. C.,** Chilling injury of germinating seeds and seedlings, *HortScience,* 17, 169, 1982.
8. **Lyons, J. M. and Breidenbach, R. W.,** Strategies for altering chilling sensitivity as a limiting factor in crop production, in *Stress Physiology of Crop Plants,* Mussell, H. and Staples, R. C., Eds., John Wiley & Sons, New York, 1979, 179.
9. **Patterson, B. D., Paull, R., and Smillie, R. M.,** Chilling resistance in *Lycopersicon hirsutum* Humb. & Bonpl., a wild tomato with a wide altitudinal distribution, *Aust. J. Plant Physiol.,* 5, 609, 1978.
10. **Patterson, B. D., Murata, T., and Graham, D.,** Electrolyte leakage induced by chilling in *Passiflora* species tolerant to different climates, *Aust. J. Plant Physiol.,* 3, 435, 1976.
11. **Muramoto, H., Hesketh, J. D., and Baker, D. N.,** Cold tolerance in a hexaploid cotton, *Crop Sci.,* 11, 589, 1971.

12. **Castleberry, R. M., Teeri, J. A., and Buriel, J. F.,** Vegetative growth responses of maize genotypes to simulated natural chilling events, *Crop Sci.,* 18, 633, 1978.

13. **Lin, S. S. and Patterson, M. L.,** Low temperature induced floret sterility in rice, *Crop Sci.,* 15, 657, 1975.

14. **McWilliam, J. R., Manokaran, W., and Kipnis, T.,** Adaptation to chilling stress in sorghum, in *Low Temperature Stress in Crop Plants: The Role of the Membrane,* Lyons, J. M., Graham, D., and Raison, J. K., Eds., Academic Press, New York, 1979, 491.

15. **Apeland, J.,** Factors affecting the sensitivity of cucumbers to chilling temperatures, *Bull. Int. Inst. Refrig.,* 46 (Annex 1), 325, 1966.

16. **Sherman, M., Paris, H. S., and Allen, J. J.,** Storability of summer squash as affected by gene B and genetic background, *HortScience,* 22, 920, 1987.

17. **Dickson, M. H. and Boettger, M. A.,** Emergence, growth, and blossoming of bean *(Phaseolus vulgaris)* at suboptimal temperatures, *J. Am. Soc. Hortic. Sci.,* 109, 257, 1984.

18. **Pollock, B. M., Roos, E. E., and Manalo, J. R.,** Vigor of garden bean seeds and seedlings influenced by initial seed moisture, substrate oxygen, and imbibition temperature, *J. Am. Soc. Hortic. Sci.,* 94, 577, 1969.

19. **Bramlage, W. J., Leopold, A. C., and Specht, J. E.,** Imbibitional chilling sensitivity among soybean cultivars, *Crop Sci.,* 19, 811, 1979.

20. **Zauberman, G., Schiffmann-Nadel, M., and Yanko, U.,** Susceptibility to chilling injury of three avocado cultivars at various stages of ripening, *HortScience,* 8, 511, 1973.

21. **Pantastico, E. B., Grierson, W., and Soule, J.,** Chilling injury in tropical fruits. I. Bananas *(Musa paradisiaca* var. Sapientum cv. Lacatan), *Proc. Trop. Reg. Am. Soc. Hortic. Sci.,* 11, 83, 1967.

22. **Hummel, R. L. and Henny, R. J.,** Variation in sensitivity to chilling injury within the Genus *Aglaonema,* *HortScience,* 21, 291, 1986.

23. **Kemp, G. A.,** Low-temperature growth responses of the tomato, *Can. J. Plant Sci.,* 48, 281, 1968.

24. **Christiansen, M. N. and Lewis, C. F.,** Reciprocal differences in tolerance to seed hydration chilling in F_1 progeny of *Gossypium hirsutum, Crop Sci.,* 13, 210, 1973.

25. **McConnell, R. L. and Gardner, C. O.,** Selection for cold germination in two corn populations, *Crop Sci.,* 19, 765, 1979.

26. **Patterson, B. D., Mutton, L., Paull, R. D., and Nguyen, V. Q.,** Tomato pollen development: stage sensitive to chilling and a natural environment for the selection of resistant genotypes, *Plant Cell Environ.,* 10, 363, 1987.

27. **Patterson, B. D.,** Genes for cold resistance from wild tomatoes, *HortScience,* 23, 794, 1988.

28. **Rick, C. M.,** Natural variability in wild species of *Lycopersicon* and its bearing on tomato breeding, *Genet. Agric.,* 30, 249, 1976.

29. **Patterson, B. D. and Payne, L. A.,** Screening for chilling resistance in tomato seedlings, *HortScience,* 18, 340, 1983.

30. **Wolf, S., Yakir, D., Stevens, M. A., and Rudich, J.,** Cold temperature tolerance of wild tomato species, *J. Am. Soc. Hortic. Sci.,* 111, 960, 1986.

31. **Staub, J. E., Lower, R. L., and Nienhuis, J.,** Correlated responses to selection for low temperature germination in cucumber, *HortScience,* 23, 745, 1988.

32. **van Arkel, H.,** New forage crop introductions for the semiarid highland areas of Kenya as a means to increase beef production, *Neth. J. Agric. Sci.,* 25, 135, 1977.

33. **Eagles, H. A.,** Cold tolerance and its relevance to maize breeding in New Zealand, *Proc. Agron. Soc. N.Z.,* 9, 97, 1979.

34. **Cannon, O. S., Gatherum, D. M., and Miles, W. G.,** Heritability of low temperature seed germination in tomato, *HortScience,* 8, 404, 1973.

35. **El Sayed, M. N. and John, C. A.,** Heritability studies of tomato emergence at different temperatures, *J. Am. Soc. Hortic. Sci.,* 98, 440, 1973.

36. **Ng, T. J. and Tigchelaar, E. C.,** Inheritance of low temperature seed sprouting in tomato, *J. Am. Soc. Hortic. Sci.,* 98, 314, 1973.

37. **Smith, P. G. and Millett, A. H.,** Germinating and sprouting responses of the tomato at low temperatures, *Proc. Am. Soc. Hortic. Sci.,* 84, 480, 1964.

38. **Whittington, W. J. and Fierlinger, P.,** The genetic control of sprouting in tomato, *Ann. Bot.,* 36, 873, 1972.

39. **De Vos, D. A., Hill, R. R., Jr., Hepler, R. W., and Garwood, D. L.,** Inheritance of low temperature sprouting ability in F_1 tomato crosses, *J. Am. Soc. Hortic. Sci.,* 106, 352, 1981.

40. **Scott, S. J. and Jones, R. A.,** Low-temperature seed germination of *Lycopersicon* species evaluated by survival analysis, *Euphytica,* 31, 869, 1982.

41. **Dickson, M. H.,** Breeding beans, *Phaseolus vulgaris* L., for improved germination under unfavorable low temperature conditions, *Crop Sci.,* 11, 848, 1971.

42. **Chaplin, G. R. and Scott, K. J.,** Association of calcium in chilling injury susceptibility of stored avocados, *HortScience,* 15, 514, 1980.

43. **Wade, N. L.,** Physiology of cool-storage disorders of fruit and vegetables, in *Low Temperature Stress in Crop Plants: The Role of the Membrane,* Lyons, J. M., Graham, D., and Raison, J. K., Eds., Academic Press, New York, 1979, 81.

44. **Bangerth, F., Dilley, D. R., and Dewey, D. H.,** Effect of postharvest calcium treatments on internal breakdown and respiration of apple fruits, *J. Am. Soc. Hortic. Sci.,* 97, 679, 1972.

45. **Scott, K. J. and Wills, R. B. H.,** Postharvest application of calcium as a control for storage breakdown of apples, *HortScience,* 10, 75, 1975.

46. **Wade, N. L.,** Effects of storage atmosphere, temperature and calcium on low-temperature injury of peach fruit, *Sci. Hortic.,* 15, 145, 1981.

47. **Ilker, Y. and Morris, L. L.,** Alleviation of chilling injury of okra, *HortScience,* 10, 324, 1975.

48. **Purvis, A. C.,** Susceptibility of 'Marsh' grapefruit to chilling injury is not related to endogenous calcium levels in flavedo tissue, *HortScience,* 20, 95, 1985.

49. **Kader, A. A. and Morris, L. L.,** Amelioration of chilling injury symptoms on tomato fruits, *HortScience,* 10, 324, 1975.

50. **Moline, H. E.,** Effects of vacuum infiltration of calcium chloride on ripening rate and chilling injury of tomato fruit, *Phytopathology,* 70, 691, 1980.

51. **Chen, N. M. and Paull, R. E.,** Development and prevention of chilling injury in papaya fruit, *J. Am. Soc. Hortic. Sci.,* 111, 639, 1986.

52. **Wang, C. Y. and Baker, J. E.,** Effects of two free radical scavengers and intermittent warming on chilling injury and polar lipid composition of cucumber and sweet pepper fruits, *Plant Cell Physiol.,* 20, 243, 1979.

53. **Senaratna, T., Mackay, C. E., McKersie, B. D., and Fletcher, R. A.,** Uniconazole-induced chilling tolerance in tomato and its relationship to antioxidant content, *J. Plant Physiol.,* 133, 56, 1988.

54. **Wise, R. R. and Naylor, A. W.,** Chilling-enhanced photooxidation — the peroxidative destruction of lipids during chilling injury to photosynthesis and ultrastructure, *Plant Physiol.,* 83, 272, 1987a.

55. **Wise, R. R. and Naylor, A. W.,** Chilling-enhanced photooxidation — evidence for the role of singlet oxygen and superoxide in the breakdown of pigments and endogenous antioxidants, *Plant Physiol.,* 83, 278, 1987b.

56. **Slater, T. F.,** *Free Radical Mechanisms in Tissue Injury,* Pion, London, 1972, 1.

57. **Hansen, E. and Mellenthin, W. M.,** Chemical control of superficial scald on Anjou pears, *Proc. Am. Soc. Hortic. Sci.,* 91, 860, 1967.

58. **Smock, R. M.,** A comparison of treatments for control of the apple scald disease, *Proc. Am. Soc. Hortic. Sci.,* 69, 91, 1957.

59. **Gough, R. E., Shutak, V. G., Olney, C. E., and Day, H.,** Effect of butylated hydroxytoluene (BHT) on apple scale, *J. Am. Soc. Hortic. Sci.,* 98, 14, 1973.

60. **Wills, R. B. H., Hopkirk, G., and Scott, K. J.,** Reduction of soft scald in apples with antioxidants, *J. Am. Soc. Hortic. Sci.,* 106, 569, 1981.

61. **Geduspan, H. S. and Peng, A. C.,** Effect of chemical treatment on the fatty acid composition of European cucumber fruit during cold storage, *J. Sci. Food Agric.,* 40, 333, 1987.

62. **Frankel, E. N.,** Lipid oxidation, *Prog. Lipid Res.,* 19, 1, 1980.

63. **Senaratna, T., McKersie, B. D., and Stinson, R. H.,** Simulation of dehydration injury to membranes from soybean axes by free radicals, *Plant Physiol.,* 77, 472, 1985.

64. **Waring, A. J., Breidenbach, R. W., and Lyons, J. M.,** *In vivo* modification of plant membrane phospholipid composition, *Biochim. Biophys. Acta,* 443, 157, 1976.

65. **Ilker, R., Waring, A. J., Lyons, J. M., and Breidenbach, R. W.,** The cytological responses of tomato seedling cotyledons to chilling and the influence of membrane modifications upon these responses, *Protoplasma,* 90, 229, 1976.

66. **Hilton, J. L., St. John, J. B., Christiansen, M. N., and Norris, K. H.,** Interaction of lipoidal materials and a pyridazinone inhibition of chloroplast development, *Plant Physiol.,* 48, 171, 1971.

67. **St. John, J. B. and Christiansen, M. N.,** Inhibition of linolenic acid synthesis and modification of chilling resistance in cotton seedlings, *Plant Physiol.,* 57, 257, 1976.

68. **St. John, J. B.,** Chemical modification of lipids in chilling sensitive species, in *Low Temperature Stress in Crop Plants: The Role of the Membrane,* Lyons, J. M., Graham, D., and Raison, J. K., Eds., Academic Press, New York, 1979, 405.

69. **Horvath, I. and van Hasselt, P. R.,** Inhibition of chilling-induced photooxidative damage to leaves of *Cucumis sativus* L. by treatment with amino alcohols, *Planta,* 164, 83, 1985.

70. **Horvath, I., Vign, L., Belea, A., and Farkas, T.,** Hardiness dependent accumulation of phospholipids in leaves of wheat cultivars, *Physiol. Plant.,* 49, 117, 1980.

71. **Horvath, I., Vigh, L., and Farkas, T.,** The manipulation of polar head group composition of phospholipids in the wheat Miranovskaja 808 affects frost tolerance, *Planta,* 151, 103, 1981.

72. **Horvath, I., Vigh, L., Woltjes, J., van Hasselt, P. R., and Kuiper, P. J. C.,** The physical state of lipids of the leaves of cucumber genotypes as affected by temperature, in *Biochemistry and Metabolism of Plant Lipids,* Wintermans, J. F. G. M. and Kuiper, P. J. C., Eds., Elsevier, Amsterdam, 1982, 427.

73. **Kinney, A. J., Clarkson, D. T., and Loughman, B. C.,** The effect of temperature on phospholipid biosynthesis in rye roots, in *Biochemistry and Metabolism of Plant Lipids,* Wintermans, J. F. G. M. and Kuiper, P. J. C., Eds., Elsevier, Amsterdam, 1982, 437.

74. **Kuiper, P. J. C.,** Lipids in alfalfa leaves in relation to cold hardiness, *Plant Physiol.,* 45, 684, 1970.

75. **Sikorska, E. and Kacperska-Palacz, A.,** Phospholipid involvement in frost tolerance, *Physiol. Plant.,* 47, 144, 1979.

76. **Sikorska, E. and Kacperska-Palacz, A.,** Frost induced phospholipid changes in cold-acclimated and non-acclimated rape leaves, *Physiol. Plant.,* 48, 201, 1980.

77. **Siminovitch, D., Singh, J., and De La Roche, I. A.,** Studies on membranes in plant cells resistant to extreme freezing. I. Augmentation of phospholipids and membrane substances without changes in unsaturation of fatty acids during hardening of black locust bark, *Cryobiology,* 12, 144, 1975.

78. **Smolenska, G. and Kuiper, P. J. C.,** Effect of low temperature on lipid and fatty acid composition of roots and leaves of winter rape plants, *Physiol. Plant.,* 41, 29, 1977.

79. **Yoshida, S. and Sakai, A.,** Phospholipid changes associated with the cold hardiness of cortical cells from poplar stem, *Plant Cell Physiol.,* 14, 353, 1973.

80. **Tajima, K. and Shimizu, N.,** Effect of membrane stabilizers and polyhydric alcohols on chilling injury of sorghum seedlings, *Jpn. J. Crop Sci.,* 46, 335, 1977.

81. **Jones, R. L., Freebairn, H. T., and McDonald, J. F.,** The prevention of chilling injury, weight loss reduction, and ripening retardation in banana, *J. Am. Soc. Hortic. Sci.,* 103, 219, 1978.

82. **Aljuburi, H. J. and Huff, A.,** Reduction in chilling injury to stored grapefruit (*Citrus paradisi* Macf.) by vegetable oils, *Sci. Hortic.,* 24, 53, 1984.

83. **Christiansen, M. N. and Ashworth, E. N.,** Prevention of chilling injury to seedling cotton with antitranspirants, *Crop Sci.,* 18, 907, 1978.

84. **Priestley, D. A. and Leopold, A. C.,** Alleviation of imbibitional chilling injury by use of lanolin, *Crop Sci.,* 26, 1252, 1986.

85. **Leopold, A. C.,** Temperature effects on soybean imbibition and leakage, *Plant Physiol.,* 65, 1096, 1980.

86. **Leopold, A. C. and Musgrave, M.E.,** Respiratory changes with chilling injury of soybean, *Plant Physiol.,* 64, 702, 1979.

87. **Perry, D. A. and Harrison, J. G.,** The deleterious effects of water and low temperature on germination of pea seed, *J. Exp. Bot.,* 21, 504, 1970.

88. **Powell, A. A. and Matthews, S.,** The damaging effect of water on dry pea embryos during imbibition, *J. Exp. Bot.,* 29, 1215, 1978.

89. **Tully, R. E., Musgrave, M. E., and Leopold, A. C.,** The seed coat as a control of imbibitional chilling injury, *Crop Sci.,* 21, 312, 1981.

90. **Schiffmann-Nadel, M., Chalutz, E., Waks, J., and Lattar, F. S.,** Reduction of pitting of grapefruit by thiabendazole during long-term cold storage, *HortScience,* 7, 394, 1972.

91. **Schiffmann-Nadel, M., Chalutz, E., Waks, J., and Dagan, M.,** Reduction of chilling injury in grapefruit by thiobendazole and benomyl during long-term storage, *J. Am. Soc. Hortic. Sci.,* 100, 270, 1975.

92. **Wardowski, W. F., Albrigo, L. D., Grierson, W., Barmore, C. R., and Wheaton, T. A.,** Chilling injury and decay of grapefruit as affected by thiabendazole, benomyl, and CO_2, *HortScience,* 10, 381, 1975.

93. **Wang, C. Y. and Anderson, R. E.,** Progress on controlled atmosphere storage and intermittent warming of peaches and nectarines, in *Controlled Atmospheres for Storage and Transportation of Perishable Agricultural Commodities,* Richardson, D. G. and Meheriuk, M., Eds., Timber Press, Beaverton, OR, 1982, 221.

94. **Chalutz, E., Waks, J., and Schiffmann-Nadel, M.,** Reducing susceptibility of grapefruit to chilling injury during cold treatment, *HortScience,* 20, 226, 1985.

95. **Wang, C. Y.,** Modification of chilling susceptibility in seedlings of cucumber and Zucchini squash by the bioregulator paclobutrazol (PP333), *Sci. Hortic.,* 26, 293, 1985.

96. **Whitaker, B. D. and Wang, C. Y.,** Effect of paclobutrazol and chilling on leaf membrane lipids in cucumber seedlings, *Physiol. Plant.,* 70, 404, 1987.

97. **Lee, E. H., Byun, J. K., and Steffens, G. L.,** Increased tolerance of plants to SO_2, chilling, and heat stress by a new GA biosynthesis inhibitor, paclobutrazol (PP333), *Plant Physiol.,* 77 (Suppl.), 135, 1985.

98. **Couture, R. M.,** PP333: a new experimental plant growth regulator from ICI, *Proc. Plant Growth Regul. Soc. Am.,* 9, 59, 1982.

99. **Graebe, J. E.,** Gibberellin biosynthesis in cell-free systems from higher plants, in *Plant Growth Substances,* Wareing, P. F., Ed., Academic Press, London, 1982, 71.

100. **Jung, J. and Rademacher, W.,** Plant growth regulating chemicals-cereal grains, in *Plant Growth Regulating Chemicals,* Vol. 1, Nickell, L. G., Ed., CRC Press, Boca Raton, FL, 1983, 253.

101. **Quinlan, J. D. and Richardson, P. J.,** Effect of paclobutrazol (PP333) on apple shoot growth, *Acta Hortic.,* 146, 96, 1983.

102. **Raese, J. T. and Burts, E. C.,** Increased yield and suppression of shoot growth and mite populations of 'd'Anjou' pear trees with nitrogen and paclobutrazol, *HortScience,* 18, 212, 1983.

103. **Steffens, G. L., Wang, S. Y., Steffens, C. L., and Brennan, T.,** Influence of paclobutrazol (PP333) on apple seedling growth and physiology, *Proc. Plant Growth Regul. Soc. Am.,* 10, 195, 1983.

104. **Wample, R. L. and Culver, E. B.,** The influence of paclobutrazol, a new growth regulator, on sunflowers, *J. Am. Soc. Hortic. Sci.,* 108, 122, 1983.

105. **Williams, M. W.,** Use of bioregulators to control vegetative growth of fruit trees and improve fruiting efficiency, *Acta Hortic.,* 146, 88, 1983.

106. **Dalziel, J. and Lawrence, D. K.,** Biochemical and biological effects of kaurene oxidase inhibitors, such as paclobutrazol, in *Biochemical Aspects of Synthetic and Naturally Occurring Plant Growth Regulators,* Monogr. No. 11, Menhenett, R. and Lawrence, D. K., Eds., British Plant Growth Regulator Group, Wantage, Oxfordshire, England, 1984, 43.

107. **Wang, S. Y., Byun, J. K., and Steffens, G. L.,** Controlling plant growth via the gibberellin biosynthesis system. II. Biochemical and physiological alterations in apple seedlings, *Physiol. Plant.,* 63, 169, 1985.

108. **Jaggard, K. W., Briscoe, P. V., and Lawrence, D. K.,** An understanding of crop physiology in assessing the effects of a plant growth regulator on sugar beet, in *Chemical Manipulation of Crop Growth and Development,* McLaren, J. S., Ed., Butterworths, London, 1982, 139.

109. **Sankhla, N., Davis, T. D., Upadhyaya, A., Sankhla, D., Walser, R. H., and Smith B. N.,** Growth and metabolism of soybean as affected by paclobutrazol, *Plant Cell Physiol.,* 26, 913, 1985.

110. **Upadhyaya, A., Sankhla, D., Davis, T. D., Sankhla, N., and Smith, B. N.,** Effect of paclobutrazol on the activities of some enzymes of activated oxygen metabolism and lipid peroxidation in senescing soybean leaves, *J. Plant Physiol.,* 121, 453, 1985.

111. **Raison, J. K.,** Alterations in the physical properties and thermal response of membrane lipids: correlations with acclimation to chilling and high temperatures, in *Frontiers of Membrane Research in Agriculture,* St. John, J. B., Berlin, E., and Jackson, P. C., Eds., Rowman & Allanheld, Totowa, NJ, 1985, 383.

112. **Wolfe, J. A.,** Chilling injury in plants — the role of membrane lipid fluidity, *Plant Cell Environ.,* 1, 241, 1978.

113. **Fletcher, R. A. and Hofstra, G.,** Triadimefon a plant multi-protectant, *Plant Cell Physiol.,* 26, 775, 1985.

114. **Asare-Boamah, N. K. and Fletcher, R. A.,** Protection of bean seedlings against heat and chilling injury by triadimefon, *Physiol. Plant.,* 67, 353, 1986.

115. **Asare-Boamah, N. K., Hofstra, G., Fletcher, R. A., and Dumbroff, E. B.,** Triadimefon protects bean plants from water stress through its effects on abscisic acid, *Plant Cell Physiol.,* 27, 383, 1986.

116. **Buchenauer, H. and Rohner, E.,** Effects of triadimefon and triadimenol on various plant species as well as on gibberellin content and sterol metabolism in shoots of barley seedlings, *Pestic. Biochem. Physiol.,* 15, 58, 1981.

117. **Weiser, C. J.,** Cold resistance and injury in woody plants, *Science,* 169, 1269, 1970.

118. **Carter, J. V. and Brenner, M.,** Plant growth regulators and low temperature stress, in *Encyclopedia of Plant Physiology,* New Series, Vol. 11, Pharis, A. P. and Reid, D. M., Eds., Springer-Verlag, New York, 1985, 418.

119. **Irving, R. M.,** Characterization and role of an endogenous inhibitor in the induction of cold hardiness of *Acer negundo, Plant Physiol.,* 44, 801, 1969.

120. **Rikin, A., Waldman, M., Richmond, A. E., and Dovrat, A.,** Hormonal regulation of morphogenesis and cold-resistance. I. Modifications by abscisic acid and by gibberellic acid in alfalfa *Medicago sativa* L. seedlings, *J. Exp. Bot.,* 26, 175, 1975.

121. **Wang, C. Y.,** Physiological and biochemical responses of plants to chilling stress, *HortScience,* 17, 173, 1982.

122. **Reaney, M. J. T., Ishikawa, M., Robertson, A. J., and Gusta, L. V.,** The induction of cold acclimation: the role of abscisic acid, in *Low Temperature Stress in Crops,* Li, P. H., Ed., CRC Press, Boca Raton, FL, 1989, chap. 1.

123. **Grierson, W.,** Chilling injury in tropical and subtropical fruit. V. Effect of harvest date, degreening, delayed storage and peel color on chilling injury of grapefruit, *Proc. Trop. Reg. Am. Soc. Hortic. Sci.,* 18, 66, 1974.

124. **Ismail, M. A. and Grierson, W.,** Seasonal susceptibility of grapefruit to chilling injury as modified by certain growth regulators, *HortScience,* 12, 118, 1977.

125. **Kawada, K., Wheaton, T. A., Purvis, A. C., and Grierson, W.,** Levels of growth regulators and reducing sugars of 'Marsh' grapefruit peel as related to seasonal resistance to chilling injury, *HortScience,* 14, 446, 1979.

126. **Brewin, N. J. and Northcote, D. H.,** Variation in the amounts of 3', 5'-cyclic AMP in plant tissue, *J. Exp. Bot.,* 24, 881, 1973.

127. **Waldman, M., Rikin, A., Dovrat, A., and Richmond, A. E.,** Hormonal regulation of morphogenesis and cold-resistance, *J. Exp. Bot.,* 26, 853, 1975.

128. **Irving, R. M. and Lamphear, F. O.,** The long day leaf as a source of cold hardiness inhibitors, *Plant Physiol.,* 42, 1384, 1967.

129. **Irving, R. M. and Lamphear, F. O.,** Regulation of cold hardiness in *Acer negundo, Plant Physiol.,* 43, 9, 1968.

130. **Reid, D. M., Pharis, R. P., and Roberts, D. W. A.,** Effects of four temperature regimens on the gibberellin content of winter wheat cv. Kharkov, *Physiol. Plant.,* 30, 53, 1974.

131. **Proebsting, E. L., Jr. and Mills, H. H.,** Gibberellin-induced hardiness responses in Elberta peach flower buds, *Proc. Am. Soc. Hortic. Sci.,* 85, 134, 1964.

132. **Edgerton, L. J.,** Some effects of gibberellin and growth retardants on bud development and cold hardiness of peach, *Proc. Am. Soc. Hortic. Sci.,* 88, 197, 1967.

133. **Wang, C. Y., Mellenthin, W. M., and Hansen, E.,** Effect of temperature on development of premature ripening in Bartlett pears, *J. Am. Soc. Hortic. Sci.,* 96, 122, 1971.

134. **Ogura, N., Hayashi, R., Ogishima, T., Abe, Y., Nakagawa, H., and Takehana, H.,** Ethylene production by tomato fruits at various temperatures and effect of ethylene on the fruits, *Nippon Nogeikagaku Kaishi,* 50, 519, 1976.

135. **Wang, C. Y.,** Relation of chilling stress to ethylene production, in *Low Temperature Stress Physiology in Crops,* Li, P. H., Ed., CRC Press, Boca Raton, FL, 1989, 177.

136. **Buescher, R. W.,** Hardcore in sweet potato roots as influenced by cultivar, curing, and ethylene, *HortScience,* 12, 326, 1977.

137. **Lipton, W. J. and Aharoni, Y.,** Chilling injury and ripening of 'Honey Dew' muskmelons stored at 25° or 5° C after ethylene treatment at 20° C, *J. Am. Soc. Hortic. Sci.,* 104, 327, 1979.

138. **Lipton, W. J., Aharoni, Y., and Elliston, E.,** Rates of CO_2 and ethylene production and of ripening of 'Honey Dew' muskmelons at a chilling temperature after pretreatment with ethylene, *J. Am. Soc. Hortic. Sci.,* 104, 846, 1979.

139. **Lipton, W. J. and Mackey, B. E.,** Ethylene and low-temperature treatments of honeydew melons to facilitate long-distance shipment, *U.S. Dep. Agric. Res. Serv. Rep.,* 10, 1984, 28.

140. **Lee, S. K. and Young, R. E.,** Temperature sensitivity of avocado fruit in relation to C_2H_4 treatment, *J. Am. Soc. Hortic. Sci.,* 109, 689, 1984.

141. **Young, R. E. and Lee, S. K.,** The effect of ethylene on chilling and high temperature injury of avocado fruit, *HortScience,* 14, 421, 1979.

142. **Chaplin, G. R., Wills, R. B. H., and Graham, D.,** Induction of chilling injury in stored avocado with exogenous ethylene, *HortScience,* 18, 952, 1983.

143. **McDonald, R. E.,** Effects of vegetable oils, CO_2, and film wrapping on chilling injury and decay of lemons, *HortScience,* 21, 476, 1986.

144. **Lyons, J. M.,** Chilling injury in plants, *Annu. Rev. Plant Physiol.,* 24, 445, 1973.

145. **Mager, J.,** The stabilizing effect of spermine and related polyamines on bacterial protoplasts, *Biochem. Biophys. Acta,* 36, 529, 1959.

146. **Smith, T. A.,** Polyamines, *Annu. Rev. Plant Physiol.,* 36, 117, 1985.

147. **Kramer, G. F. and Wang, C. Y.,** Relationship of polyamine levels to chilling injury in Zucchini squash, *Plant Physiol.,* 86 (Suppl.), 51, 1988.

148. **Wang, C. Y.,** Influence of low oxygen atmosphere on polyamines in chilled Zucchini squash, *HortScience,* 23, 831, 1988.

149. **Kramer, G. F. and Wang, C. Y.,** Correlation of reduced chilling injury and oxidative damage with increased polyamine levels in Zucchini squash, *Physiol. Plant.,* 1989, in press.

150. **Guye, M. G., Vugh, L., and Wilson, J. M.,** Polyamine titre in relation to chilling sensitivity in *Phaseolus* sp., *J. Exp. Bot.,* 37, 1036, 1986.

151. **Kushad, M. M. and Yelenosky, G.,** Evaluation of polyamine and proline levels during low temperature acclimation of citrus, *Plant Physiol.,* 84, 692, 1987.

152. **Nadeau, P., Delaney, S., and Chouinard, L.,** Effects of cold hardening on the regulation of polyamine levels in wheat (*Triticum aestivum* L.) and alfalfa (*Medicago sativa* L.), *Plant Physiol.,* 84, 73, 1987.

153. **Galston, A. W. and Kaur-Sawhney, R.,** Polyamines and senescence in plants, in *Plant Senescence: Its Biochemistry and Physiology,* Thomson, W. W., Nothnagel, E. A., and Huffaker, R. C., Eds., American Society for Plant Physiology, Rockville, MD, 1987, 167.

154. **Slocum, R. D., Kaur-Sawhney, R., and Galston, A. W.,** The physiology and biochemistry of polyamines in plants, *Arch. Biochem. Biophys.,* 235, 283, 1984.

155. **Altman, A. and Bachrach, U.,** Involvement of polyamines in plant growth and senescence, in *Advances in Polyamine Research,* Vol. 3, Caldarera, C. M., Zappia, V., and Bachrach, U., Eds., Raven Press, New York, 1981, 365.

156. **Galston, A. W.,** Polyamines as modulators of plant development, *BioScience,* 33, 382, 1983.

157. **Galston, A. W. and Kaur-Sawhney, R.,** Polyamines as endogenous growth regulators, in *Plant Hormones and Their Role in Plant Growth and Development,* Davies, P. J., Ed., Martinus Nijhoff, Boston, 1987, 280.

158. **Flores, H. E., Young, N. D., and Galston, A. W.,** Polyamine metabolism and plant stress, in *Cellular and Molecular Biology of Plant Stress,* Key, J. L. and Kosuge, T., Eds., Alan R. Liss, New York, 1985, 93.

159. **Wang, S. Y. and Steffens, G. L.,** Effect of paclobutrazol on water stress-induced ethylene biosynthesis and polyamine accumulation in apple seedling leaves, *Phytochemistry,* 24, 2185, 1985.

160. **Turner, L. B. and Stewart, G. R.,** The effect of water stress upon polyamine levels in barley (*Hordeum vulgare* L.) leaves, *J. Exp. Bot.,* 37, 170, 1986.

161. **Smith, T. A. and Sinclair, C.,** The effect of acid feeding on amine formation in barley, *Ann. Bot.,* 31, 103, 1967.

162. **Young, N. D. and Galston, A. W.,** Putrescine and acid stress, *Plant Physiol.,* 71, 767, 1983.

163. **Flores, H. E. and Galston, A. W.,** Polyamines and plant stress: activation of putrescine biosynthesis by osmotic shock, *Science,* 217, 1259, 1982.

164. **Richards, F. J. and Coleman, R. G.,** Occurrence of putrescine in potassium deficient barley, *Nature (London),* 170, 460, 1952.

165. **Coleman, R. G. and Richards, F. J.,** Physiological studies in plant nutrition. XVIII. Some aspects of nitrogen metabolism in barley and other plants in relation to potassium deficiency, *Ann. Bot.,* 20, 393, 1956.

166. **McDonald, R. E. and Kushad, M. M.,** Accumulation of putrescine during chilling injury of fruits, *Plant Physiol.,* 82, 324, 1986.

167. **Galston, A. W., Altman, A., and Kaur-Sawhney, R.,** Polyamines, ribonuclease and the improvement of oat leaf protoplasts, *Plant Sci. Lett.,* 11, 69, 1978.

168. **Kaur-Sawhney, R. and Galston, A. W.,** Interaction of polyamines and light on biochemical processes involved in leaf senescence, *Plant Cell Environ.,* 2, 189, 1979.

169. **Kaur-Sawhney, R., Altman, A., and Galston, A. W.,** Dual mechanisms in polyamine mediated control of ribonuclease activity in oat leaf protoplasts, *Plant Physiol.,* 62, 158, 1978.

170. **Cohen, S. S.,** *Introduction to the Polyamines,* Prentice-Hall, Englewood Cliffs, NJ, 1971, chap. 4.

171. **Srivastava, S. K. and Rajbabu, P.,** Effect of amines and guanidines on ATPase from maize scutellum, *Phytochemistry,* 22, 2675, 1983.

172. **Altman, A., Kaur-Sawhney, R., and Galston, A. W.,** Stabilization of oat leaf protoplasts through polyamine mediated inhibition of senescence, *Plant Physiol.,* 60, 570, 1977.

173. **Shih, L. M., Kaur-Sawhney, R., Fuhrer, J., Samanta, S., and Galston, A. W.,** Effects of exogenous 1,3-diaminopropane and spermidine on senescence of oat leaves, *Plant Physiol.,* 70, 1592, 1982.

174. **Altman, A.,** Retardation of radish leaf senescence by polyamines, *Physiol. Plant.,* 54, 189, 1982.

175. **Ormrod, D. P. and Beckerson, D. W.,** Polyamines as antiozonants for tomato, *HortScience,* 21, 1070, 1986.

176. **Wang, C. Y.,** Changes of polyamines and ethylene in cucumber seedlings in response to chilling stress, *Physiol. Plant.,* 69, 253, 1987.

177. **Smith, T. A.,** The function and metabolism of polyamines in higher plants, in *Plant Growth Substances,* Wareing, P. F., Ed., Academic Press, New York, 1982, 463.

178. **Ballas, S. K., Mohandas, N., Marton, L. J., and Shohet, S. B.,** Stabilization of erythrocyte membranes by polyamines, *Proc. Natl. Acad. Sci. U.S.A.,* 80, 1942, 1983.

179. **Roberts, D. R., Dumbroff, E. B., and Thompson, J. E.,** Exogenous polyamines alter membrane fluidity — a basis for potential misinterpretation of their physiological role, *Planta,* 167, 395, 1986.

180. **Srivastava, S. K. and Smith, T. A.,** The effect of some oligoamines and guanidines on membrane permeability in higher plants, *Phytochemistry,* 21, 997, 1982.

181. **Drolet, G., Dumbroff, E. B., Legge, R. L., and Thompson, J. E.,** Radical scavenging properties of polyamines, *Phytochemistry,* 25, 367, 1986.

182. **Kitada, M., Igaraski, K., Hirose, S., and Kitagawa, H.,** Inhibition by polyamines of lipid peroxide formation in rat liver microsomes, *Biochem. Biophys. Res. Commun.,* 87, 388, 1979.

183. **Trappel, A. L.,** Lipid peroxidation and fluorescent molecular damage to membranes, in *Pathology of Cell Membranes,* Vol. 1, Trump, B. and Arstila, A., Eds., Academic Press, New York, 1975, 145.

184. **Adams, D. O. and Yang, S. F.,** Methionine metabolism in apple tissue, *Plant Physiol.,* 60, 892, 1977.

185. **Adams, D. O., Wang, S. Y., and Lieberman, M.,** Control of ethylene production in apple cell suspension cultures, *Plant Physiol.,* 65 (Suppl.), 41, 1980.

186. **Wang, S. Y., Adams, D. O., and Lieberman, M.,** Recycling of 5'-methylthioadenosine-ribose carbon atoms into methionine in tomato tissue in relation to ethylene production, *Plant Physiol.,* 70, 117, 1982.

187. **Kushad, M. M., Yelenosky, G., and Knight, R.,** Interrelationship of polyamine and ethylene biosynthesis during avocado fruit development and ripening, *Plant Physiol.,* 87, 463, 1988.

188. **King, A. I., Reid, M. S., and Patterson, B. D.,** Diurnal changes in the chilling sensitivity of seedlings, *Plant Physiol.,* 70, 211, 1982.

189. **Rikin, A., Alsmon, D., and Gitler, C.,** Chilling injury in cotton (*Gossypium hirsutum* L.): prevention by abscisic acid, *Plant Cell Physiol.,* 20, 1537, 1979.

190. **Rikin, A., Atsmon, D., and Gitler, C.,** Chilling injury in cotton *Gossypium hirsutum* L.): effects of antimicrotubular drugs, *Plant Cell Physiol.,* 21, 829, 1980.

191. **Semeniuk, P., Moline, H. E., and Abbott, J. A.,** A comparison of the effects of ABA and an antitranspirant on chilling injury of coleus, cucumber, and dieffenbachia, *J. Am. Soc. Hortic. Sci.,* 111, 866, 1986.

192. **Sasson, N. and Bramlage, W. J.,** Effects of chemical protectants against chilling injury of young cucumber seedlings, *J. Am. Soc. Hortic. Sci.,* 106, 282, 1981.

193. **Rikin, A. and Richmond, A. E.,** Amelioration of chilling injury in cucumber seedlings by abscisic acid, *Physiol. Plant.,* 38, 95, 1976.

194. **Tseng, M. J. and Li, P. H.,** Mefluidide protection of severely chilled crop plants, *Plant Physiol.,* 75, 249, 1984.

195. **Tseng, M. J., Zhang, C. L., and Li, P. H.,** Quantitative measurements of mefluidide protection of chilled corn plants, *J. Am. Soc. Hortic. Sci.,* 111, 409, 1986.

196. **Zhang, C. L., Li, P. H., and Brenner, M. L.,** Relationship between mefluidide treatment and abscisic acid metabolism in chilled corn leaves, *Plant Physiol.,* 81, 699, 1986.

197. **Zhang, L. X., Li, P. H., and Tseng, M. J.,** Amelioration of chilling injury in rice seedlings by mefluidide, *Crop Sci.,* 27, 531, 1987.

198. **Cummins, W. R., Kende, H., and Raschke, K.,** Specificity and reversibility of the rapid stomatal response to abscisic acid, *Planta,* 99, 347, 1971.

199. **Orton, P. J. and Mansfield, T. A.,** The activity of abscisic acid analogues as inhibitors of stomatal opening, *Planta,* 121, 263, 1974.

200. **Mittleheuser, C. J. and van Steveninck, R. F. M.,** Stomatal closure and inhibition of transpiration induced by (RS)± abscisic acid, *Nature (London),* 221, 281, 1969.

201. **Uehara, Y., Ogawa, T., and Shibata, K.,** Effects of ABA and its derivatives on stomatal closing, *Plant Cell Physiol.,* 16, 543, 1975.

202. **Markhart, A. H., III,** Amelioration of chilling-induced water stress by abscisic acid-induced changes in root hydraulic conductance, *Plant Physiol.,* 74, 81, 1984.

203. **James, R. J. and Mansfield, T. A.,** Suppression of stomatal opening in leaves treated with abscisic acid, *J. Exp. Bot.,* 21, 714, 1970.

204. **Wright, M.,** The effect of chilling on ethylene production, membrane permeability and water loss of leaves of *Phaseolus vulgaris, Planta,* 120, 63, 1974.

205. **Wilson, J. M.,** Interaction of chilling and water stress, in *Crop Reactions to Water and Temperature Stresses in Humid, Temperate Climates,* Raper, C. D., Jr. and Kramer, P. J., Eds., Westview Press, Boulder, CO, 1983, chap. 9.

206. **Rikin, A., Blumenfeld, A., and Richmond, A. E.,** Chilling resistance as affected by stressing environments and abscisic acid, *Bot. Gaz.,* 137, 307, 1976.

207. **Christiansen, M. N. and Ashworth, E. N.,** Prevention of chilling injury to seedling cotton with anti-transpirants, *Crop Sci.,* 18, 907, 1978.

208. **Wilson, J. M.,** The mechanism of chill and drought hardening of *Phaseolus vulgaris* leaves, *New Phytol.,* 76, 257, 1976.

209. **Bornmann, C. H. and Jansson, E.,** *Nicotiana tabacum* callus studies. X. ABA increases resistance to cold damage, *Physiol. Plant.,* 48, 491, 1980.

210. **Markhart, A. H., III, Fiscus, E. L., Naylor, A. W., and Kramer, P. J.,** Effect of temperature on water and ion transport in soybean and broccoli systems, *Plant Physiol.,* 64, 83, 1979.

211. **Markhart, A. H., III, Fiscus, E. L., Naylor, A. W., and Kramer, P. J.,** Effect of abscisic acid on root hydraulic conductivity, *Plant Physiol.,* 64, 611, 1979.

212. **Markhart, A. H. III,** Chilling injury: a review of possible causes, *HortScience,* 21, 1329, 1986.

213. **Rikin, A., Atsmon, D., and Gitler, C.,** Quantitation of chill-induced release of a tubulin-like factor and its prevention by abscisic acid in *Gossypium hirsutum* L., *Plant Physiol.,* 71, 747, 1983.

214. **Rikin, A. and Richmond, A. E.,** Factors affecting leakage from cucumber seedlings during chilling stress, *Plant Sci. Lett.,* 14, 263, 1979.

215. **McColloch, L. P.,** Chilling injury and alternaria rot of bell peppers, *U.S. Dep. Agric. Mark. Res. Rep.,* 536, 1, 1962.

216. **Wheaton, T. A. and Morris, L. L.,** Modification of chilling sensitivity by temperature conditioning, *Proc. Am. Soc. Hortic. Sci.,* 91, 529, 1967.

217. **Harding, P. L., Soule, M. J., and Sunday, M. B.,** Storage studies on Marsh grapefruit, *U.S. Dep. Agric. Agric. Res. Serv. Mark. Res. Rep.,* AMS-202, 1957.

218. **Hetherington, S. E., Smillie, R. M., Hardacre, A. K., and Eagles, H. A.,** Using chlorophyll fluorescence *in vivo* to measure the chilling tolerances of different populations of maize, *Aust. J. Plant Physiol.,* 10, 247, 1983.

219. **Stamp, P.,** Seedling development of adapted and exotic maize genotypes at severe chilling stress, *J. Exp. Bot.,* 38, 1336, 1987.

220. **Campbell, T. A.,** Responses of Stokes aster achenes to chilling, *J. Am. Soc. Hortic. Sci.,* 109, 736, 1984.

221. **Hatton, T. T. and Cubbedge, R. H.,** Preconditioning Florida grapefruit to prevent or reduce chilling injury in low-temperature storage, *HortScience,* 15, 423, 1980.

222. **Chan, H. T., Jr.,** Alleviation of chilling injury in papayas, *HortScience,* 23, 868, 1988.

223. **McDonald, R. E., Hatton, T. T., and Cubbedge, R. H.,** Chilling injury and decay of lemons as affected by ethylene, low temperature, and optimal storage, *HortScience,* 20, 92, 1985.

224. **Hirose, T.,** Effects of pre- and interposed warming on chilling injury respiratory rate and membrane permeability of cucumber fruits during cold storage, *J. Jpn. Soc. Hortic. Sci.,* 53, 439, 1985.

225. **Abe, K. and Chachin, K.,** Influence of conditioning on the occurrence of chilling injury and the changes of surface structure of eggplant fruits, *J. Jpn. Soc. Hortic. Sci.,* 54, 247, 1985.

226. **Picha, D. H.,** Postharvest fruit conditioning reduces chilling injury in watermelons, *HortScience,* 21, 1407, 1986.

227. **Harris, P. and James, A. T.,** The effect of low temperatures on fatty acid biosynthesis in plants, *Biochem. J.,* 112, 325, 1969.

228. **Ketchie, D. O.,** Fatty acid in the bark of Halehaven peach as associated with hardiness, *Proc. Am. Soc. Hortic. Sci.,* 88, 204, 1966.

229. **Tabacchi, M. H., Hicks, J. R., Ludford, P. M., and Robinson, R. W.,** Chilling injury tolerance and fatty acid composition in tomatoes, *HortScience,* 14, 424, 1979.

230. **De La Roche, I. A., Andrews, C. J., Pomeroy, M. K., Weinberger, P., and Kates, M.,** Lipid changes in winter wheat seedlings at temperatures inducing cold hardiness, *Can. J. Bot.,* 50, 2401, 1972.

231. **Willemot, C., Hope, H. J., Williams, R. J., and Michaud, R.,** Changes in fatty acid composition of winter wheat during frost hardening, *Cryobiology,* 14, 87, 1977.

232. **Gerloff, E. D., Richardson, T., and Stahmann, M. A.,** Changes in fatty acids of alfalfa roots during cold hardening, *Plant Physiol.,* 41, 1280, 1966.

233. **Kuiper, P. J. C.,** Lipids in alfalfa leaves in relation to cold hardiness, *Plant Physiol.,* 45, 684, 1970.

234. **Canvin, D. T.,** The effect of temperature on the oil content and fatty acid composition of the oils from several oil seed crops, *Can. J. Bot.,* 43, 63, 1965.

235. **Harris, H. C., McWilliam, J. R., and Mason, W.,** Influence of temperature on oil content and composition of sunflower, *Aust. J. Agric. Res.,* 29, 1203, 1978.

236. **De La Roche, I. A. and Andrews, C. J.,** Changes in phospholipid composition of a winter wheat cultivar during germination at 2° C and 24° C, *Plant Physiol.,* 51, 468, 1973.

237. **Wilson, J. M. and Crawford, R. M. M.,** The acclimatization of plants to chilling temperatures in relation to the fatty acid composition of leaf polar lipids, *New Phytol.,* 73, 805, 1974.

238. **Guinn, G.,** Changes in sugars, starch, RNA, protein and lipid-soluble phosphate in leaves of cotton plants at low temperatures, *Crop Sci.,* 11, 262, 1971.

239. **Guinn, G.,** Chilling injury in cotton seedlings: changes in permeability of cotyledons, *Crop Sci.,* 11, 101, 1971.

240. **Smith, C. W. and McWilliams, E. L.,** Amelioration of chilling injury in *Maranta leuconeura* and *Scindapsus pictus* by preconditioning, *HortScience,* 14, 439, 1979.

241. **Lyons, J. M. and Breidenbach, R. W.,** Chilling injury, in *Postharvest Physiology of Vegetables,* Weichmann, J., Ed., Marcel Dekker, New York, 1987, 305.

242. **Lange, A. and Minchin, P. E. H.,** Phylogenetic distribution and mechanism of translocaton inhibition by chilling, *J. Exp. Bot.,* 37, 389, 1986.

243. **Kidd, F. and West, C.,** The cause and control of superficial scald of apples, *G.B. Dep. Sci. Ind. Res. Food Invest. Board Rep.,* 111, 1935.

244. **Smith, W. H.,** Control of superficial scald in stored apples, *Nature (London),* 183, 760, 1959.

245. **Brooks, C. and McColloch, L. P.,** Some storage diseases of grapefruit, *J. Agric. Res.,* 52, 319, 1936.

246. **Davis, P. L. and Hofmann, R. C.,** Reduction of chilling injury of citrus fruits in cold storage by intermittent warming, *J. Food Sci.,* 38, 871, 1973.

247. **Cohen, E., Shuali, M., and Shalom, Y.,** Effect of intermittent warming on the reduction of chilling injury of 'Villa Franka' lemon fruits stored at cold temperature, *J. Hortic. Sci.,* 58, 593, 1983.

248. **Hruschka, H. W.,** Physiological breakdown in cranberries — inhibition by intermittent warming during cold storage, *Plant Dis. Rep.,* 54, 219, 1970.

249. **Anderson, R. E. and Penney, R. W.,** Intermittent warming of peaches and nectarines stored in a controlled atmosphere or air, *J. Am. Soc. Hortic. Sci.,* 100, 151, 1975.

250. **Ben-Arie, R., Lavee, S., and Guelfat-Reich, S.,** Control of woolly breakdown of 'Elberta' peaches in cold storage by intermittent exposure to room temperature, *J. Am. Soc. Hortic. Sci.,* 95, 801, 1970.

251. **Buescher, R. W. and Furmanski, R. J.,** Role of pectinesterase and polygalacturonase in the formation of woolliness in peaches, *J. Food Sci.,* 43, 264, 1978.

252. **Smith, W. H.,** Control of low-temperature injury in the Victoria plum, *Nature (London),* 159, 541, 1947.

253. **Hruschka, H. W., Smith, W. L., and Baker, J. E.,** Reducing chilling injury of potatoes by intermittent warming, *Am. Potato J.,* 46, 38, 1969.

254. **Lieberman, M., Craft, C. C., and Wilcox, M. S.,** Effect of chilling on the chlorogenic acid and ascorbic acid content of Porto Rico sweetpotatoes, *Proc. Am. Soc. Hortic. Sci.,* 74, 642, 1959.

255. **Marcellin, P. and Baccaunaud, M.,** Effect of a gradual cooling and an intermittent warming on the cold storage life of tomatoes, *Bull. Int. Congr. Refrig.,* XV, 6, 1979.

256. **Cohen, E.,** Commercial use of long-term storage of lemon with intermittent warming, *HortScience,* 23, 400, 1988.

257. **Tatsumi, Y., Iwamoto, M., and Murata, T.,** Electrolyte leakage from the discs of *Cucurbitaceae* fruits associated with chilling injury, *J. Jpn. Soc. Hortic. Sci.,* 50, 114, 1981.

258. **Christiansen, M. N., Carns, H. R., and Slyter, D. J.,** Stimulation of solute loss from radicles of *Gossypium hirsutum* L. by chilling, anaerbiosis and low pH, *Plant Physiol.,* 46, 53, 1970.

259. **Eaks, I. L. and Morris, L. L.,** Respiration of cucumber fruits associated with physiological injury at chilling temperatures, *Plant Physiol.,* 33, 308, 1956.

260. **Eaks, I. L.,** Effect of chilling on respiration and volatiles of California lemon fruit, *J. Am. Soc. Hortic. Sci.,* 105, 865, 1980.

261. **Wilson, J. M. and Crawford, R. M. M.,** Leaf fatty acid content in relation to hardening and chilling injury, *J. Exp. Bot.,* 25, 121, 1974.

262. **Paull, R. E., Patterson, B. D., and Graham, D.,** Chilling injury assays for plant breeding, in *Low Temperature Stress in Crop Plants: The Role of the Membrane,* Lyons, J. M., Graham, D., and Raison, J. K., Eds., Academic Press, New York, 1979, 507.

263. **Chaplin, G. R., Wills, R. B. H., and Graham, D.,** Objective measurement of chilling injury in the mesocarp of stored avocados, *HortScience,* 17, 238, 1982.

264. **Smillie, R. M. and Nott, R.,** Assay of chilling injury in wild and domestic tomatoes based on photosystem activity of the chilled leaves, *Plant Physiol.,* 63, 796, 1979.

265. **Oquist, G.,** Effects of low temperature on photosynthesis, *Plant Cell Environ.,* 6, 281, 1983.

266. **Melcarek, P. K. and Brown, G. A.,** Effects of chilling stress on prompt and delayed chlorophyll fluorescence from leaves, *Plant Physiol.,* 60, 822, 1977.

267. **Potvin, C.,** Effects of leaf detachment on chlorophyll fluorescence during chilling experiments, *Plant Physiol.,* 78, 883, 1985.

268. **Smillie, R. M., Nott, R., Hetherington, S. E., and Oquist, G.,** Chilling injury and recovery in detached and attached leaves measured by chlorophyll fluorescence, *Physiol. Plant.,* 69, 419, 1987.

269. **Greaves, J. A. and Wilson, J. M.,** Assessment of the nonfreezing cold sensitivity of wild and cultivated potato species by chlorophyll fluorescence analysis, *Potato Res.,* 29, 509, 1986.

270. **Kemps, T. L., Isleib, T. G., Herner, R. C., and Sink, K. C.,** Evaluation of techniques to measure chilling injury in tomato, *HortScience,* 22, 1309, 1987.

271. **Abbott, J. A. and Massie, D. R.,** Delayed light emission for early detection of chilling in cucumber and bell pepper fruit, *J. Am. Soc. Hortic. Sci.,* 110, 42, 1985.

272. **Abbott, J. A., Krizek, D. T., Semeniuk, P., Moline, H. E., and Mirecki, R. M.,** Refreshed delayed light emission and fluorescence for detecting pretreatment effects on chilling injury in coleus, *J. Am. Soc. Hortic. Sci.,* 112, 560, 1987.

273. **Strehler, B. L. and Arnold, W. A.,** Light production by green plants, *J. Gen. Physiol.,* 34, 809, 1951.

274. **Wang, C. Y.,** Use of 1-aminocyclopropane-1-carboxylic acid level as an index for chilling exposure at various temperatures, *Plant Physiol.,* 72 (Suppl.), 43, 1983.

275. **Field, R. J.,** The role of 1-aminocyclopropane-1-carboxylic acid in the control of low temperature induced ethylene production in leaf tissue of *Phaseolus vulgaris* L., *Ann. Bot.,* 54, 61, 1984.

276. **Chen, Y. Z. and Patterson, B. D.,** Ethylene and 1-aminocyclopropane-1-carboxylic acid as indicators of chilling sensitivity in various plant species, *Aust. J. Plant Physiol.,* 12, 377, 1985.

277. **Wang, C. Y. and Adams, D. O.,** Ethylene production by chilled cucumbers (*Cucumis sativus* L.), *Plant Physiol.,* 66, 841, 1980.

278. **Wang, C. Y. and Adams, D. O.,** Chilling-induced ethylene production in cucumbers (*Cucumis sativus* L.), *Plant Physiol.,* 69, 424, 1982.

279. **Chan, H. T., Sanxter, S., and Couey, H. M.,** Electrolyte leakage and ethylene production induced by chilling injury of papayas, *HortScience,* 20, 1070, 1985.

280. **Yao, K., Yu, L., and Zhou, S.,** A study of storage temperature and chilling injury on sweet pepper, *Acta Hortic. Sin.,* 13, 119, 1986.

281. **Lipton, W. J. and Wang, C. Y.,** Chilling exposures and ethylene treatment change the level of 1-aminocyclopropane-1-carboxylic acid in 'Honey Dew' melons, *J. Am. Soc. Hortic. Sci.,* 112, 109, 1987.

282. **Lipton, W. J., Peterson, S. J., and Wang, C. Y.,** Solar radiation influences solar yellowing, chilling injury and ACC accumulation in 'Honey Dew' melons, *J. Am. Soc. Hortic. Sci.,* 112, 503, 1987.

283. **Kaku, S. and Iwaya-Inoue, M.,** Monitoring primary response to chilling stress in etiolated *Vigna radiata* and *V. mungo* seedlings using thermal hysteresis of water proton NMR relaxation times, *Plant Cell Physiol.,* 29, 1063, 1988.

284. **Kaku, S. and Iwaya-Inoue, M.,** Estimation of chilling sensitivity and injury in gloxinia leaves by the thermal hysteresis of NMR relaxation times of water protons, *Plant Cell Physiol.,* 28, 509, 1987.

285. **Spalding, D. H. and Reeder, W. F.,** Low-oxygen high-carbon dioxide controlled atmosphere storage for control of anthracnose and chilling injury of avocados, *Phytopathology,* 65, 458, 1975.

286. **Vakis, N., Grierson, W., and Soule, J.,** Chilling injury in tropical and subtropical fruits. III. The role of CO_2 in suppressing chilling injury of grapefruit and avocados, *Proc. Trop. Reg. Am. Soc. Hortic. Sci.,* 14, 89, 1970.

287. **Hatton, T. T., Cubbedge, R. H., and Grierson, W.,** Effects of prestorage carbon dioxide treatments and delayed storage on chilling injury of 'Marsh' grapefruit, *Proc. Fla. State Hortic. Soc.,* 88, 335, 1975.

288. **Hatton, T. T., Smoot, J. J., and Cubbedge, R. H.,** Influence of carbon dioxide exposure on stored mid- and late-season 'Marsh' grapefruit, *Proc. Trop. Reg. Am. Soc. Hortic. Sci.,* 16, 49, 1972.

289. **Iwata, T. and Yoshida, T.,** Chilling injury to fruits of Japanese apricot *(Prunus mume)* and prevention measures, *Stud. Inst. Hortic. Kyoto Univ.,* 9, 135, 1979.

290. **Anderson, R. E., Parson, C. S., and Smith, W. L.,** Controlled atmosphere storage of eastern-grown peaches and nectarines, *U.S. Dep. Agric. Mark. Res. Rep.,* 836, 1969.

291. **Paull, R. E. and Rohrback, K. G.,** Symptom development of chilling injury in pineapple fruit, *J. Am. Soc. Hortic. Sci.,* 110, 100, 1985.

292. **Mencarelli, F., Lipton, W. J., and Peterson, S. J.,** Responses of 'zucchini' squash to storage in low-O_2 atmospheres at chilling and nonchilling temperatures, *J. Am. Soc. Hortic. Sci.,* 108, 884, 1983.

293. **Mencarelli, F.,** Effect of high CO_2 atmospheres on stored zucchini squash, *J. Am. Soc. Hortic. Sci.,* 112, 985, 1987.

294. **Wang, C. Y. and Ji, Z. L.,** Effect of low oxygen storage on chilling injury and polyamines in zucchini squash, *Sci. Hortic.,* 39, 1, 1989.

295. **Wilson, J. M.,** Drought resistance as related to low temperature stress, in *Low Temperature Stress in Crop Plants: The Role of the Membrane,* Lyons, J. M., Graham, D., and Raison, J. K., Eds., Academic Press, New York, 1979, 47.

296. **Morris, L. L. and Platenius, H.,** Low temperature injury to certain vegetables after harvest, *Proc. Am. Soc. Hortic. Sci.,* 36, 609, 1939.

297. **Pantastico, E. B., Soule, J., and Grierson, W.,** Chilling injury in tropical and subtropical fruits. II. Limes and grapefruits, *Proc. Trop. Reg. Am. Soc. Hortic. Sci.,* 12, 171, 1968.

298. **Brooks, C. and McColloch, L. P.,** Some storage diseases of grapefruit, *J. Agric. Res.,* 52, 319, 1936.

299. **McColloch, L. P.,** Chilling injury and Alternaria rot of bell peppers, *U.S. Dep. Agric. Mark. Res. Rep.,* 563, 1962.

300. **Scott, K. J. and Chaplin, G. R.,** Reduction of chilling injury in avocados stored in sealed polyethylene bags, *Trop. Agric.(Trinidad),* 55, 87, 1978.

301. **Scott, K. J. and Gandanegara, S.,** Effect of temperature on the storage life of bananas held in polyethylene bags with ethylene absorbent, *Trop. Agric. (Trinidad),* 51, 23, 1974.

302. **Grierson, W.,** Chilling injury in tropical and subtropical fruits. IV. The role of packaging and waxing in minimizing chilling injury of grapefruit, *Proc. Trop. Reg. Am. Soc. Hortic. Sci.,* 15, 76, 1971.

303. **Wardowski, W. F., Grierson, W., and Edwards, G. J.,** Chilling injury of stored limes and grapefruit as affected by differentially permeable packaging films, *HortScience,* 8, 173, 1973.

304. **Ben-Yehoshua, S., Kobiler, I., and Shapiro, B.,** Effects of cooling versus seal-packaging with high-density polyethylene on keeping qualities of various citrus cultivars, *J. Am. Soc. Hortic. Sci.,* 106, 536, 1981.

305. **Purvis, A. C.,** Relationship between chilling injury of grapefruit and moisture loss during storage: amelioration by polyethylene shrink film, *J. Am. Soc. Hortic. Sci.,* 110, 385, 1985.

306. **Adamicki, F.,** Effects of storage temperature and wrapping on the keeping quality of cucumber fruits, *Acta Hortic.,* 156, 269, 1984.

307. **Rij, R. E. and Ross, S. R.,** Effects of shrink wrap on internal gas concentrations, chilling injury and ripening of 'Honey Dew' melons, *J. Food Quality,* 11, 175, 1988.

308. **Hobson, G. E.,** The short term storage of tomato fruit, *J. Hortic. Sci.,* 56, 363, 1981.

309. **Hobson, G. E.,** Low-temperature injury and the storage of ripening tomatoes, *J. Hortic. Sci.,* 62, 55, 1987.

310. **Chace, W. G., Jr.,** Controlled atmosphere storage of Florida citrus fruits, *Proc. 1st Int. Citrus Symp.,* 3, 1365, 1969.

311. **Kramer, G. F. and Wang, C. Y.,** Reduction of chilling injury in zucchini squash by temperature management, *HortScience,* 1989, in press.

Index

INDEX